D1187825

SATELLITE COMMUNICATIONS SYSTEMS

SATELLITE COMMUNICATIONS SYSTEMS

G MARAL
ECOLE NATIONALE SUPÉRIEURE DES TÉLÉCOMMUNICATIONS
Département 'Télécommunications et Systèmes Aérospatiaux'
TOULOUSE
FRANCE

and

M BOUSQUET
ECOLE NATIONALE SUPÉRIEURE DE L'AÉRONAUTIQUE ET DE L'ESPACE
Département 'Electronique'
TOULOUSE
FRANCE

Translated by S. David,
Insight Consultancy

A Wiley–Interscience Publication

JOHN WILEY & SONS

Chichester · New York · Brisbane · Toronto · Singapore

Originally published under the title
Les Systèmes de Télécommunications
par Satellites by G. Maral,
M. Bousquet and J. Pares
© Masson, Paris 1975, 1982

Copyright © 1986 by John Wiley & Sons Ltd.

Reprinted March 1987

Library of Congress Cataloging in Publication Data:

Maral, Gérard.
   Satellite communications systems.

   Translation of: Les systèmes de télécommunications par satellites.
   'A Wiley—Interscience publication'.
   Bibliography: p.
   Includes index.
   1. Artificial satellites in telecommunication.
I. Bousquet, Michel. II. David, S. III. Title.
TK5104.M3613 1986   621.38′0422   84-24433

ISBN 0 471 90220 9

British Library Cataloguing in Publication Data:

Maral, G.
   Satellite communications systems.
   1. Artificial satellites in telecommunication
   I. Title   II. Bousquet, M.   III. Les Systèmes
   de télécommunications par satellites. *English*
621.38′0422   TK5104

ISBN 0 471 90220 9

Typeset by Mathematical Composition Setters Ltd.
Printed and Bound in Great Britain.

CONTENTS

PREFACE

Satellite communications is one of the most impressive spin-offs from the space programmes and has made a major contribution to, indeed totally altered, the patterns of international communications. Communications by satellite evolved from the simple technology of 'Early Bird' launched in 1965 to the highly sophisticated present-day satellites.

A large number of technical papers and books have been published on the subject of satellite communications. However, the available literature deals only with specific topics mainly related to communications techniques, and little has been written on the other various aspects of a satellite system, even though they are essential to fulfill the mission. Therefore the need was obvious for a practical book offering an overview of all aspects of satellite communications systems, which in addition examined these subjects in technical depth.

The aim of this book is to offer such an overview to students, engineers or scientists willing to enter the field. It can also be considered as a useful handbook to practising engineers. The book displays a detailed description of the various parts and interfaces of a satellite communications system, together with the constraints to be considered and the corresponding issues. Techniques commonly used and advanced concepts are covered.

The material in this book is presented in both undergraduate and first-year graduate courses at the Ecole Nationale Supérieure des Télécommunications, at the Ecole Nationale Supérieure de l'Aéronautique et de l'Espace and at the University of Toulouse (France). It has also been well appreciated by a large number of engineers involved in the field who attended continuing engineering education and in-house courses that we have conducted in France and abroad.

The book is divided into nine chapters. Chapter 1 presents the evolution and use of satellite communications systems from the most naive ones, at the beginning of the space era, to the most technically advanced present-day systems. Chapter 2 deals with transmission of information between the satellite and earth stations, i.e. analyses the factors which condition the link budget, and describes modulation techniques and their performances. Chapter 3 discusses the various aspects of multiple access (FDMA, TDMA and CDMA) by several earth stations to one satellite. Beam switching is examined in the context of multibeam satellites. The simplest non-regenerative satellite repeater and the more complex regenerative ones with on-board processing are both discussed. Chapter 4 deals with the various topics related to the geometry of the system such as orbits, distance between satellite and earth stations, coverage areas, earth stations antenna pointing angles, eclipses and solar interferences. Chapter 5 aims at a description of the main environmental factors which condition the design and operation of the satellite on its orbit during its lifetime. First a description of the space environment is given, and then the effects of this environment on the satellite itself. Chapter 6 is devoted to the various subsystems which constitute a geostationary communications satellite. It is common practice to distinguish the communications subsystems (repeater and antennas) from the bus or platform (attitude and orbit control, propulsion, telemetry tracking and command, thermal control, structure, electric power supply). For each subsystem the purpose, a description of the possible solutions and the performances according to the state-of-the-art technology are presented. Chapter 7 explains the constraints and the procedures of the launching and positioning of a geostationary satellite. This chapter offers also detailed characteristics and performances of current launch vehicles including STS and ARIANE. Chapter 8 is devoted to earth station technology. Chapter 9 analyses the technics used to evaluate the reliability of a satellite communications system and ensure the required availability.

We have made an extensive use of the important material presented in the various technical journals from the IEEE, AIAA, and others, and in papers presented at various conferences and conventions by members of aerospace companies or space agencies. We have attempted to give adequate credit to all authors of the above cited material, and we apologize in advance to any author for unintended omissions. We wish to thank the many excellent and assiduous technical staff members of the various companies (Matra, Alcatel-Thomson, Aerospatiale) and agencies (CNES, CNET, ESA) with whom we have been in contact for many years for helping us to understand a little more about satellite communications, and for influencing our teaching in this area.

G. MARAL
M. BOUSQUET

Acknowledgment

Initially the French publisher Masson published a book entitled *Les systèmes de communications par satellites* by J. Pares and V. Toscer. This book was mainly a contribution by J. Pares, a former student of the Ecole Polytechnique and a graduate of the Ecole Nationale Supérieure des Télécommunications. At that time J. Pares was with the Direction Générale de l'Armement of the Ministry of Defence and in charge of advanced studies in satellite communications systems. The original material came from the course he gave at the Ecole Nationale Supérieure des Techniques Avancées, Paris. The great success of the book encouraged the publisher to issue a new edition. In the meantime J. Pares had left the Ministry of Defence for Siemens S. A., where he is Head of the Communications Division for France. As J. Pares was no longer directly involved in the field of satellite communications, the publisher asked us to update the material. We maintained the initial organization of the first edition while adding an original contribution to cover the rapid development of the field. This second edition entitled *Les Systèmes de Télécommunications par Satellites* by G. Maral, M. Bousquet and J. Pares was duly published. Subsequently, John Wiley decided to have the book translated and to publish it in English. We took this opportunity to update and upgrade most of the material again, while still retaining the original organization set up by J. Pares. We wish to express to J. Pares our gratitude for leading the way and for his contribution to the success of this new version.

G. Maral
M. Bousquet

CHAPTER 1   Introduction

Communications involve the transfer of information between a source and a user. Terrestrial communications face long distance communications constraints because they either use guided media—wirelines, coaxial cables, and optical fibre cables—which have in common the fact that they require a physical path between terminals or wireless transmissions such as microwave radio relays which due to propagation problems must be in line of sight. Modern satellite communications originate from Clarke's idea to install radio relays on geostationary satellites, thus allowing for transmission of radio microwave signals over large distances (Clarke, 1945). The explosive growth of communications satellites and the perceived potential of the medium for novel applications has generated intense interest in both government and private sectors. This chapter presents the evolution and use of satellite communications systems from the simplest ones, at the beginning of the space era in the late fifties, to the most technically advanced present day systems, and also takes a look into the future.

1.1   ORIGINS OF SATELLITE COMMUNICATIONS SYSTEMS

Satellite communications are the outcome of research in the field of radio communications with the aim of achieving the greatest coverage and capacity, at the lowest cost. The Second World War encouraged rapid development of missile and microwave technology. The joint application of expertise gained in these two technologies heralded an era of radio communications by satellite. The services provided complemented those supplied until then exclusively by terrestrial networks (radio or cables).

Satellite communications systems are divided into two parts:

(1) Space segment (includes the satellite and the means on Earth necessary for launching and station keeping),
(2) Earth segment (earth stations containing transmitters and receivers for transmission and reception of signals from satellites).

Whereas communications on the Earth's surface benefited from advances

1

made in microwaves, the space segment required development in many various fields such as: launchers, orbit propulsion motors, attitude control, structure, power supply, electronic components, etc.

The space age began in 1957 with the launching of the first artificial satellite (SPUTNIK). The following years were marked by various experiments involving space communications—a Christmas greeting was sent by President Eisenhower in 1958 via the SCORE satellite; and 1960 saw a reflector satellite ECHO in orbit. The satellite COURIER was able in 1960 to record a message which could be played back later and during 1962 the active communications satellites (repeaters) TELSTAR and RELAY were in use. Then the first geostationary satellite SYNCOM was launched in 1963.

In 1965 the first commercial geostationary satellite, Intelsat I (EARLY BIRD) inaugurated a long INTELSAT series. The same year the first Russian communications satellite in the MOLNYA series was launched. Both these systems are in use to this day and in addition most regions and countries have or intend to use communications by satellite.

1.2   SPACE RADIOCOMMUNICATION SERVICES

Radio Regulations originate from the World Administrative Radio Conference (WARC), held under the auspices of the ITU (International Telecommunication Union), a governing organization for public telecommunications administrations throughout the world. WARC allocates frequencies for radio communications according to the different types of user. However, sharing the same band between services, particularly between 'terrestrial services' and 'satellite services' performing the same functions, is a potential source of interference.

The principal space radiocommunication services are:

Fixed-Satellite Service for communication between earth stations at specified fixed points via one or more satellites (e.g. INTELSAT).

Mobile-Satellite Service provides communication between mobile earth stations and one or more space stations, or between mobile earth stations via one or more space stations. Earth stations can be situated on-board ships (Maritime Mobile-Satellite Service, for example, MARISAT and MARECS); on-board aircraft (Aeronautical Mobile-Satellite Service, for example experiments with the ATS satellite; project AEROSAT was deferred); on-board terrestrial vehicles (Land Mobile-Satellite Service). This service may also be used to detect and locate distress signals from survival craft stations and emergency position-indicating radio beacon stations (e.g. SARSAT and SARGOS).

Broadcasting-Satellite Service allows sound and vision to be received by individuals or communities via satellite, the feeder link being part of the fixed satellite service. Experiments were carried out with the following satellites: ATS 6; CTS Canadian; BSE Japanese. Examples of on-going projects are: TDF/TVSAT, and TELE-X in Europe; STC, USSB, DBSC in the USA; ARABSAT in Arabian countries and AUSSAT in Australia.

Earth Exploration-Satellite Service involves observation of the Earth for various purposes, e.g. meteorological (TIROS 1, METEOSAT); geodesy (TRANSIT); earth resources (LANDSAT, SPOT, SEASAT); data collection (ARGOS).

Space Research Service where spacecraft or other objects in space are used for scientific or technical research.

Space Operation Service concerned exclusively with the operation of spacecraft (tracking, telemetry, telecommand).

Radiodetermination-Satellite Service for the purpose of determining the position and velocity of an object using one or more space stations (e.g. NAVSTAR).

Amateur-Satellite Service for radio amateur use carrying out technical investigations and learning about intercommunication (e.g. OSCAR, ARSENE).

Intersatellite service providing links between artificial earth satellites.

1.3 CHARACTERISTICS OF A SATELLITE COMMUNICATIONS SYSTEM

This book examines principally satellite communications systems related to fixed and mobile satellite services. The fixed satellite service has been in use longest, and is the most demanding in terms of capacity, i.e. the number of links required. For instance Figure 1.1 shows that the intercontinental traffic demand has been increasing exponentially at a rate of about 20 per cent per year.

Figure 1.2 shows the structure of a satellites communication system. Though the principles used are similar to those of microwave links, satellite communication systems can be distinguished by three characteristics:

(1) The signal has to travel a long distance without amplification and because of this, the early reflector satellites (ECHO) were replaced by active satellites with on-board amplification of the received signals
(2) Equipment is installed in an unmanned automatic craft and is subject to extreme environmental stresses.
(3) Repair is currently considered to be impossible after launching. Though Space Shuttle has demonstrated that repair of a satellite is possible, it must be stressed that this is presently applicable to low orbiting satellites only.

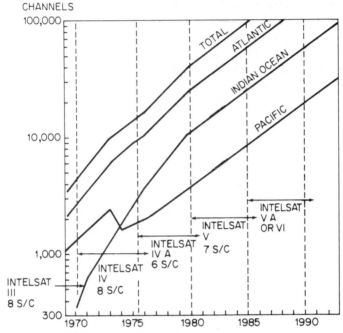

Figure 1.1   INTELSAT traffic development. From Rusch and Cuccia (1980). *Copyright American Institute of Aeronautics and Astronautics and reprinted with permission of the AIAA.*

Initially, satellite communications systems were designed to ensure links between a small number of earth stations with very large antennae (up to 32 m in diameter). The current trend is towards systems of one or several satellites, a group of several thousand receive-only (RO) earth stations with small antennae (a few metres).

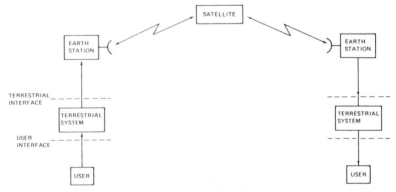

Figure 1.2   Structure of a satellite communication system. *Reproduced with permission from Van Trees (1979).*

Whereas terrestrial systems on the whole make use of a mesh network of sub-stations, a satellite communications system is a star-shaped network in which the satellite forms a nodal point. This has made necessary the further development of specific signal processing techniques and multiple access to a single satellite using several transmitting stations. The fact that the satellite is at the nodal point of a communications system also points out the vulnerability of the system in the event of satellite failure and makes it necessary to improve the reliability of the satellite. This reliability must be guaranteed over the lifetime of the satellite which is in the order of seven to ten years.

The frequencies used, intially in the C-band for commercial communications (6 GHz uplink/4 GHz downlink frequencies) have been extended by the WARC above 10 GHz. Various systems are using 14/11 GHz (K or Ku-band) (e.g. INTELSAT V, OTS, ECS, TELECOM 1, ANIK B, etc.) and experiments are in progress using 30/20 GHz (e.g. the Japanese satellite CS).

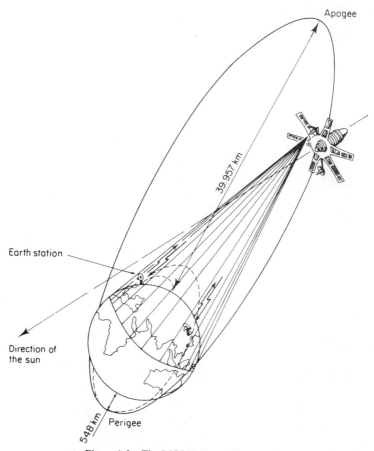

Figure 1.3   The MOLYNA satellite system

Satellite communications systems can be sub-divided according to their type of orbit:

(1) Satellites with a relatively undefined trajectory launched at low or medium altitude and making several revolutions per day around the Earth. Imprecisions in placing it in orbit and drift are not corrected and several satellites are needed to establish communication links. The number of satellites required is determined in terms of the probability of at least one link being available for a given percentage of time
(2) Satellites with a prescribed orbit equipped with a trajectory corrective system permitting the placing of the satellite on the determined orbit and maintaining it. A case of particular interest is that of an elliptical orbit at about 64 ° inclination to the Equator and having an orbital period of half the period of the rotation of the Earth: the axis of the ellipse of the orbit does not revolve around the centre of the Earth and the apogee is always situated above the same geographical area. The MOLNYA system, started by the Russians during 1965, is a system of this type (Figure 1.3).

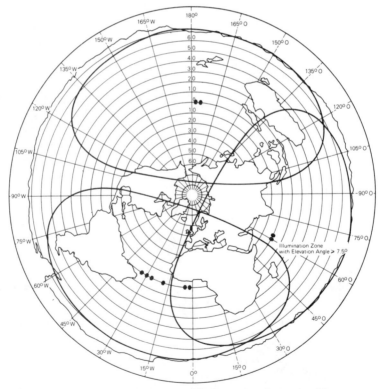

Figure 1.4   Global coverage with three geostationary satellites

(3) Geosynchronous satellites launched on a circular orbit and having a period equal to that of the rotation of the Earth. Amongst these satellites those which orbit the Equator in the same direction as the Earth's rotation are geostationary. For a terrestrial observer, a geostationary satellite is therefore immobile in the sky. A geostationary satellite permits coverage of a zone equal to about 40 per cent of the Earth's surface and its immobility in the sky makes it possible for the antennae to point in a fixed direction. A single geostationary satellite is able to link-up earth stations up to 17 000 km apart, and global coverage of the Earth excluding the polar regions can be achieved with three geostationary satellites (Figure 1.4).

An important consideration in the geosynchronous satellite communications system is the time delay of about 0.25 s, experienced when linking two earth stations. This is particularly noticeable during telephone conversations using a satellite link. The International Telegraph and Telephone Consultative Committee (CCITT) has established a standard of a maximum delay of 0.6 s so that communications involving two successive hops with this type of satellite are not recommended. In the future intersatellite links may avoid this inconvenience. Such links, which will not be made through the atmosphere, can be established at millimetre or optical wavelengths, possibly using lasers.

With the exception of the MOLNYA system, which is used for specific purposes by the Soviet Union (coverage of high latitude regions and high latitude of the launch site), the development of various satellite communications systems worldwide has seen geostationary satellites take on a particular importance.

1.4   EXAMPLES OF SATELLITE COMMUNICATIONS SYSTEMS

Currently, 20 per cent of the world's long distance traffic is handled by satellites. The early trend was for a rapid development of international systems; however, more recently several countries (e.g. Australia, Brazil, Canada, France, Germany, India, Indonesia, USA, etc.) have opted for domestic systems.

The following description of various communications satellite systems, while not exhaustive, offers pertinent examples.

1.4.1   International systems

1.4.1.1   The INTELSAT system (Alper and Pelton, 1984)

The International Telecommunications Satellite Organization has more than

Table 1.1 INTELSAT COMMERCIAL SATELLITES

	INTELSAT I	INTELSAT II	INTELSAT III	INTELSAT IV
Launches	6 April 1965	26/10/66 (F1) (U)	18/09/68 (F1) (U)	25/01/71 (F2) A 19°W
(U) = Unsuccessful		11/01/67 (F2)	18/12/68 (F2)	19/12/71 (F3) A 21.5°W
A = Atlantic Ocean	('Early Bird')	22/03/67 (F3)	05/02/69 (F3)	22/01/72 (F4) P 179°E
P = Pacific Ocean		27/09/67 (F4)	21/05/69 (F4)	13/06/72 (F5) I 57°E
I = Indian Ocean			14/01/70 (F6)	23/08/73 (F7) A 1°W
MCS = Maritime			22/04/70 (F7)	21/11/74 (F8) P 174°E
Communication			23/07/70 (F8) (U)	20/02/75 (F6) (U)
Sub-system				22/05/75 (F1) IA 18.5°W
Mission Life (years)	1.5	3	5	7
Mass (kg) Launch/ orbital	68/38	162/86	293/151	1415/700
Dimensions:				
Length (cm)	72	142	141	238
Height (cm)	59	67	202	528
Installed Power (Watts) (Initial/ Final)	45/33	83/75	160/125	600/400
Antennae	11° × 360°	12° × 360°	20° × 20°	Global Receive 2
				Global transmit 2
				Spot beam E/R 2
EIRP (dBW)	11.5	15.5	23	Global 22.5
				Spot beam 33.7
Transponders				
Frequency (GHz) up/down link	6/4	6/4	6/4	6/4
(Watts)	6	18	11	6
Bandwidth (MHz)	25	125	225	36
Number	2	1	2	12
Telephone Circuits Capacity	240 or	240 or	1200+	4000+
Television channels	1	1	1	2
Multiple Access	No	Yes	Yes	Yes
Total effective bandwidth (MHz)				
6/4 GHz	50	125	450	432
14/11 GHz	—	—	—	—
Attitude control	Spin	Spin	Dual Spin	Dual Spin

100 members and is based currently in Washington DC. It is responsible for providing communication links to its members. Administratively, INTELSAT hires out the given service to its members. One of the important members of INTELSAT is COMSAT (Communications Satellite Corporation) which represents the USA and had a management service contract with INTELSAT until 1979 and has the largest investment (about 20 per cent). It is also notable that most of the important Eastern European countries are not members of INTELSAT though the People's Republic of China is a member (0.33 per cent share in 1981).

Table 1.1 (continued)

	INTELSAT IV-A		INTELSAT V		INTELSAT VI	
Launches (U) = Unsuccessful A = Atlantic Ocean P = Pacific Ocean I = Indian Ocean MCS = Maritime Communication Sub-system	25/01/71 (F2) A 329°E 29/01/76 (F2) A 57°E 26/05/77 (F4) A 338.5°E 29/09/77 (F5) (U) 06/01/78 (F3) 179°E 31/03/78 (F6) 174°E		6/12/80 (F2) A 34.5°W 23/05/81 (F1)I 57°E Autumn 81 (F3) A 24.5°W Spring 82 (F4) A 27.5°W 28/9/82 (F5-MCS) I 63°E 19/5/83 (F6-MCS) A 18.5°W 18/10/83 (F7 MCS) I 60°E 4/3/84 (F8 MCS) A 53°W 10/6/84 (F9) (U)		Planned 1986/7	
Mission Life (years)	7		7		10	
Mass (kg) Launch/ orbital	1515/793		1928/1037		3748(AR4)/2232 13806 (STS)	
Dimensions: Length (cm) Height (cm)	238 693		1558 643		364 1183	
Installed Power (Watts) (Initial/ Final)	700/500		1800/1200		2800/2200	
Antennae	Global Hemisphere Spot beam		Global Hemisphere Zonal Spot beam	2 2 2 2	Global Hemisphere Zonal Spot beam	2 2 4 2
EIRP (dBW)	Global Hemisphere spot beam	22 26 29	Global Hemisphere Zonal Spot beam	23.5 29 29 41–44	Global Hemisphere Zonal Spot beam	26.5 31.0 31.0 41–44
Transponders Frequency (GHz) up/down link	6/4		14/12 and 6/4		14/11 and 6/4	
(Watts)	6.5		4.5–8.5–10		1.8 to 16	
Bandwidth (MHz)	36		36–72–77–241		36–41–72–77–150	
Number	20		27		50	
Telephone Circuits Capacity	6000 +		12 000 +		36 000 +	
Television channels	2		2		2	
Multiple Access	Yes		Yes		Yes	
Total effective bandwidth (MHz) 6/4 GHz	720		1357		2386	
14/11 GHz	—		780		886	
Attitude control	Dual Spin		3-axis		Dual Spin	

Table 1.1 shows the main characteristics of various satellites in the INTELSAT series. The increase in capacity, from one generation to the next, is achieved by an increase in the power available, the directivity of the antennas and the bandwidth used. Once the authorized frequency spectrum is completely full, increases in capacity can be achieved using frequency reuse by separate beams for East and West coverage and by orthogonal polarization (INTELSAT IV-A, V, VA and VI). These developments have necessitated increases in the size and weight of each successive satellite in the INTELSAT series (see Figure 1.5).

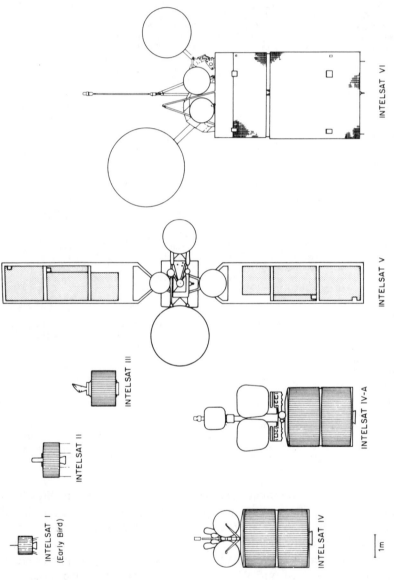

Figure 1.5   INTELSAT satellites. *Reproduced with permission from Rusch and Dwyer (1978)*

The INTELSAT system simultaneously handles telephone traffic, telegraphy messages, data, TV signals and facsimile information through more than 250 earth stations spread across more than 130 countries.

In the space of twenty years the cost of a telephone circuit has reduced by a factor larger than 7. INTELSAT presently imposes a monthly charge of $390 per international telephone half-circuit (the INTELSAT Board of Governors decided on constant cost since December 1982). The actual cost of a circuit by satellite is higher than that which INTELSAT alone charges when one includes the cost of the earth segment (sometimes called the 'terrestrial tail') which belongs to private companies or administrations. It should be noted that the two parts are owned and managed by different organizations (INTELSAT for the space segment and various organizations for the earth stations and ground links) and this administrative barrier may hamper the potential development and technical evolution of the system.

1.4.1.2 The INMARSAT system (Lundberg, 1984)

The International Maritime Satellite Organization, INMARSAT, was started with a relatively small membership but expanded to have a wide international membership, and had over 40 members by mid-1983 including Eastern European countries such as the USSR. It is similarly financed to INTELSAT and has its headquarters in London and in its initial years used transponders on various satellites available from INTELSAT, the European Space Agency (ESA) with its MARECS satellite, and the MARISAT Joint Venture providing the MARISAT satellites. In 1982 INTELSAT V (F–5) was the first of a series launched with a MCS (Maritime Communications Subsystem) specifically for INMARSAT.

INMARSAT's primary mission is to provide a service to ships and platforms. By July 1984, more than 2600 ships were equipped with ship earth stations. The service includes low rate (2.4 kbit/s) and high rate (up to 56 kbit/s) data, voice and emergency services. During mid-1983 INMARSAT launched an ambitious plan to procure a world-wide satellite system for most remote mobile platforms and appeared to be offering the potential of an air traffic communications and control service. This second generation system is due to come into service in 1988. All delivered satellites are required to be compatible not only with the US launch vehicles and the European Ariane, but also with the Proton launcher of the USSR. Table 1.2 gives the main parameters of the INMARSAT system.

1.4.1.3 The European system (Howell, 1980; Amadesi and Dharmadasa, 1984; Payet *et al.*, 1984)

The European international organization which provides telecommunication

Table 1.2 THE **INMARSAT** SYSTEM'S CHARACTERISTICS

S/C name(s)	MARISAT	MARECS	INTELSAT-5-MCS[a]
Prime contractor	Hughes Aircraft	British Aerospace	Ford Aerospace
Launch dates	19 February 1976; 9 June 1976; 14 October 1976	A: 20 December 1981 B: 9 November 1984	F5: 28 September 1982 F6: 19 May 1983 F7: 18 October 1983 F8: 4 March 1984
Launcher(s)	Delta 2914	Ariane	Atlas Centaur and Ariane
Launch site	ETR, Florida	CSG, Kourou, French Guiana	ETR and CSG
Design life (yrs)	5	7	7
Orbital positions	15°W; 176.5°E; 73°E	26°W; 177.5°E	63°E; 18.5°W; 60°E; 53°W
Transfer orbit			
Mass (kg)	655	1006 (A); 1014 (B)	1870
Primary power (W)	330	1000	1800 (EOL)
Frequencies	250–400 MHz 1537–1541 MHz 1638.5–1642.5 MHz 4195–4199 MHz 6420–6424 MHz	1538.8–1542.5 MHZ 1638.6–1644.5 MHz 4188.5–4200.5 MHz 6416.9–6425 MHz	1535–1542.5 MHz 1636.5–1644 MHz 4192.5–4200 MHz 6417.5–6420 MHz
No. of active transponders	5 (3 UHF; 1 L-band; 1 C-band)	1 C TOL 1 L TOC	2 (1 C–L; 1 L–C) of 7.5 MHz bandwidth each
EIRP (dBW)	UHF (28 dBW—wide band; 23 dBW—narrow band); C (18.8 dBW); L (26 dBW) (Medium Power Band)	C (15.7 dBW) L (24.7 dBW)	C-band (20 dBW)[b] L-band (32 dBW)[c]
'Ground' stations	3 coast earth stations: 12.8 mΩ ~200 ship terminals: 1.2 mΩ	~15 coast earth stations ~2500 ship terminals	Same stations as for MARECS
Capacity	Three UHF channels (one wide band and two narrow band), plus one two-way voice circuit and 44 telex channels while US was using UHF (due to power limitations) This rose to nine voice and 110 telex channels when Navy closed use of its UHF channels	46 voice channels. Each channel can carry 25 to 30 telex messages instead of voice	35 voice channels
Comments			

[a] Information on number of active transponder frequencies, power relates to MCS package only.
[b] Typical single carrier edge of coverage saturated EIRP.
[c] Typical multi-carrier edge of coverage saturated EIRP.

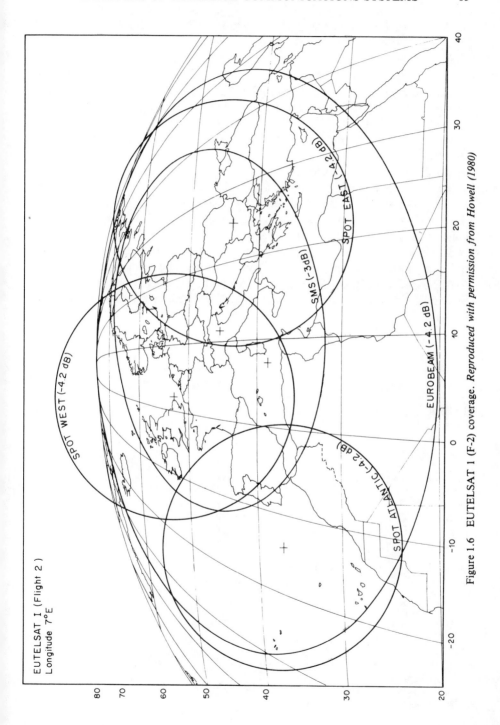

Figure 1.6 EUTELSAT 1 (F-2) coverage. *Reproduced with permission from Howell (1980)*

services by satellite is named EUTELSAT. By 1983 EUTELSAT consisted of 20 member states (signatories) who are represented through their Telecom administrations (including Post when these are combined and called PTT).

The EUTELSAT satellites are used for main route telephony in Europe, TV distribution services, and exchange of TV programmes within the European Broadcasting Union (EBU). In addition, special services are provided (Satellite Multiservice Systems: SMS) for the purpose of data transmission and teleconferencing required by the international business community, using small dish terminals.

The European programme started with an experimental satellite—OTS— and moved to the operational phase with the launching in June 1983 of EUTELSAT 1(F-2). EUTELSAT 1(F-2) with its SMS payload was launched in August 1984. Figure 1.6 shows the coverage area. EUTELSAT is now considering the implementation of a three-satellite space segment, as opposed to the two-satellite configuration originally envisaged. The main characteristics of this system are shown in Table 1.3. From the beginning of the next decade the EUTELSAT 1 satellite will be replaced gradually by a newly designed ECS-A, of capacity intermediate to that of EUTELSAT 1 and a second generation satellite.

Table 1.3 EUTELSAT SYSTEM CHARACTERISTICS

Telephony and EBU/TV		Business services	
Up-link frequency band	14–14.5 GHz	Up-link frequency band	14–14.08 GHz
Down-link frequency band	10.95–11.2 GHz 11.45–11.7 GHz	Down-link frequency	12.5–12.58 GHz
Satellite Transmitter power per transponder	20 W	*Satellite* Transmitter power per transponder	20 W
Receiver G/T (edge of coverage)	-5.3 db/K	Receiver G/T (edge of coverage)	-2 dB/K
EIRP at edge of coverage		EIRP at edge of coverage	39.8 dBW
(Spot beam)	40.8 dBW		
(Eurobeam)	34.8 dBW		
Earth station G/T (clear sky)	39 dBK^{-1}	*Earth station* G/T (clear sky)	30.4/27.4 dBK^{-1a}
EIRP	87.5 dBW	EIRP	71.4 dBW[b]
HPA	Max 2 kW	HPA	250 to 500 W[c]

[a] Two standards are defined
[b] Maximum EIRP carrier (for the highest bit rate and reception with the lower G/T).
[c] Depending on the number and type of carriers.

1.4.1.4 Other systems

The USSR and its allies (Bulgaria, Cuba, Czechoslovakia, East Germany, Hungary, Mongolia, Poland and Rumania) set up in November 1971 a consortium called INTERSPUTNIK. The original signatories have since been joined by Afghanistan, Laos, South Yemen, Syria, and Vietnam. In addition, several other countries, such as Algeria, Iraq, and North Korea, which are not full signatory members, used the INTERSPUTNIK facilities for communications with the signatory members. INTERSPUTNIK operates both MOLNYA satellites orbiting on an elliptical inclined orbit (Figure 1.3) and geostationary ('statsionar') satellites (RADUGA, EKRAN AND GORIZONT). This network is intended for telephony, telegraph, radio, data and telex communications, and distribution or direct broadcasting of television using the SECAM III B system. Table 1.4 shows the main characteristics of the INTERSPUTNIK system.

The Arab Satellite Communications (ARABSAT) Organization was formed in 1976 as a consequence of the decision of the League of the Arab States to set up a satellite communications system dedicated to the Arab region, when it became obvious that building a terrestrial network for this region would be both too expensive and time-consuming (Al-Mashat, 1982). A contract between the ARABSAT organization and the Aerospatiale of France was concluded in May 1981 for the production and delivery of communications spacecraft, and the provision of related services. The first ARABSAT satellite was launched in February 1985 by the Ariane launcher. The system is to provide a variety of communications services such as telephone, low and medium data transmission, multiplexed telex/telegraphy, radio and television distribution, community TV reception. It operates in the C-band (6/4 GHz) using twenty-five 33 MHz bandwidth channels. A specific channel is dedicated to community television which operates at 6 GHz in the up-link and at 2.5 GHz (S-band) in the down-link. The satellite lifetime is at least seven years. Earth stations for the C-band communications services are: major earth stations with 11-m diameter antennae, implemented in large metropolitan areas; transportable earth stations with antenna diameter of 1.6 m, used for emergency communications; earth stations with 4.5-m diameter antennae used for TV reception in remote areas. For community television, down-link reception is obtained from small receive only S-band earth stations with antennae of 3 m diameter.

The PALAPA system is used by the ASEAN (Association of South-East Asian Nations) and includes more than 200 stations in total. This system will be described later, with domestic systems, as it was initially planned to be a domestic system for the use of Indonesia, and became an international system because of initial excess capacity, which enabled ASEAN to lease transponders to Indonesia, Philippines and Thailand, (Bratahalim and Steady, 1984).

Table 1.4 THE INTERSPUTNIK SYSTEM CHARACTERISTICS

S/C Names(s)	MOLNYA-1 ('Lightning')	MOLNYA-2	MOLNYA-A
First Launch	March 23, 1956	November 24, 1971	November 21, 1974
Launcher	Vostok	Vostok	Vostok
Design Life (yrs)	~ 2	1 ~ 2	1 ~ 2
Launch Site	Plesetsk or Baykonur	Baykonur	Plesetsk
No. Launched (by mid-1984)	60	17	22
Orbit	12 hr, 16 min, i = 62.8°	i = 65.4°	
Orbital Positions			
Mass (kg)	1000	NA	1600
Frequencies	(1) 1000/800 MHz (2) 3.4–4.1 GHz down	6.2–5.7 GHz up 3.4–3.9 GHz down	6–6.2 GHz up 3.6–3.9 GHz down
No. of Transponder and output Power	1 × 20 W 3 × 40 W	NA NA	3 × 30 W? 1 × 40 W
EIRP (dBW) (max.)	NA	NA	32
Mission	Telephone, Telegraph, Radio, TV Distribution	Telephone, Telegraph, Radio, TV Distribution	TV Distribution, ± 800 phone channels
Ground Stations	12 m Φ Orbita Station Also used for mobile transmission with the Mars terminals	12 m Φ Orbitat Station	12 m Φ Orbita Station
Comments	One MOLNYA 1-S was launched into GEO July 1974 Launch rate: about every 3 years. Now used only for military	Replaced by RADUGA & GORIZONT	Links with Arctic regions

Table 1.4   (continued)

S/C Name(s)	RADUGA (Rainbow), STATSIONAR	EKRAN ('Screen'), STATSIONAR-T	GORIZONT ('Horizon'), STATSIONAR
First Launch	December 22, 1975	October 26, 1976	December 19, 1978
Launcher	Proton-D	Proton-D	Proton-D
Design Life (yrs)	2 ~ 3	2	2 ~ 3
Launch Site	Baykonur	Baykonur	Baykonur
No. Launched (by mid-1984)	15	22	10
Orbit		Quasi-stationary	
Orbital Positions	35°E, 45°E, 85°E, 128°E	99°E	14°W, 53°E, 90°E, 140°E
Mass (kg)	1940	1970	2120
Frequencies	(1) 6.1–6.2 GHz up 3.6–3.9 GHz down (2) 8 GHz up 7.2–7.7 GHz down	6.2 GHz up 702–726 MHz down	7.2–7.7 GHz up R3.6–3.9 GHz down
No. of Transponder and output Power	(1) 4 × 40 W (2) 2 × 40 W	1 (single klystron) 200 W	1 × 40 W 5 × 15 W
EIRP (dBW) (max.)	(1) 31	45–55	40–45
Mission		Direct Broadcase	Data and Telex
Ground Stations	12 m Φ Orbita Stations	More that 1500 terminals, covering some 9 million km^2 (40% of USSR territory). Yagi antennas are used	2.5 m Φ (Moskva TV receive only System) 25 m Φ & 12 m Φ Orbita Stations (Voice & Data) Launched for Olympics in Moscow
Comments	Also serves military pupuses and is used for maritime links	R90 helical transmit array antennas used on spacecraft Broadcast to Central Siberia: 1500 stations	Broadcast to Algeria, Cuba, Bulgaria, Hungary, E-Germany, Poland, Mongolia, Afghanistan, Vietnam, Laos. Can be received in Europe (cable)

The Nordic countries (Sweden, Denmark, Finland, Iceland and Greenland) have agreed on a direct broadcasting satellite (NORSAT) for television. Sweden, Norway and Finland (NOTELSAT) have decided to go ahead with a satellite system called TELE-X (Backlund, 1984). TELE-X is a multi-mission experimental/preoperational telecommunications satellite which will provide new services for data and video communication and direct TV to users in Sweden, Norway and Finland. The prime contractor for the spacecraft is Aerospatiale (France), and Saab-space (Sweden) is co-prime contractor. Ericsson (Sweden) had an overall responsibility for the payload. Thomson (France) assists Ericsson and will also supply the repeaters. Table 1.5 summarizes the main characteristics of the satellite. The TELE-X platform is based on the platform used in the Franco-German TV satellite programme TVSAT/TDF-1. The data/video transponders are equipped with 220-W travelling wave the tube amplifiers (TWTA). Two transponders can be

Table 1.5 THE TELE-X SATELLITE

Data/video repeater	
Number of operating transponders	2
Redundancy	3 for 2
Frequency band	14/12 GHz
Bandwidth	40 MHz (NBT)
	86 MHz (NBT)
G/T	11 dBK^{-1}
EIRP	>59 dBW
TV repeater	
Number of operating transponders	2
Redundancy	3 for 2
Frequency band	18/12 GHZ
Bandwidth	27 MHz
G/T	7 dBK^{-1}
EIRP	65 dBW

Orbital position: 5° east
Life time: 5–7 years
Mass
  at launch: 2130 kg
  dry: 1020 kg
Power generation: 3200 W
Attitude control: three-axis
Random pointing errors (3 sigma)   receive antenna $<0.075°$
                                    transmit antenna $<0.05°$
Dimensions:
  Satellite body   L = 2.4 m, W = 1.65 m, H = 2.4 m
  Solar arrays,   span = 19 m

operated simultaneously, one narrow band transponder (NBT), with capacity of 500×64 kbit/s or 20×2.048 Mbit/s or a combination of these channels commensurate with power and bandwidth and one wide band transponder (WBT), with capacity of 25×2.048 Mbit/s or 6×8.848 Mbit/s or 2×34.368 Mbit/s or 1×139.264 Mbit/s. The TV transponders are like those of TVSAT/TDF-1 with additional efforts to minimize baseband distortion of the TV signals for the most advanced transmission method (C-MAC). The TWTA operates at saturation and delivers an RF output power of about 220 W. The earth segment consists of many different types of stations. The data station with one 64 kbit/s channel capacity is in the one extreme. The transportable video/data station with simultaneous transmission capability of up to four carriers at data rates up to 34 Mbit/s is on the other extreme. The antenna diameter of all types of data/video stations is in the range of 1.5–2.5 m, depending on location.

The TELE-X system has an initial capacity of handling 500 earth stations and 3 calls/s. The capacity is extendable to 5000 earth stations and 30 calls/s. The earth segment comprises also a Feeder Link Station with a capability of transmitting three TV channels at 18 GHz using a 8-m diameter antenna. The home terminals will have dishes of 0.5–0.9 m in diameter.

1983 saw the emergence of private systems offering international services in competition with established international organizations. ORION and ISI (International Satellite Inc.) are examples of potential Intelsat competitors. On a regional basis, in mid-1984 a private company EBS (European Business satellite) announced it was procuring a satellite system to provide European business services by 1988. In addition, national systems such as the French TELECOM I and the proposed British UNISAT offered regional services and the potential for transatlantic traffic.

1.4.2   National systems

1.4.2.1   The United States

The United States has several different programmes that provide national links. Table 1.6 lists these different systems, as of December 1984.

The WESTAR network has been operational since 1974. Its satellites were produced by the Western Union Telegraph Company, manufactured by Hughes Aircraft Company and used for telephone, broadcasting of television and high quality audio programme material, and data services. Figure 1.7 shows the coverage of WESTAR. Some of the transponders of this network were leased to the American Satellite Corporation (ASC). These satellites have 12 or 24 C-band (6/4 GHz) transponders each of 36 MHz bandwidth.

The RCA Corporation initially leased transponders on WESTAR and ANIK, then in 1975 and 1976 constructed and launched its own satellites

Table 1.6   U.S. SYSTEMS (DECEMBER 1984)

Owner	Name of system
Western union	WESTAR
RCA Americom	SATCOM
ATT	COMSTAR
	TELSTAR
SBS	SBS (USA-SAT)
GTE	SPACENET
	GSTAR
Hughes	GALAXY

(SATCOM 1 and SATCOM 2). These satellite each weight approximately 400 kg at beginning of orbit life (BOL), use three-axis stabilization and have a capacity of 24 channels of 36 MHz bandwidth in the C-band (6/4 GHz) with frequency reuse by cross-polarization. SATCOM 1 operation terminated in June 1984 while SATCOM 2 is still in service. Two additional C-band satellites were launched in December 1981 and January 1982. In October 1982 the first all-solid-state satellite, SATCOM 5, was launched, soon followed by SATCOM 1R and SATCOM 2R. These RCA Advanced Satcom spacecraft contain 24 8-W solid state transponder channels at C-band (Freeling and Weinrich, 1984; Braun and Keigler, 1984).

The American Telegraph and Telephone Company (ATT) leased four satellites manufactured by Hughes Aircraft Company at the request of Comsat General Corporation. These COMSTAR satellites offer 24 repeaters with

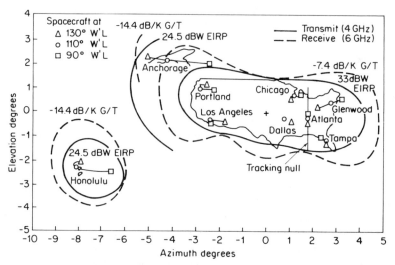

Figure 1.7   WESTAR satellite coverage. *Reprinted with permission from Sion (1978) Copyright (1978) Pergaman Press Ltd*

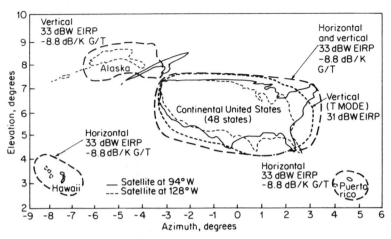

Figure 1.8   COMSTAR satellite coverage. *Reprinted with permission from Sion (1978). Copyright (1978) Pergaman Press Ltd*

36 MHz bandwidth and use twice the C-band (6/4 GHz) by means of frequency reuse by cross-polarization. Figure 1.8 shows the coverage of a COMSTAR satellite. The TELSTAR satellites are due to replace the ageing COMSTAR satellites, starting in 1983. These are improved 24 transponder C-band satellites which incorporate with both travelling-wave tube amplifiers (TWTA) and solid-state power amplifiers (SSPA). (Benden *et al.*, 1982).

The Satellite Business Systems (SBS) satellite was in orbit in 1980. This project was funded by a consortium of Comsat General, IBM, and Aetna Life and Casualty Inc. It is designed to provide services such as telephone, video conferencing and data transmission for a large number of almost entirely business users. SBS-1 and SBS-2 (or USASAT) operate in the K-band (14/12 GHz) and use digital transmission. They displayed, at the time of their launching, novel features such as time division multiple access, varying-with-demand transmitted burst format and digital speech interpolation (DSI). Satellite capacity is 10 channels at a bit rate of 48 Mbit/s giving a total of 480 Mbit/s. Capacity can be 14 000 telephone circuits or 8000 links for data transmission at 54 Kbit/s (Stamminger and Stein, 1980; Schnipper, 1980). Figure 1.9 depicts the coverage of these satellites (Barrila and Zitzmann, 1977). In SBS-3 two of the ten transponders provide 40 W of transmitting power each and are leased to COMSAT to provide programme distribution services to the NBC television network. In SBS-4 the output of five transponders can be concentrated via spot beams on the east or west coasts of the USA and may be used for direct broadcast television service. An increased coverage is also anticipated using the same satellite but reducing the number of digital channels (Ludwig, 1980). SBS-5, planned for launch in 1986, will have 14 transponders, the extra four with 110 MHz of bandwidth.

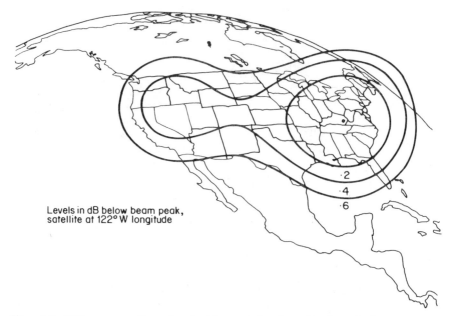

Levels in dB below beam peak,
satellite at 122° W longitude

Figure 1.9   SBS coverage. *Reproduced with permission from Barnla and Zitzmann (1977). Copyright © 1977 IEEE.*

GTE offers satellite services through two satellite systems: SPACENET and GSTAR (Waylass, 1982; Jacobs and Pourmand, 1984). The SPACENET system consists of three hybrid satellites operating in both the C-band (6/4 GHz) and Ku-band (14/12 Ghz). The C-band transponder configuration consists of twelve 36 MHz transponders using 8.5-W SSPAs, and six 72 MHz transponders using 16-W TWTAs. The Ku-band configuration is six 72-MHz transponders using 16-W TWTAs. These satellites, manufactured by RCA Astro Electronics, are designed to meet the needs of video and audio distributors, specialized common carriers, business users and government agencies requiring high quality long distance communications services. The first SPACENET satellite was launched in May 1984. The GSTAR system consists of two satellites, also manufactured by RCA Astro Electronics, and operating at Ku-band (14/12 GHz). Several antennae patterns and transponder combinations can be generated by ground command. Each satellite offers sixteen 54-MHz transponders. Two transponders using 27-W TWTAs are dedicated to the 50-state coverage. The 14 other transponders utilize 20-W TWTAs and can be switched to regional beams (east and west) and to the continental US (CONUS) beam. The GSTAR system is designed for voice, data and video communications, and direct-to-home broadcasting.

A Hughes subsidiary, Hughes Communications Inc., operates a C-band

(6/4 GHz) system, called GALAXY, which consists of three satellites. This system serves the business community, cable programmers, and broadcasters.

Owing to the increase in demand, the above cited American Satellite Company's services decided in 1981 to operate two wholly owned commercial communication satellites. ASC's first generation satellites, manufactured by RCA Astro Electronics, will operate in both the C-band (6/4 GHz) and Ku-band (14/12 GHz). Transponders on-board the spacecraft will include both 36 MHz (equipped with SSPAs) and 72 MHz bandwidth transponders with total RF power of 300 W. Launch dates are scheduled for September 1985 and 1986. These satellites will provide voice, data, facsimile and video conferencing communication services to business and government agencies.

NASA is presently actively pursuing a substantial and innovative satellite communications technology development programme, whose output should be the Advanced Communications Technology Satellite (ACTS)(Sivo, 1983; Nickelson, 1984). This satellite is designed to develop electronically hopped spot-beam technology with on-board message switching. Fixed and hopped-spot beams will deliver their signals to a baseband processor which will contain demodulation, buffering, forward error correction, memory, baseband switching and remodulation circuits. ACTS will operate in the Ka-band (30/20 GHz) and will be launched in 1988.

Another system, developed for NASA by TRW, and called Tracking and Data Relay Satellite System (TDRSS), has provided since the end of 1983 both Ku-band (14/12 GHz) and S-band (2 GHz) communication and tracking services to low earth-orbiting user spacecrafts; (Holmes, 1978; Landon and Raymond, 1982). TDRSS consists of two geostationary relay satellites 130° apart in longitude and a ground terminal located at White Sands, New Mexico (Figure 1.10). Each satellite consists of a total of seven antennae among which are two independently mechanically steerable 4.9-m reflector antennae operating simultaneously at both S-band and Ku-band. These provide single access high capacity (up to 300 Mbit/s) communications links. Each spacecraft also consists of a S-band phased array antenna of 30 helical elements which forms and steers beams under ground control for up to 20 multiple access users.

For the needs of the US Department of Defense, a worldwide satellite communications system, called Defense Satellite Communication System (DSCS) was set up in 1966. The system operates in the X-band (8/7 GHz). To begin with, the system was based on 26 small (45 kg) communications satellites. Each satellite relayed voice, imagery, computerized digital data and teletype transmissions. The DSCS II generation, for a total of sixteen 619 kg (BOL) satellites manufactured by TRW starting in 1971, offered a substantial increase in capacity and mechanically steerable spot-beam antennae for intensified coverage to link small portable ground stations into the communication

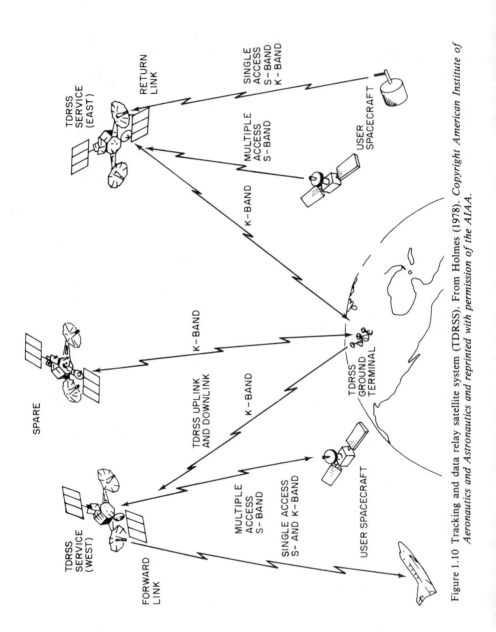

Figure 1.10 Tracking and data relay satellite system (TDRSS). From Holmes (1978). *Copyright American Institute of Aeronautics and Astronautics and reprinted with permission of the AIAA.*

system. In 1977, four additional DSCS II satellites were produced for launch in 1979 and later on, and offered a two fold increase in effective isotropic radiated power (EIRP) through the use of 40-W TWTAs in place of 20-W TWTAs used in the first twelve DSCS II satellites. The DSCS III generation, under development and production by General Electric since 1982, consists of ten 10-year lifetime satellites weighing 1125 kg (BOL), each with six channels (Donovan, *et al.*, 1983). Three electronically steerable multiple beam antennae (two transmit and one receive) are used for earth-coverage, area-coverage and spot beams. The receive antennae have 61 beams with the capability of producing steerable nulls in the earth-coverage mode to counter-act jamming. Two transmit and two receive horns provide earth-coverage. In addition, a steerable transmit dish antenna, associated with a 40-W TWTA, provides a spot beam with increased radiation power for users with small receivers.

1.4.2.2 Canada

Canada has been operating domestic satellite systems since 1973, with the first of the ANIK satellite series. Some 130 earth stations, ranging from 1.2 m to 30 m, are operated by TELESAT. They provide a variety of sound and tele-vision broadcasting, data, message and record services, including a medium density telephone service to communities in the high Arctic using 8 m or 10 m antennae; (Smart, 1983; Nickelson, 1984). The ANIK D satellites are gradually replacing the former A and B series. ANIK D satellites were built by SPAR Aerospace Ltd in cooperation with the Hughes Aircraft Co. ANIK D satellites weigh 1217 kg (BOL) and are designed for a 10-year lifetime. They offer 24 channels with 36 MHz of bandwidth at C-band (6/4 GHz) using 11-W TWTAs (Smalley, 1983). ANIK C-3, launched in November 1982, is the first satellite of the C series operating at Ku-band (14/12 GHz). Four down-link spot beams share sixteen 54 MHz bandwidth channels equipped with 15-W TWTAs. Each channel has a basic capability for one 90 Mbit/s digital message carrier or two

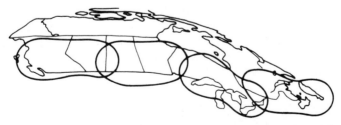

Figure 1.11   Anik-C satellite coverage. From Lester (1980). *Copyright American Institute of Aeronautics and Astronautics and reprinted with permission of the AIAA*

Table 1.7 CANADIAN ANIK PROGRAMME. FROM LESTER (1980). *Copyright American Institute of Aeronuatics and Astronautics and reprinted with permission of the AIAA*

Satellite	Launch date	Frequency (GHz)	Satellite channels	Channel bandwidth (MHz)	EIRP
ANIK A-1	November 1972	6/4	12	36	24 dBW
A-2	April 1973				(2 × 8 deg beam)
A-3	May 1975				
ANIK B-1	December 1978	6/4	12	36	34 dBW
					(2 × 8 deg beam)
		14/12	4	72	46.5 dBW
					(2 × 2 deg beam)
ANIK C-1	June 1984	14/12	16	54	48 dBW
C-2	June 1983				(1 × 2 deg beam)
C-3	November 1982				
ANIK D-1	August 1982	6/4	24	36	36 dBW
D-2	Scheduled late 1984				(2 × 8 deg beam)

analogue television signals. Figure 1.11 shows the coverage of the ANIK C satellite. Table 1.7 summarizes the characteristics of the satellites of the ANIK series.

1.4.2.3 Indonesia

Indonesia is a growing country with a population of about 150 million inhabitants, stretching about 5000 km along the equator and containing 13 677 islands. Interconnecting those islands by conventional means, using microwave links and submarine cables, would have been an extremely time-consuming and expensive task. It is therefore not surprising that Indonesia decided very early to embark upon the establishment of a national telecommunications satellite system. In February 1975, the Indonesian National Telecommunications Trading Company (PERUMTEL) awarded a contract to Hughes Aircraft Co. for the construction of two satellites, a master control station and nine earth stations, and to Ford Aerospace and ITT, for each to provide an additional 15 earth stations. The PALAPA satellite communication system was inaugurated in August 1976, with the PALAPA A satellite. Figure 1.12 shows the coverage of the PALAPA A satellite. The replacement PALAPA B satellites ordered in 1979 were launched by the Space Shuttle in June 1983 and in February 1984. Due to upper stage failure, this latter launch was unsuccessful but the satellite was recovered during a subsequent flight in November 1984 by the Shuttle's crew. The PALAPA system provides telephony and telex services together with TV distribution. The system has grown from the initial 40 earth stations to nearly 230 in 1984 (Bratahalim and Steady, 1984). Table 1.8 gives the main parameters of the PALAPA system. As quoted above (Section 1.4.3), some of the PALAPA transponders are leased to the ASEAN.

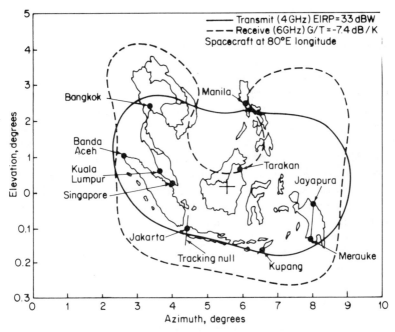

Figure 1.12 PALAPA satellite coverage. *Reprinted with permission from Sion (1978). Copyright (1978) Pergaman Press Ltd.*

Table 1.8 THE PALAPA SYSTEM'S PARAMETERS

S/C Name(s)	PALAPA-A	PALAPA-B
Prime contractor	Hughes	Hughes
Launch dates	8 August 1976; 10 March 1977	18 June 1983 3 February 1984 (failure)
Launcher	Delta 2914	Shuttle
Reserved orbital slot(s)	83°E, 77°E	108°E, 113°E, 118°E
Design life (yrs)	7	8
Initial mass in GEO (kg)	297	628
Array power (W, BOL/EOL)	307/250	1108/831
Frequencies (GHz)	6/4	6/4
No. of transponders (active/total)	12/12	24/30
Transponder bandwidth (MHz)	36	36
Output Power/transponder (W)	5	10
EIRP (dBW)	32 (Indonesia) 27 (ASEAN)	34 (Indonesia) 32 (ASEAN)
G/T at edge of coverage (dBK^{-1})	−7	−5
Capacity	6000 voice circuits or 12 TV channels, or combination	12 000 voice circuits or 24 TV channels, or combination

1.4.2.4 Brazil

Brazil has about 80 per cent of its population living in a 500-km strip along the Atlantic coast. The remaining sparsely populated area consists of hundreds of small cities and seven state territory capitals, several hundreds of kilometres distant from the main centres. It appeared that the most cost-effective way of providing telecommunication services to these isolated cities would be by using a satellite system, and hence the Brazilian State Company, EMBRATEL, responsible for satellite services started renting, in 1975, transponders from the INTELSAT network. Based on the forecast traffic demands for the beginning of 1985, Brazil decided to implement its own dedicated communication satellite network. A first BRAZILSAT satellite was launched by the ARIANE launcher in February 1985. This satellite consists of 24 channels operating in the C-band. The payload design is similar to that of the ANIK D satellite (Berridge, 1983)

Figure 1.13 illustrates a look comparison of several domestic satellites cited above, which were manufactured by North American firms.

1.4.2.5 The Soviet Union

The Soviet Union began its National Satellite Communications Network in 1965 with ORBITA using several MOLNYA satellites. This network permitted transmission from two earth stations at Moscow and Vladivostok, for telephone traffic, radio, telegraph, TV and facsimiles to about one hundred earth stations. The network further extended using RADOUGA, EKRAN and GORIZONT geostationary satellites (see section 1.4.1.4).

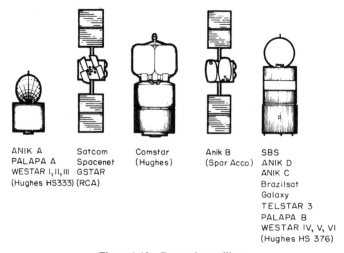

ANIK A
PALAPA A
WESTAR I, II, III
(Hughes HS333)

Satcom
Spacenet
GSTAR
(RCA)

Comstar
(Hughes)

Anik B
(Spar Acco.)

SBS
ANIK D
ANIK C
Brazilsat
Galaxy
TELSTAR 3
PALAPA B
WESTAR IV, V, VI
(Hughes HS 376)

Figure 1.13   Domestic satellites

1.4.2.6 France

The French administration decided in February 1979 to establish the TELECOM 1 domestic satellite network (Lombard *et al.* 1983). The satellite is designed to provide:

(1) Intra-company business services (data transmission with bit rates from 2.4 kbit/s to 2 Mbit/s, videoconferencing, high speed facsimile, telephony, etc.) and videotransmission within continental France, using the Ku-band (14/12 GHz);
(2) Telephony and television distribution between mainland and overseas territories, using the C-band (6/4 GHz);
(3) Communications for governmental use in the X-band (8/7 GHz).

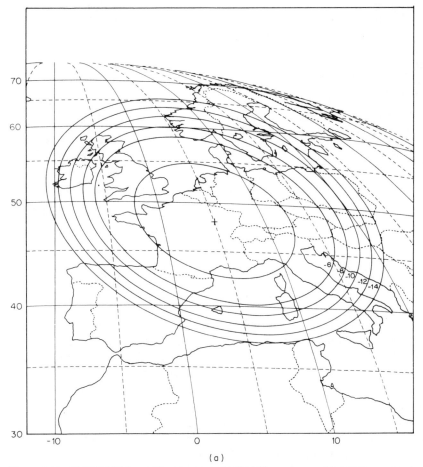

(a)

Figure 1.14   TELECOM 1 satellite coverage: (a) 12 GHz coverage; (b) 4 GHz coverage.

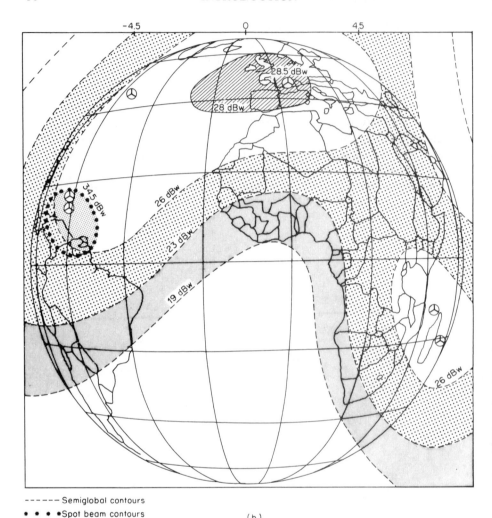

—————— Semiglobal contours
● ● ● ●Spot beam contours

(b)

Figure 1.14(b)

The satellite is manufactured by Matra Space Branch, weighs 660 kg and has an expected lifetime of seven years. The 14/12 GHz payload consists of six 36-MHz bandwidth channels equipped with 20-W TWTAs. Each channel allows for either a 25-Mbit/s capacity, or one video-transmission link. The antenna coverage is a $2°$ by $1.4°$ spot beam over continental France, obtained with a 0.8 m × 1.2 m reflector antenna. The 6/4 GHz payload consists of four channels, two 40-MHz bandwidth channels and two 120-MHz bandwidth channels equipped with 8.5-W TWTAs. The receive antenna is a horn with global coverage. Three channels are connected to a semi-global shaped beam transmitting antenna. The remaining channel is connected to a spot-beam antenna illuminating French Guiana ˙and the Antilles. The 8/7-GHz payload offers two channels of 40-MHz bandwidth with global coverage (Blachier *et al.*,1982). The first TELECOM 1 satellite was launched in August 1984 by the ARIANE 3 launcher. Figure 1.14 illustrates the TELECOM 1 down-link coverage.

In addition, France collaborates with the Federal Republic of Germany on a direct broadcasting satellite system programme. The first satellite (TVSAT) should be launched in 1986 and will be followed by the second one (TDF1). Figure 1.15 shows the coverage of the French TDF1 satellite (Georgy, 1984). Table 1.9 gives the main characteristics of both the German TVSAT and the French TDF1 satellite.

1.4.2.7 West Germany

West Germany worked closely with France on the 'Symphonie' and TVSAT/TDF programme and in 1983 announced the Deutsche Fermelde Satellite (DFS) due to be launched in 1987. The satellite is designed for TV programme distributed with five 44-MHz bandwidth channels at Ku-band (14/12 GHz), country-wide network for new services with two 44-MHz bandwidth channels, allowing for transmission of 60 Mbit/s at Ku-band, three point-to-point connections for digital telephony, data and TV programme exchange with three 90-MHz bandwidth channels at Ku-band and one 90-MHz bandwidth channel at 30/20 GHz (Schindler, 1984).

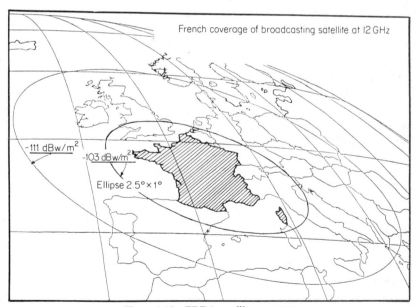

Figure 1.15   TDF 1 satellite coverage.

Table 1.9   TDF1/TV-SAT SATELLITE CHARACTERISTICS

Lifetime: 7 years
Reliability: 0.75
Dry mass: 1000 kg
Mass at launch: 2060 kg
Launcher: Ariane 2
Electric power (end-of-life, summer solstice): 3230 W (two wings of four
  1.6 m × 3.6 m panels); growth potential to 6 kW (twice six panels allowing for
  five operating channels)
Dimensions (m): 2.4 × 1.64 (19.23 with solar array extended), total height: 6.4
Number of channels: 5, 4 simultaneously operating
Transmitted power per channel: about 230 W
Uplink frequency: 17.3–17.7 GHz (TDF1)
                  17.7–18.1 GHz (TV-SAT)
Receiving antenna coverage: $0.7° × 0.7°$
Downlink frequency: 11.7–12.1 GHz
Polarization: RHC (TDF1), LHC (TV-SAT)
Transmitting antenna coverage: $2.5° × 0.98°$ (TDF1), $1.62° × 072°$ (TV-SAT)

1.4.2.8   The United Kingdom

The United Kingdom whilst leading a European consortium with British
Aerospace as a prime contractor (later partnered by Matra Space Branch of
France in a joint subsidiary called SATCOM International) for the European
Space Agency satellites OTS and ECS, has channelled its resources via the
International Agencies (e.g. INMARSAT, ESA, INTELSAT, NATO, etc.)
rather than own and build its own telecommunications satellite network. The
exception would have been the multimission satellite project UNISAT which
was announced in mid-1983 by United Satellites Ltd (a joint venture between
British Aerospace, Marconi and British Telecom). The satellite would have
offered two direct broadcast channels equipped with 240-W TWTAs, to be
received by ground receivers with 0.9-m dishes. These channels would
have been operated by the BBC. The satellite would have also contained a
communication payload with six 36-MHz bandwidth channels operating in the
K-band (14/12 GHz) for business services, data and voice. Two
receive/transmit beams would have been provided, the east beam encompass-
ing most of Europe and the west beam covering most of the eastern parts of
the USA and Canada (Reed, 1983).

1.4.2.9   Japan

Japan has been very active in the field, starting with experimental satellites
such as the Broadcasting Satellite for Experimental Purpose (BSE), launched
in 1978, and the Experimental Communication Satellites (CS and ECS),
launched respectively in 1977 and 1979, and turning afterwards to fully operational systems.

Nippon Telegraph and Telephone Public Corporation (NTT) initiated the world's first Ka-band (30/20 GHz) domestic satellite communications system for commercial use, using the CS-2 satellites (CS-2a, launched in February 1983; CS-2b, launched in August 1983). The launch vehicles used were N-11 rockets developed by the National Space Development Agency of Japan (NASDA). Each satellite has eight channels with six for the Ka-band (30/20 GHz) and two for the C-band (6/4 GHz). The satellite employs a horn-reflector antenna that produces a shaped beam over the main island of Japan in the Ka-band and a coverage of the whole territory of Japan in the C-band. The system is designed for digital communications at 65 Mbit/s or 100 Mbit/s between large earth stations of the public communications network, and also allows small transportable earth stations to access easily to the public communications network from any place in Japan. Table 1.10 gives the main parameters of the system (Tanaka, 1984).

Table 1.10 JAPAN'S DOMESTIC SATELLITE COMMUNICATION SYSTEM

Main performance of CS-2

Diameter	2.18 m	
Height	3.29 m	
Weight	350 kg (when launched)	
Design life	3 years (5-year objectives)	
Stabilization	Spin stabilized	

	C-band	K-band
Number of transponder units	2	6
Receiver noise figure	Less than 6.2 dB	Less than 12 dB
Transmitter output power	More than 34.5 dBm	More than 34.0 dBm
Beacon oscillator output power		More than 15 dBm
Antenna coverage	All Japanese islands	Main Japanese island
Antenna gain	More than 25 dB	More than 33 dB

NTT's satellite communication system

Frequency band	30/20 GHz		6/4 GHz	
Type of communication	Fixed–Fixed	Fixed–Transportable	Fixed–Fixed	Fixed–Transportable
Communication mode	TDMA 8 stations (max.)	Point to point 2 stations	TDMA 8 stations (max.)	Point to point 2 stations
Modulation	65-Mbit/s BPSK	FM	100-Mbit/s QPSK	FM

NTT's earth station parameters

Frequency band	30/20 GHz		6/4 GHz	
Type of station	Fixed	Transportable	Fixed	Transportable
Antenna (type)	11.5-m diam. (offset Cassegrain)	2.7-m diam.	11.5-m diam. (Cassegrain)	3-m diam.
HPA	300-W klystron	200-W klystron	450-W TWT	400-W klystron
LNA	220 K	300 K	57 K	100 K
G/T	41 dBK^{-1}	27 dBK^{-1}	32 dBK^{-1}	18 dBK^{-1}
EIRP	91 dBW	79 dBW	80 dBW	67 dBW

The first operational broadcasting satellite of Japan (BS-2) was launched in January 1984. The satellite has the capacity of two 27-MHz bandwidth channels to transmit colour television signals, using 100-W travelling wave tubes. The satellite's expected lifetime is five years. In most parts of mainland Japan, high quality pictures can be received using home receivers equipped with a 75-cm diameter antenna. Receiving antennae of 1.6 m are to be installed in remote sites such as the Okinawa Islands. The fabrication of direct broadcast home receivers is under way at a cost of 400 dollars for the front end and 450 dollars for the indoor unit. It is estimated that there will be more than 500 000 sets by the end of BS-2 life (Nakamura, 1984)

1.4.2.10 India

The first indian domestic satellite INSAT-1 (INSAT-1A) launched on 10th April 1982 had to be desactivated on 6th September 1982 due to failure in orbit and was replaced by the INSAT-1B satellite successfully launched from the Space Shuttle in September 1983. This satellite is intended for long distance telecommunications (telephony, data, facsimile, etc., meteorological earth observation and data relay, direct TV broadcasting to community TV sets in rural areas and networking of terrestrial TV transmitters, and finally regional and national networking of radio transmitters.

1.4.2.11 Other countries

Italy is planning an ambitious experimental ITALSAT satellite with six individual spot beams operating at Ka-band (30/20 GHZ) and covering Italy. Each spotbeam is served by a regenerative repeater working at 147 Mbit/s, with a total capacity of 884 Mbit/s. Interspotbeam connectivity will be provided by a synchronous baseband switching matrix. Along with the multibeam regenerative payload, the satellite includes a conventional global Italian coverage payload and a propagation payload with two beacon generators at 40 GHz and 50 GHz, both with European coverage (Marconicchio and Valdoni, 1983). ITALSAT is scheduled to be ready for launch in the middle of 1988.

The Australian AUSSAT (Nowland, 1983), Colombian SATCOL (Romero, 1982), Mexican MORELOS (Sanchez-Ruiz and Elbert, 1984), are further examples of national systems planned or operating. In addition, many countries have leased INTELSAT transponder capacity for domestic services. By 1983, there were more than 24 countries leasing channel capacity for domestic use, which varied from television distribution to trunk and thin route telephony traffic.

1.5 FUTURE DEVELOPMENTS

The above description of existing and planned communication satellite systems shows obviously a formidable development during the last 20 years. Early satellites had a mass of about 40 kg and an on-board available power of a few watts, while the present day satellites weigh up to about 2000 kg (BOL) and can offer more than 2000 W. This trend is shown in Figure 1.16 (Koelle, 1984).

This continuous increase has been a result of three factors: technology advances, availability of larger launchers, and growing demand for satellite

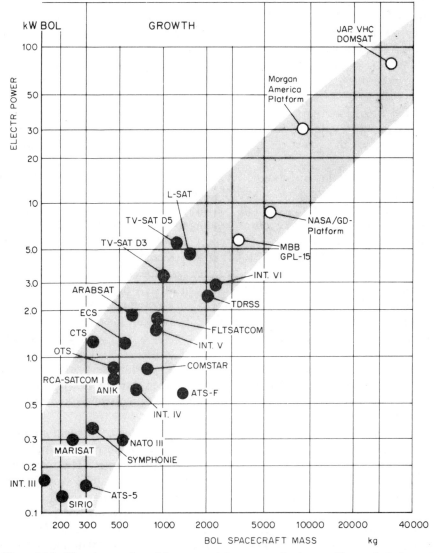

Figure 1.16   Historic growth and future trend for communication satellites and platforms in terms of mass and power. From Koelle and Kleinau (1980). *Copyright American Institute of Aeronautics and Astronautics and reprinted with permission of the AIAA*

services all over the world. As the demand grew it also diversified from trunk telephony, provided initially by international satellite networks, to new services (video communications, data transfer, direct broadcast) often based on digital facilities, and first offered within regional or domestic systems. The ground segment similarly underwent a change, from the large and expensive earth stations with 30-m diameter antennae to smaller and cheaper earth stations with antenna diameters in the range of 3 to 10 m for telecommunication applications, and 0.6 to 1.2 m for direct broadcast applications. As the organization of the satellite became more complex, earth stations simplified and increased in number.

What can we foresee in the future? The simplest approach to this 'guessing game' is to imagine the future as an extrapolation of the present trend. The extrapolation of present spacecraft configurations would lead to multiple payload satellites, offering a wide diversity of services, with mass in the range of 3000 to 5000 kg, launched by either the Shuttle or ARIANE 5. The present sate of the art of 10 years' design lifetime would probably be extended to twelve years.

A first step in this direction is illustrated by the European Space Agency OLYMPUS programme (also known as the L-SAT: Large Satellite) which consists of a body-fixed stabilized platform suitable for accommodation of payload masses up to 600 kg, corresponding to a total lift-off mass of 3500 kg, and delivering up to 6.5 kW of power (Bonhomme *et al*, 1984). Other examples are the MBB-GEOPLATFORM (Koelle, 1984) and the NASA geostationary communications platform programme (Ramler *et al*, 1984).

The above scenario is appealing as, even though the cost per spacecraft increases with size, the specific cost (cost per kg, or better per channel-year) decreases with growing mass. Market studies indicate that beyond 1995 the demand for such satellites should be in the order of about 20 to 40 per year. However, this scenario is constrained by two factors: the maximum capacity of the available launch vehicles, and the geostationary orbit crowding which will lead to intolerable interference between satellite systems. To face the ever increasing demand for more communication channels and services, new configurations must be envisaged.

Modular platform assemblies with automatic docking of payload modules: they consist of a common service module providing the power supply and other support functions for a number of payload modules. The total mass would be in the range 5000 to 10 000 kg. The payload modules would incorporate apogee propulsion, rendezvous and docking systems.

Large platforms with in-orbit assembly: such platforms are permanent stations with servicing and repair (modules exchange) operations. The mass ranges from 20 000 to 100 000 kg, span is of the order of 100 m and power is 30 to 250 kW. Figure 1.17 illustrates an example of such a large communications platform.

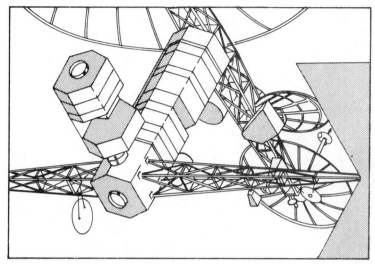

Figure 1.17 Large platform concept. *Reprinted with permission from Cuccia (1978). Copyright (1978) Pergaman Press Ltd.*

The advantages of geostationary platforms would be a higher capacity per orbital slot by combining the functions and operations provided by many satellite systems onto one orbiting platform, lower costs per unit of capacity, and improved communications networks as a result of interconnection facilities. However, a significant penalty of geostationary platforms is that, to achieve a small number of launches, it is necessary to put large payload capacities into orbit well in advance of the time they are needed. The penalty resulting from unused capacity may offset the economic benefits of scale. So the concept of clustering a number of satellites has been proposed as a substitute for geostationary platforms (Visher, 1979).

Cluster concept: it consists of placing a group of six to twelve geostationary satellites in orbit, connected via radiofrequency or optical intersatellite links to a central switching satellite, and spaced about 100 km apart. Each satellite would be equipped with a large diameter antenna producing several spot beams, with beamwidth of about $0.1°$, allowing transmission to the earth stations with low power transmitters. The links with the switching satellite would ensure the required interconnectivity between all satellites. The system would be deployed progressively through successive launches as the need arose. It could use relatively small satellites as a result of low power requirements, and in the case of a failure of one of them the faulty satellite could be replaced at a reasonably low cost. The main difficulties are: station keeping the individual satellites within the cluster as relative motion may entail the loss of antenna tracking and the shortage of the intersatellite links, and the implementation of the switching device in the central satellite.

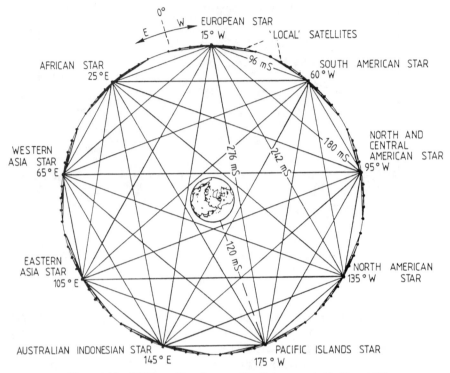

Figure 1.18   STAR satellites intersatellite link network (Golden, 1982).

Finally, a possible future integrated organization of the geostationary orbit has been proposed, based on a primary STAR network of nine satellites (or platforms) and an additional network composed of LOCAL satellites (Golden, 1982). All satellites would be equipped with intersatellite links providing a flexible interconnectivity. Figure 1.18 illustrates this rather futuristic concept of a 'wired sky'.

REFERENCES

AL-MASHAT, A. (1982) The Arab satellite communications system, *AIAA 9th CSSC*, San Diego, pp. 187–191.

ALPER, J. and J. N. PELTON. *The INTELSAT Global Satellite System*, Progress in Astronautics and Aeronautics **93**, 1984.

AMADESI, P. and D. DHARMADASA (1984) ECS-A: An evolution in the European regional satellite communication system, *GLOBECOM 84*, Atlanta, Paper 36.7.

BACKLUND, L. (1984) TELE-X, the first step in a satellite communication system for the Nordic countries, *AIAA 10th CSSC*, Orlando, pp. 299–309.

BARNLA, J. D. and F. R. ZITZMANN (1977) Digital communications satellite system of SBS, *IEEE Electronics and Aerospace System Conference*, Sept. 26–28, pp. 7.2a–7.21.

BENDEN, W. J., G. CAMPBELL, E. T. HARKLESS and T. F. MCMASTER (1982) Telstar 3 spacecraft design summary, *AIAA 9th CSSC*, San Diego, pp. 653–663.

BERRIDGE, B. M., and N. M. G. FREITAS (1983) The design of the Brazilian domestic satellite system, SCC83, Ottawa, pp. 20.5.1–20.5.4.

BLACHIER, B., J. BOUGUET, J. D'HOLLANDER and J. L. SOULA (1982) Telecom I payload, *AIAA 9th CSSC*, Orlando, pp. 456—474.

BONHOMME, R., B. L. HERDAN and R. STEELS (1984) Development and application of new technologies in the ESA Olympus programme, *AIAA 10th CSSC*, Orlando, pp. 249–262.

BRATAHALIM, S. and G. STEADY (1984) PALAPA B and beyond, *ICC84*, Amsterdam, pp. 83–88.

BRAUN, W. H. and J. E. KEIGLER (1984) RCA satellite networks: high technology and low user cost, *Proc. IEEE*, **72**, 1483–1505.

CLARKE, A. (1945) Extraterrestrial relays, *Wireless World*, Oct., pp. 305–308.

CUCCIA, C. L. (1978) The technology of large multifunction communication satellites in the post Intelsat V era. Astronautics for space and human progress. *29th Intern. Astron. Congress, Dubrovnik*, pp. 367–380.

DOAN, D. L., A. R. RAAB and R. ASTHON (1980) Design of Anik D communication antenna, *IEE Antenna and Propagation Int. Symposium*, Paper AP 3–6, 4 pp.

DONOVAN, R., R. KELLEY and K. SWIMM (1983) Evolution of the DSCS Phase III satellite through the 1990s, *ICC83*, Boston, pp. C1.6.1–1.6.9.

FREELING, R. and W. WEINRICH (1984) RCA advanced Satcom, the first all solid state communication satellite, *AIAA 10th CSSC*, Orlando, pp. 501–589.

GEORGY, J. (1984) The French broadcasting satellite system TDF 1, *ICC84*, Amsterdam, pp. 1084–1087.

GOLDEN, E. (1982) The wired sky, *AIAA, 9th CSSC*, San Diego, pp. 174–180.

HOLMES, W. M. (1978) The tracking and data relay satellite system, *AIAA, 7th CSSC*, San Diego, Paper 78–554, 6 pp.

HOWELL, T. F. (1980) Communications mission and system aspects of European regional satellite system, *ESA Journal*, **4**, 227–246.

JACOBS, A. and B. POURMAND (1984) A digital satellite communications system, *AIAA 10th CSSC*, Orlando, pp. 722–730.

KOELLE, D. E. (1984) GEO space platform economics, *AIAA 10th CSSC*, Orlando, pp. 241–248.

KOELLE, D. E. and W. KLEINAU (1980) A third generation communication satellite concept, *AIAA 8th CSSC*, Orlando, Paper 80–0505.

LANDON, R. B. and H. G. RAYMOND (1982) Ku-band satellite communication via TDRSS, *AIAA 9th CSSC*, San Diego, pp. 741–756.

LESTER, R. M. (1980) Telesat Canada plans for new satellite systems, *J. Spacecraft*, **17** (2), 75–78.

LOMBARD, D., F. RANCY and D. ROUFFET (1983) Telecom 1, a national communication satellite for domestic and business services, SCC83, Ottawa, pp. 17.4.1–17.4.4.

LUDWIG, L. G. (1980) Satellite system for direct broadcast of television, *IEEE Eascon 80*, pp. 103–106.

LUNDBERG, O. (1984) The INMARSAT system and its future, *Space Communications and Broadcasting*, **2**(3), pp. 215–227.

MARCONICCHIO, F. and F. VALDONI (1983) The Italsat programme, *Space Communications and Broadcasting*, **1**(2), pp. 199–204.

NAKAMURA, Y. (1984) The operational broadcasting satellite system for Japan, *ICC84*, Amsterdam, pp. 1094–1097.

NICKELSON, R. L. (1984) Domestic satellite communications overview, *Space Communications and Broadcasting*, **2**(3), pp. 205–214.

NOWLAND, W. L. (1983) AUSSAT—A milestone in Australia's communication history, *Space Communications and Broadcasting*, **1**(1), pp. 73–89.

PAYET, G., J. C. RAISON and D. DHARMADASA (1984) The future of the Eutelsat system, *ICC84*, Amsterdam, Paper 3.8

RAMLER, J., R. DURRETT and G. C. MARSHALL (1984) NASA's geostationary communications platform program, *AIAA 10th CSSC*, Orlando, pp. 613–621.

REED, A. G. (1983) Unisat—UK direct TV broadcasting satellite, *SCC83*, Ottawa, pp. 17.5.1–17.5.2.

ROMERO, D. D. (1982) SATCOL—A domestic satellite for Colombia, *AIAA 9th CSSC*, San Diego, pp. 429–435

RUSCH, R. J. and D. G. DWYRE (1978) Intelsat V spacecraft design, *Acta Astronautica*, **5**, pp. 173–188.

RUSCH, R. J. and C. L. CUCCIA (1980) A projection of the development of high capacity communications satellites in the 1980's, *AIAA, 8th CSSC*, Orlando, Paper 80–0544, 6 pp.

SANCHEZ-RUIZ, M. E., and B. R. ELBERT (1984) Mexico's first domestic satellite, *AIAA 10th CSSC*, Orlando, pp. 310–318.

SCHINDLER, O. (1984) The German communications satellite system, *ICC84*, Amsterdam (oral presentation).

SCHNIPPER, H. (1980) The SBS system and services, *IEEE Communications Magazine*, **18**(5), pp. 12–15.

SION, E. (1978) Hughes domestic communicaitons satellite systems, *Acta Astronautica*, **5**, pp. 189–218.

SIVO, J. (1983) Advanced Communications Satellite Systems, *IEEE J. Selected Areas in Communication*, **1**, pp. 580–588.

SMALLEY, A. R. (1983) Overview of ANIK D satellite and services, *SCC83*, Ottawa, pp. 7.2.1–7.2.4.

SMART, F. H. (1983) Overview of the ANIK C satellite and services, *SCC83*, Ottawa, pp. 7.1.1–7.1.5.

STAMMINGER, R. and J. A. STEIN (1980) Business satellite developments, *IEE Eascon 80*, pp. 95–101.

TANAKA, R. (1984) 30/20 GHz domestic satellite communication system in the public communication network in Japan: design and operation, *Proc. IEEE*, **72**, pp. 1637–1644.

THUE, M. (1981) *Radiocommunications spatiales*, Cours ENST Option TSA, Toulouse.

VAN TREES, H. L. (1979) Satellite communications, *IEEE* Press Wiley.

VISHER, P. S. (1979) Satellite clusters, *Satellite Communications*, **3**(9), pp. 22–27.

WAYLASS, C. J. (1982) The Spacenet satellite, *AIAA 9th CSSC*, San Diego, pp. 730–735.

CHAPTER 2   Communication techniques

This chapter deals with transmission of information between the satellite and earth stations in the context of geostationary satellites, and examines the essential aspects of telecommunications such as the link budget, modulation techniques and characteristics of the transmission medium.

2.1   LINK BUDGET

The link budget is a calculation of the signal/noise power ratio at the receiving side of a transmission link considering the transmission medium and the transmitter/receiver characteristics. This calculation is achieved by totalling the signal power gains and losses and taking into consideration the contribution of noise at the receiver input.

2.1.1   Received signal power

The received signal power is a function of the transmitted power, taking into account the distance between transmitter and receiver and the characteristics of the transmitting and receiving antennae.

2.1.1.1   Fundamentals

Figure 2.1 illustrates the link. If the transmitter T_x were equipped with an isotropic antenna, the transmitted power P_T from a sphere with centre T_x and radius R, would be evenly distributed across the sphere. The power flux would then be equal to

$$\frac{P_T}{4\pi R^2}$$

As the transmitting antenna has a gain of G_T in the receiver direction, the power flux density at the receiver is given by

$$\frac{G_T P_T}{4\pi R^2}$$

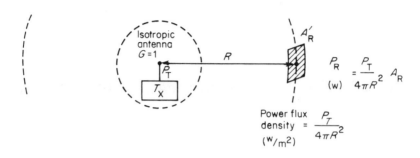

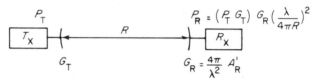

Figure 2.1　Link parameters

The product, $G_T P_T$, is the *Equivalent Isotropic Radiated Power* (EIRP) in the absence of transmission losses.

The receiving antenna, with an effective aperture area of A_R', captures a signal power equal to

$$\left(\frac{G_T P_T}{4\pi R^2}\right) A_R'$$

Therefore the power at the receiver input is:

$$P_R = P_T G_T \frac{A_R'}{4\pi R^2} \tag{1}$$

The receiving antenna gain G_R is related to the equivalent surface area A_R' by the relationship;

$$G_R = \frac{4\pi A_R'}{\lambda^2} \tag{2}$$

where λ is the wavelength. The ratio of the received power to the transmitted power is hence given by:

$$\frac{P_R}{P_T} = G_T G_R \frac{\lambda^2}{(4\pi)^2 R^2} \tag{3}$$

The power attenuation expressed in decibels is $a_{dB} = 10 \log_{10} P_T/P_R$, therefore, we get:

$$a_{dB} = 22 + 20 \log_{10}\left(\frac{R}{\lambda}\right) - G_T - G_R \tag{4}$$

where G_T and G_R represent the antenna gains in decibels.

Attenuation of the signal against frequency varies such that:

$$L_{FS} = 22 + 20 \log_{10}\left(\frac{R}{\lambda}\right) \qquad (5)$$

which is the 'free space loss', (expressed in dB) that is the signal power attenuation between two isotropic antennae in free space. Free space loss varies with frequency: the higher the frequency, the higher the free space loss. However, this increased loss is compensated by the increase in the antenna gain with increased frequency, for a given antenna aperture area. For a geostationary satellite free space loss ranges from 195 to 213 dB for frequencies between 4 and 30 GHz assuming a subsatellite earth station.

2.1.1.2 Real situation

Expressions (3) and (4) are theoretical and apply to an ideal situation; in practice, various losses must be taken into account as follows. Expression (3) should be written as:

$$P_R/P_T = G_T G_R/L \qquad (6)$$

where L expresses the total loss: free space loss L_{FS} and additional losses L_A.

$$L = L_{FS} \times L_A$$

where

$$L_A = L_{FTX} \times A_{AG} \times A_{RAIN} \times L_{POL} \times L_{POINT} \times L_{FRX} \qquad (7)$$

L_{FTX} represents losses between the transmitter output and the transmitting antenna (wiring, duplexers, filters, etc.),

A_{AG} represents attenuation by the atmosphere and ionosphere,

A_{RAIN} represents attenuation due to precipitations and clouds,

L_{POL} represents losses caused by polarization mismatch between the transmitting and receiving antennae (See Table 2.1). For example, if the wave polarization on the transmitting antenna axis is circular, it becomes increasingly elliptical towards the limit of the main lobe of the antenna corresponding to the 3 dB beamwidth,

L_{POINT} represents losses caused by antenna depointing (earth station near the coverage boundary, errors in pointing, misalignment of the radio-electrical axis with the geometrical axis, imperfect satellite stabilization, etc.)

L_{FRX} represents losses between the receiving antenna and the receiver input (wiring, duplexer, etc).

2.1.1.3 Antenna efficiency—gain and beamwidth

The gain of an antenna (gain relative to an isotropic antenna) is expressed in terms of the actual surface area A by the equation:

$$G = \frac{4\pi\eta A}{\lambda^2} \qquad (8)$$

where η is the efficiency of the antenna. Sixty per cent is usual, and good antennae can achieve 70 per cent. A is the actual surface area of the antenna, and the product ηA corresponds to the effective aperture A' in formula (1). If the antenna is circular, with diameter D, the gain is expressed as follows:

$$G = \eta \left(\frac{\pi D}{\lambda}\right)^2 \qquad (9)$$

The above expression represents the antenna gain in its radioelectric axis direction. The gain is then maximum and equal to $G = G_{\text{MAX}}$ as given by (9). Figure 2.2 shows the variation of gain when one departs from the axis by an angle α. For small depointing angles, the gain is given by

$$G(\alpha)_{\text{dB}} = G_{\text{MAX,dB}} - 12\left(\frac{\alpha}{\theta_{\text{3dB}}}\right)^2 \qquad (10)$$

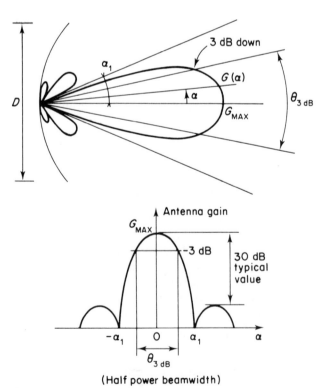

(Half power beamwidth)

Figure 2.2   Antenna radiation pattern.

Table 2.1 LOSSES CAUSED BY POLARIZATION MISMATCH L_{POL}.

			Linear	Circulaire Left	Right
				Circulaire	
			Linear	Left	Right
		Linear	$20 \log \cos \psi(*)$	3 dB	3 dB
Polarization of the received wave		Left	3 dB	0	∞
	Circulaire	Right	3 dB	∞	0

Column "Polarization of the receiving antenna" spans Linear/Left/Right.

* ψ is the angle of the two directions of polarization.

where θ_{3dB} is the halfpower beamwidth of the antenna. It corresponds to a 3 dB drop-off in the antenna gain as a result of a depointing angle α equal to $\theta_{3dB}/2$. θ_{3dB} in degrees is given by $\theta_{3dB} = 70(\lambda/D)$.

With an antenna giving a 65 per cent efficiency the gain is;

$$G = \frac{0.65\pi^2(70)^2}{\theta_{3dB}^2} = \frac{32\,000}{\theta_{3dB}^2}$$

Hence the smaller θ_{3dB}, the higher the gain.

2.1.1.4 Example

Assuming a link between a geostationary satellite and an earth station [35 786 km at a frequency of 12 GHz ($\lambda = 2.5$ cm)]; applying the following

Table 2.2 POWER LINK BUDGET.

		Positive factors	Negative factors
Power required at receiver input	P_R		-116 dBW
Receiver losses	L_{FRX}	0 dB	
Receiving antenna gain	G_R		-40 dB
Polarization losses	L_{POL}	1 dB	
Free space loss	$L_{FS} = 22 + 20 \log D/\lambda$	206 dB	
Depointing loss (total T_x/R_x)	L_{point}	3 dB	
Transmitting antenna gain	G_p		-42 dB
Transmitter losses	L_{FTX}	1 dB	
Total		211 dB	-198 dBW
Transmitter power			$+13$ dBW (20 W)

antenna characteristics:

  60 per cent efficiency
  1.3 m diameter transmitting dish 60 (42 dB gain);
  1 m receiving dish (40 dB gain);

and assuming that with the selected modulation technique, the service quality asked for requires that the carrier power at the receiver input be -116 dBW.

The link budget for the signal power can be established following the guidelines of Table 2.2 and the above equations. The goal is to calculate the required transmitter power, given in decibels by:

$$P_T = P_R - G_E - G_R + L$$

2.1.2   Noise temperature

The principal role of the receiver is to amplify the signal to a usable level. Receiver noise, external background noise and interference from other transmitters corrupt this amplified signal.

2.1.2.1   Receiver noise temperature

Noise generated by the receiver is characterized by its *noise figure* or its *effective input noise temperature*.

The *noise figure* is the ratio of the maximum usable noise power at the output of the receiver N_{out} to the maximum noise power that there would be if there were no noise source other than the generator connected to the receiver input at standard reference temperature $T_0 = 290$ K, i.e. GkT_0B_{IF}, where G is the maximum usable power gain of the receiver, B_{IF} is the *noise equivalent bandwidth of the receiver* (considered to be the bandwidth of the intermediate frequency amplifier) and k the Boltzmann constant ($k = 1.38 \times 10^{-23}$ J K^{-1} and $10 \log k = -228.6$ dBW K^{-1} Hz^{-1}).

$$F = \frac{N_{out}}{GkT_0B_{IF}} \tag{11}$$

The noise figure, ratio of two power terms, can be expressed in decibels. As it is inconvenient to operate at noise figures of less than 3 dB, for low noise figures it is better to consider the *effective input noise temperature* of the receiver, T_R. This is the noise temperature of a source at the input of a *noiseless* receiver, which would produce the same contribution to the receiver output noise as the internal noise of the actual system itself.

If the actual source has a noise temperature of T_0 at the input, the maximum noise power at the output is a given by

$$N_{out} = GkT_0B_{IF} + GkT_RB_{IF} = Gk(T_0 + T_R)B_{IF}$$

Table 2.3 RELATION BETWEEN EFFECTIVE INPUT NOISE TEMPERATURE AND NOISE FIGURE.

T_R(K)	7	35	75	290	865	2610	28 710
F(dB)	0.1	0.5	1	3	6	10	20

which gives the following relation between F and T_R

$$F = 1 + \frac{T_R}{T_0} \tag{12}$$

Table 2.3 shows the relation between effective input noise temperature and noise figure. As a first approximation at low noise temperatures, (less than 100 K) $T_R = 70 F_{dB}$.

Earth station receivers have noise temperatures between 10 and 200 K, and receivers on satellites have noise temperatures in the order of 1000 K.

2.1.2.2 Antenna noise temperature

All matter emits radiant energy. When picked up by an antenna, this radiation is superimposed on the usable signal as background noise. If N_0 is the power spectral density of such noise (expressed in watt/Hz) the antenna temperature T_A (expressed in kelvin) is such that:

$$N_0 = kT_A$$

where k is the Boltzmann constant.

The antenna temperature is affected by:

(1) The temperature and absorptance of external radiators.
(2) The antenna gain and its orientation relative to these external radiators.

A blackbody absorbs all the radiant energy that falls upon it, and hence its radiant energy depends solely on its temperature at a given frequency. With real radiators the energy received is partly reflected and the radiant energy is therefore lower than that of a blackbody of the same size and the same temperature. Therefore, one considers for the radiator an effective noise temperature, lower than the actual radiator temperature, defined as that of a blackbody of the same size which would radiate with the same power spectral density in the given frequency band.

When a receiving antenna is pointed towards a satellite, there are two main sources of noise to be considered:

(1) Background noise from the sky.
(2) Noise from the Earth's own radiation.

(1) *Background sky noise* The main contribution to background sky noise above 2 GHz comes from the non-ionized region of the atmosphere, which being an absorbent medium, is a noise source. Figure 2.3 illustrates the

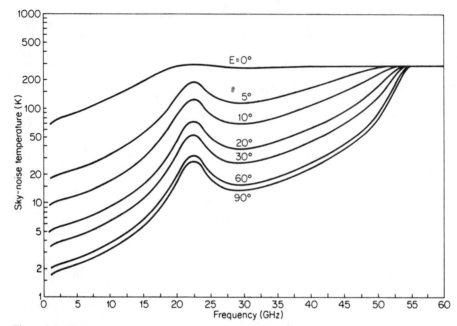

Figure 2.3 Sky noise temperature for clear air and 7.5 g/m³ of water vapour concentration (E is the elevation angle). *Reproduced with permission from CCIR (1982).*

clear sky noise temperature versus frequency assuming an infinitely narrow beam at various elevation angles E, a ground temperature of 20 °C and 7.5 g/m³ of water vapour concentration (CCIR, Vol. V, Rep. 720-1, 1982). The 2.7 K background cosmic noise contribution has not been included. With elevation angles higher than 5° the total sky noise in clear weather is less than 40 K for frequencies between 1 and 15 GHz, which are presently the most used frequencies for commercial satellite communications. The increase in noise temperature between 20 and 30 GHz, should be particularly noted. The noise due to absorption in rain and cloud must be added to the clear sky noise temperature (see section 2.2). The sky noise due to extraterrestrial sources contributes also to the total noise. For geostationary satellites the Sun is of particular importance. Its noise temperature (quiet sun) ranges from 5×10^4 K at 4 GHz to 10^4 K at 12 GHz.

Considering a radiofrequency source of apparent angular diameter α with noise temperature T_0 within the given frequency band, the additional noise temperature T, for an antenna having an equivalent noise beamwidth θ is given by:

$$T = T_0 \alpha^2/\theta^2 \quad \text{if } \theta > \alpha \quad \text{and}$$
$$T = T_0 \quad\quad\quad \text{if } \theta \leqslant \alpha \tag{13}$$

For example, with a 2 m diameter antenna, θ is about 1° at 12 GHz.

Pointing towards the Sun ($T = 10^4$ K and $\alpha = 32$ minutes), the additional temperature is about 2×10^3 K. If it were pointed towards the Moon ($T = 200$ K and $\alpha = 32$ minutes) the temperature would be 50 K. Contributions from other extraterrestrial sources can be ignored for geostationary satellites. The Sun appears, therefore, to be a major disturbing source. The times when Sun–Satellite conjunction occurs is discussed in Chapter 4.

(2) *Noise from Earth radiation* Although the earth station antenna normally points skywards, noise from the Earth contributes to the overall antenna noise temperature as it introduces itself through all side and back lobes. The noise temperature of Earth is approximately equal to its actual temperature (about 290 K). The earth contribution to antenna noise temperature depends on the elevation of the side and back lobes. Each lobe's contribution is determined by $T = G(d\Omega/4\pi)T_G$, where G is the average gain of the lobe with a total beamwidth of $d\Omega$ and T_G is the ground temperature. Practical values range from 10 K for large Cassegrain reflector antennae to 100 K for small dish antennae.

2.1.2.3 Total noise temperature at receiver input

Assume a receiver of effective input noise temperature T_R, linked to an antenna with noise temperature T_A by a circuit with an attenuation L at physical temperature θ_L.

The total noise temperature at receiver input is given by:

$$T = T_A/L + \theta_L(1 - 1/L) + T_R \tag{14}$$

The total noise power N to take into account therefore is:

$$N = kTB_{IF} \tag{15}$$

where k is the Boltzmann constant and B_{IF} is the system's noise equivalent bandwidth as determined by the bandwidth of the intermediate frequency amplifier. As can be seen the feeder connecting the antenna to the receiver is of great concern, especially with low noise earth stations, as every tenth of a decibel attenuation represents not only a signal loss, but also contributes a 7 K increase in noise temperature.

2.1.2.3.1 Earth station receivers Earth stations with large antennae display an antenna noise temperature T_A of about 30 K. It is therefore desirable to have receiver noise temperatures of the same order. This is best achieved by using parametric amplifiers. Figure 2.4 illustrates typical equivalent noise temperatures of various devices.

2.1.2.3.2 Satellite receivers The receiving antennae on board satellites have beamwidths less than or equal to the angle at which the earth is viewed from

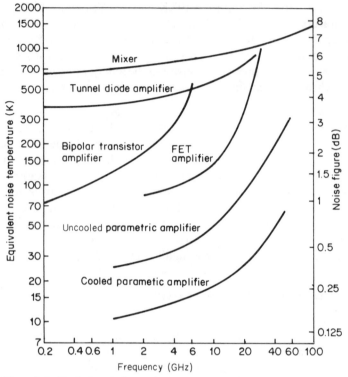

Figure 2.4   Typical equivalent noise temperature and noise figure of various devices

the satellite (17.4° for a geostationary satellite) and the antenna noise temperature is then equal to the noise temperature of the Earth in the given frequency range, that is of about 290 K.

2.1.2.4   Figure of merit of a receiving station

Earth and satellite receiving stations are often described in terms of their figure of merit G/T (dB K^{-1}). This is the ratio of antenna power gain to the total noise temperature at the receiver input. For example, the G/T of the INTELSAT V satellite receiver with global coverage using a 6/4 GHz link is -18.6 dB K^{-1}, and for standard A earth receiving stations the figure of merit is better than 40.7 dB K^{-1} at 4 GHz (for a more complete description of INTELSAT earth station standards, see Chapter 8).

2.1.3   Carrier-to-noise power ratio at receiver input

The ratio of received carrier power C, to the noise power N received within the receiver bandwidth, (B_{IF} being the noise equivalent bandwidth), is obtained from formulae (6) and (15):

$$C/N = P_T G_T G_R / L k T B_{IF} \qquad (16)$$

$N_0 = N/B_{IF}$ is the noise power spectral density (W/Hz)

$$C/N_0 = (P_T G_T)(\lambda/4\pi R)^2 \times (G_R/T) \times 1/k \times 1/L_A \qquad (17)$$

where L_A corresponds to the additional losses not considered as free space loss, $k = 1.38 \times 10^{-23}$ JK^{-1}.

Expressed in dB, the carrier power to noise power spectral density ratio is given by:

$$\left(\frac{C}{N_0}\right)_{dBHz} = 10 \log P_T G_T - 20 \log \frac{4\pi R}{\lambda} + 10 \log \frac{G_R}{T} - 10 \log L_A - 10 \log k \qquad (18)$$

EIRP (dBW)	Free space loss (dB)	Figure of merit of receiving station (dBK^{-1})	Additional losses (dB)	-228.6 dBW K^{-1} Hz^{-1}

2.2   INFLUENCE OF THE PROPAGATION MEDIUM

In free space, spherical radio waves only undergo free space loss as defined in section 2.1.1.1. Atmospheric and ionospheric propagation are subject to the following phenomena: absorption; diffusion (or diffraction); refraction; rotation of the polarization plane of the electromagnetic wave.

These phenomena depend on the path length, so the effects are greater for small elevation angles. Absorption and diffusion are mainly caused by the lower layers of atmosphere. The absorption of radio wave energy by atmosphere goes along with a specific emission which results in a noise power increase at the receiving antenna. Refraction is generated in the upper layers of atmosphere (troposphere). Depolarization occurs when the radiowave travels through the ionosphere, or within the atmosphere when raining.

2.2.1   Atmospheric effects

The atmosphere has a number of 'radio windows' where attenuation of radio waves is small. Transmission is particularly good at frequencies below 10 GHz (3 cm). Absorption by atmospheric gases is caused by the existence of absorption lines of oxygen and water vapour in the atmosphere. Oxygen has an isolated absorption line at 118.74 GHz and a series of close lines between 50 and 70 GHz which act as a continuous absorption band. Waver vapour has three absorption lines at the frequencies 22.3 GHz, 183.3 GHz and 323.8 GHz.

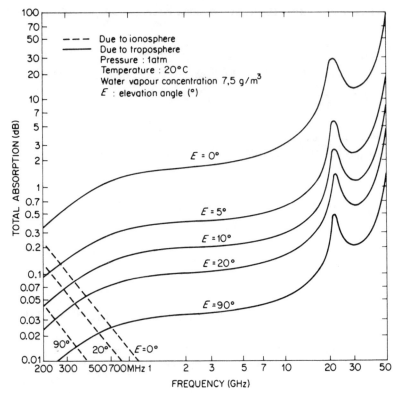

Figure 2.5   Total attenuation of radio waves by atmospheric gases versus frequency
for various elevation angles, E

Figure 2.5 shows the total attenuation by atmospheric gases versus
frequency for various elevation angles.

Absorption caused by precipitation Absorption by rain depends on
frequency, rainfall rate, diameter and distribution of raindrops. Given a rainfall rate exceeded for a required percentage of time, the resulting attenuation
of radio waves A_{RAIN} can be estimated by multiplying the specific attenuation
due to rain, γ_R expressed in terms of dB/km, by the effective path length, L_e
expressed in km:

$$A_{RAIN} = \gamma_R L_e$$

In the absence of experimental data, statistical characteristics of rainfall
intensity which is exceeded at a given point for a given percentage of an
average year can be obtained from maps presented in the CCIR Report 564–2,
1982. Figure 2.6 from the CCIR, Report 563–1, 1978, represents a simplified
map of rain climatic zones where only five zones have been considered. Figure
2.7 displays, for each one of these zones, the rain intensity in terms of

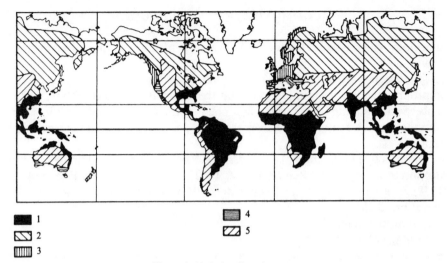

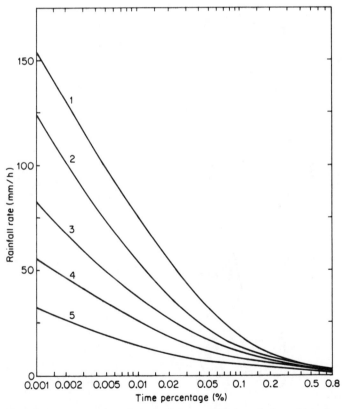

Figure 2.6 Rain climatic zones

Figure 2.7 Percentage of time of an average year during which the
indicated rainfall rate is exceeded for each of the zones of
Figure 2.6

mm/hour which is exceeded for the indicated percentage of time of an average year.

The specific attenuation γ_R is obtained from Figure 2.8, by entering the appropriate values of rain intensity and frequency (CCIR, Rep 721–1, 1982). This specific attenuation is to be multiplied by the effective path length to obtain the attenuation of radio waves exceeded for the considered percentage of time.

The effective path length, L_e expressed in km, is estimated using the method proposed in the CCIR, Report 564–2, 1982. Figure 2.9 (CCIR, Rep. 564–1, 1978) gives a rapid estimation, which depends only on the rainfall rate and the elevation angle.

System planning often requires the attenuation value exceeded for the time percentage of any month, that is the worst month, of a year. The relationship between the time percentage of an average year and that of the worst month is not an obvious one. An approximate relation has been proposed (CCIR,

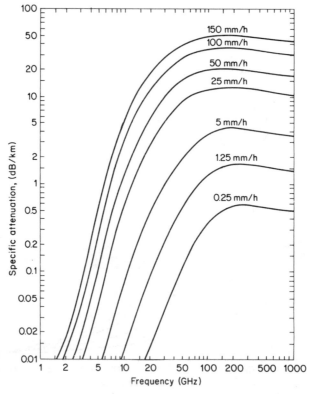

Figure 2.8   Specific attenuation due to rain. *Reproduced with permission from Olsen* et al (1978). Copyright © (1978) IEEE.

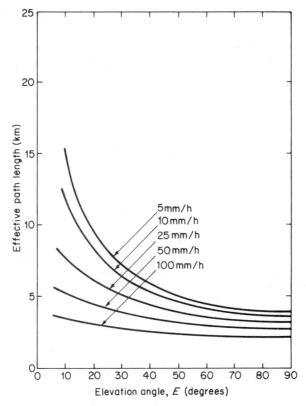

Figure 2.9 Effective path length. *Reproduced with permission from CCIR (1978).*

Rep 564–2, 1982):

$$p = 0.29\, p_w^{1.15}$$

where p represents the annual time percentage, and p_w is the worst month time percentage. The attenuation exceeded for the annual time percentage p, calculated above, may then be taken as the attenuation value exceeded for p_w per cent of the worst month.

Diffusion by precipitation can cause coupling between the signals transmitted by the antennae of earth stations and terrestrial microwave stations, and thus plays a major role in the reciprocal interference of the two services.

Contribution to sky noise temperature The clear sky noise temperature displayed in Figure 2.3 only includes the contribution of atmospheric gases. Rainfall acts as an additive source of noise related to the attenuation according to the following:

$$T_{rain} = T_m(1 - 1/A_{RAIN})$$

where A_{RAIN} is the rainfall attenuation and T_m is the effective temperature of the attenuative medium. T_m may be taken as 260 K (CCIR, Rep. 564–2, 1982) or calculated from the empiric expression

$$T_m = 1.12\ T_{amb} - 50\ K$$

which is sometimes adopted to allow use of the readily available ambient temperature, T_{amb}, at the terminal location (Arnbak, 1983).

Tropospheric scintillation and refraction   Diminution of the refraction index with altitude involves an increase in the apparent elevation angle of the radio path from the satellite. In addition, the fluctuation in the refraction index involves variations in the apparent position of a satellite as seen from an earth station. For an elevation angle of 5°, the shift is 0.2°, and it varies very little in terms of the water vapour concentration (0.04°). It is negligible for frequencies below 30 GHz and for small antennae.

Transpolarization due to precipitation   Transmitted energy with a given polarization transforms partly into orthogonal polarization during its travel through the atmosphere. This phenomenon causes a quality loss due to interference between links where frequency is reused by orthogonal polarization to increase link capacity without requiring more bandwidth (Kobayashi 1977; CCIR 722–1, 1982).

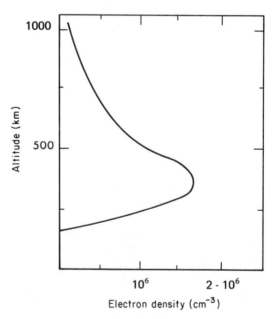

Figure 2.10   Vertical distribution of the electron density within the ionosphere

2.2.2 Ionospheric effects

The ionosphere is a zone of high density of electrons (Figure 2.10) which extends over an altitude of 80–1000 km. Ionization in the upper atmosphere is caused by solar radiation, so the density of electrons is stronger by day, and during periods of great solar activity (Figure 2.11).
Radio waves in the ionosphere are subject to:

(1) Attenuation.
(2) Refraction.
(3) Depolarization (Faraday effect: rotation is inversely proportional to the square of frequency).
(4) Propagation delay variation (delay is inversely proportional to the square of frequency).

These effects are generally negligible above 2 GHz.

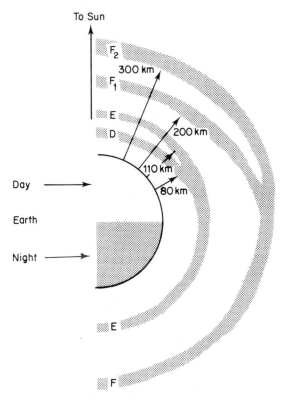

Figure 2.11   Regions of the ionosphere

2.2.3 Conclusion

Atmosphere has a small effect on the link quality at frequencies between 2 GHz and 10 GHz and for high elevation angles. At higher frequencies, the major effects are absorption by atmospheric gases and especially rain attenuation. The signal/noise power ratio degrades due to the combined effects of attenuation of the signal and increase of the noise power.

At lower frequencies, ionospheric effects play a leading part as they are frequency dependent according to a inverse square law.

As the radio wave path-length through the atmosphere is higher for a slant path, the above effects are more pronounced at small elevation angles of the earth station antenna. A commonly adopted minimum value of the elevation angle is therefore 5 to $10°$.

2.3 SIGNAL PROCESSING

The transfer of information is achieved by transmission of signals which can be digital (telegraphy, data, etc.) or analogue (radio, television, telephone, etc.). However, analogue signals may be converted into digital form before transmission. It is assumed that the original baseband signal has a bandwidth of 0 to f_m.

Transmission of baseband signals requires modulation of a carrier. This is achieved by altering one of the carrier parameters (amplitude, frequency, phase). In analogue modulation the selected parameter is changed in accordance with baseband signal amplitude. Digital modulation is achieved by altering the selected parameter in a binary or M-ary manner (M-ary modulation).

Many baseband/modulation combinations are used. Typical examples are given below:

Single Channel Per Carrier/Frequency Modulation (SCPC/FM)   The carrier is frequency modulated by one voice or TV channel. With voice channel, use is often made of compandors to reduce non-linear distortion and to compensate for signal level difference between loud and soft talkers. Voice activation is a mean to achieve a bandwidth/power advantage and consists of turning off the carrier during pauses in speech.

Frequency Division Multiplex/Frequency Modulation (FDM/FM)   Several channels are combined within the total available baseband signal bandwidth in such a way that each channel occupies a given frequency sub-band. The resulting multiplexed signal is the baseband signal used to frequency modulate the carrier. Use of companding and speech interpolation for FDM voice channels on satellite links has been shown to increase capacity. Speech interpolation consists in assigning the same channel to different speakers on a voice activated basis, thus allowing a larger number of conversations to be routed on a given satellite link.

Single Channel Per Carrier/Phase Shift Keying Modulation (SCPC/PSK) The carrier is phase modulated by digital information. The digital information is original data or a quantized form of analogue signals.

Time Division Multiplex/Phase Shift Keying Modulation (TDM/PSK) Several channels are combined on a time sharing basis. Each channel occupies a time slot within a time frame and occupies the total available bandwidth within its time slot. Analogue signals are quantized prior to transmission and the resulting multiplex consists of a stream of bits used to modulate the carrier. Digital speech interpolation (DSI), a digital version of speech interpolation techniques, is often used to provide a capacity gain of the order of 2.

At the receiver, the carrier is demodulated, and the individual signals are eventually recovered from the multiplexed baseband signal. The demodulator performance and the related service quality depend on the carrier-to-noise power ratio at the demodulator input, i.e. at the output of the IF amplifier, as obtained from (17) or (18).

2.3.1 Transmission of analogue signals

The service quality is defined by the signal/noise power ratio after demodulation: S/N.

At the intermediate frequency amplifier output, before demodulation, the modulated signal occupies a frequency band B_{IF} which depends on f_m and the modulation technique used. The carrier/noise power ratio is C/N or C/N_0 (Hz) as given by (17) or (18).

With satellite communications, the modulation scheme which is most used is frequency modulation (FM), although a recent trend indicates the return of amplitude modulation, with its derived single side band (SSB) application, which allows for narrower bandwidth operation than does FM (Brown *et al.*, 1983; Laborde and Freedenberg, 1984). However SSB does not offer the bandwidth power tradeoff that FM does.

In frequency modulation the mean power of a modulated carrier is constant and equal to the carrier power C; The modulated carrier occupies a bandwidth equal to $2f_m(m_f + 1)$ according to Carson's rule, where m_f is the modulation index (ratio of the peak frequency deviation to the maximum baseband signal frequency, f_m). To achieve appropriate matching between the modulated carrier bandwidth and the receiver bandwidth, B_{IF} is defined as

$$B_{IF} = 2f_m(m_f + 1) \qquad (19)$$

If $m(t)$ is the baseband signal, then

$$S/N = 3m_f^2 C/N_0 f_m E\{m^2(t)\} \tag{20}$$

where $E[m^2(t)]$ represents the average power of the baseband signal.

When the modulating signal is sinusoidal, $m(t) = \cos \omega_m(t)$, and $E[m^2(t)] = 1/2$. Then

$$S/N = C/N_0 f_m \cdot 3/2 \cdot m_f^2 \tag{21}$$

and so

$$S/N = 3(m_f + 1)m_f^2 C/N \tag{22}$$

The curves giving S/N in terms of C/N ($N = N_0 B_{IF}$) are shown in Figure 2.12 for a modulation index between 1 and 20. Below threshold (C/N around 10 dB with conventional demodulators, 6 dB with extended threshold demodulators) S/N decreases rapidly with C/N.

2.3.1.1 Single Channel Per Carrier/Frequency Modulation (SCPC/FM) telephony

2.3.1.1.1 Characteristics of a telephone channel. A telephone channel

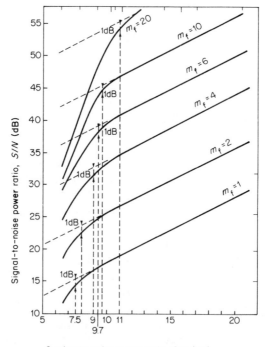

Figure 2.12   Signal-to-noise power ratio after frequency demodulation versus carrier-to-noise power ratio for various modulation index

occupies a bandwidth from 300 to 3400 Hz. The maximum amount of energy transmitted is around 800 Hz and 99 per cent of the energy is below 3000 Hz.

The average power level of a telephone channel measured at a point of zero relative level (a point where the test signal, which consists of a test tone at 800 Hz, has a power of 1 mW when measured in a 600 Ω load) is considered to be equal to − 15 dBm0 (CCITT, Rec. G−223) assuming that the channel is active on the average, 0.25 of the time.

2.3.1.1.2   Performance objectives (CCIR, *Rec. 353−4 1982*)   Noise power at a point of zero relative level in any telephone channel should not exceed:

10 000 pW0p, psophometrically weighted, 1 minute mean power, for more than 20 per cent of any month.
50 000 pW0p, psophometrically weighted, 1 minute mean power, for more than 0.3 per cent of any month.
1 000 000 pW0, unweighted (with an integrating time of 5 ms) for more than 0.01 per cent of any year.

'Psophometrically weighted power' means that the noise power has been measured using a psophometric filter which takes into account the filtering effect of the human ear on noise. Figure 2.13 represents the psophometric filter gain and shows the resulting reduction of noise power, and the related improvement on signal/noise ratio.

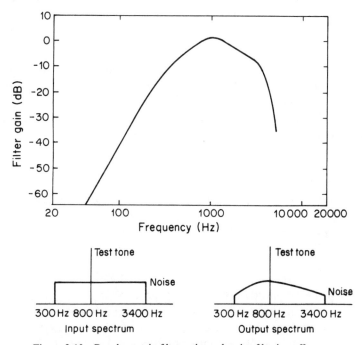

Figure 2.13   Psophometric filter gain and noise filtering effect

Using equation (21) and considering the test signal as the modulating signal, the noise power expressed in pW0p at a point of zero relative level is given by:

$$N_{pW0p} = 10^9 \left(\frac{S}{N}\right)^{-1} \qquad (23)$$

where:

$$S/N = 3\,[(\Delta F_r)^2/(f_{max}^3 - f_{min}^3)]\,pw(C/N_0)$$

ΔF_r = r.m.s. frequency deviation due to test tone
f_{max} = upper telephone channel frequency = 3400 Hz
f_{min} = lower telephone channel frequency = 300 Hz
p = improvement by preemphasis (4 dB) and companding, if any
w = psophometric weighting factor (2.5 dB)

The required radiofrequency bandwidth is given by:

$$B_{IF} = 2(\Delta F_p + f_{max})$$

where:

ΔF_p = peak frequency deviation (Hz) = $\Delta F_r g L$
L = loading factor, equal to the ratio of the r.m.s. frequency deviation produced by an active speech signal to the r.m.s. frequency deviation ΔF_r produced by the test tone = 0.25
g = 'peak to r.m.s.' factor = 12.6 (assuming 3% of the users suffer clipping)

2.3.1.2   Frequency Division Multiplex/Frequency Modulation (FDM/FM)

2.3.1.2.1   Characteristics of the frequency multiplex baseband signal.
Frequency multiplex is the combining of separate telephone channels in the frequency spectrum, with a 4 kHz bandwidth reserved for each telephone channel. Table 2.4 gives the capacity and bandwidth of telephony multiplex used for satellite communications.

For a large number of telephone channels amplitude distribution depends on:

(1) The number of telephone channels—n.
(2) The average power level of a telephone channel (-15 dBm0).

The average power S_{av} of a multichannel telephony frequency division multiplex, for n large enough, is given by:

$$10 \log S_{av} = -15 + 10 \log n \text{ (dBm0)} \qquad (24)$$

For large n the amplitude distribution follows a Gaussian law and the power spectral density of the multiplex is constant within the occupied bandwidth. A telephony multiplex can hence be considered as white Gaussian noise with mean power given by (24). Recommendation G223 of the CCITT extends (24)

Table 2.4 MULTICHANNEL TELEPHONY FREQUENCY DIVISION MULTIPLEX. (CCIR, Rec 481–1, 1982). *Reproduced by permission of CCIR.*

System capacity (number of channels)	Limits of band occupied by telephone channels (kHz)	Centre frequencies of noise measuring channels (kHz)
12	12–60	66
24	12–108	116
36	12–156	172
48	12–204	224
60	12–252	277
72	12–300	331
96	12–408	448
132	12–552	607
192	12–804	884
252	12–1052	1157
312	12–1300	1499
432	12–1796	1976
612	12–2540	2794
792	12–3284	3612
972	12–4028	4430
1092	12–4892	5381
1332	12–5884	6300
1872	12–8120	8932

to values of n smaller than 240:

$$10 \log S_{av} = \begin{array}{l} -15 + 10 \log n \text{ for } n \geqslant 240 \\ -1 + 4 \log n \text{ for } 12 < n < 240 \end{array} \qquad (25)$$

2.3.1.2.2 Signal/noise power ratio in the highest telephone channel of a multichannel telephony multiplex The ratio of test tone signal power to the psophometrically weighted noise power ratio in the highest telephone channel of a multichannel telephony multiplex is given by:

$$(S/N) = (m^2 B/b) pw(C/N_0) \qquad (26)$$

The above expression takes into account the improvement by emphasis ($p = 4$ dB) and the weighting factor related to psophometric weighting ($w = 2.5$ dB). b is the bandwidth of the telephony channel and is equal to 3100 Hz, m_f is the r.m.s. modulation index equal to $\Delta F_r/f_m$, where ΔF_r is the r.m.s. frequency deviation due to the test tone and f_m is the maximum frequency of the baseband multiplex given by Table 2.4.

Using equation (26), the noise power N, expressed in pW0p, at a point of relative zero level is given by:

$$10 \log N = 83.5 - C/N_0 + 10 \log b - 20 \log m_f \qquad (27)$$

where C/N_0 as given by (18) is expressed in dBHz. ΔF_r can be obtained from the peak frequency deviation ΔF_{p} which defines the required bandwidth given by Carson's rule:

$$B_{\mathrm{IF}} = 2(\Delta F_{\mathrm{p}} + f_{\mathrm{m}}) \tag{28}$$

hence

$$\Delta F_{\mathrm{p}} = B_{\mathrm{IF}/2} - f_{\mathrm{m}}$$

and

$$\Delta F_{\mathrm{r}} = \Delta F_{\mathrm{p}}/gL \tag{29}$$

g represents the 'peak-to-r.m.s' factor of frequency deviation and is equal to 10 dB, while L is the loading factor defined from the average power S_{av} of the telephony multiplex as given by (25). As frequency deviation is proportional to signal amplitude one has:

$$20 \log L = 10 \log S_{\mathrm{av}}$$

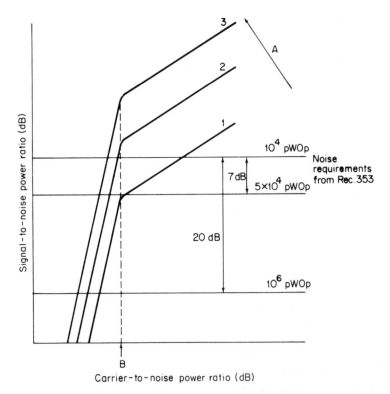

Figure 2.14 Signal-to-noise power ratio versus carrier-to-noise power ratio in FDM/FM telephony systems. *Reproduced with permission from CCIR (1978).*

that is

$$L = 10^{(-15 + 10\log n)/20} \quad \text{for } n \geqslant 240$$
$$L = 10^{(-4\log n)/20} \quad \text{for } 12 < n < 240 \tag{30}$$

Figure 2.14 shows S/N as a function of $C/N = C/N_0 B_{\mathrm{IF}}$. One must select a value of C/N which ensures S/N above the horizontal lines corresponding to the specified values of noise during the stipulated percentages of time.

The curves of Figure 2.14 represent three situations (CCIR, Rep. 708–1, 1982).

Curve 1: bandwidth limited systems.
Curve 2: power and bandwidth limited systems.
Curve 3: power limited systems.

At frequencies where rain losses are negligible, the trade-off between power and bandwidth which results from changing the modulation index value, is easily computed. At frequencies where rain losses are important (for instance, at 30/20 GHz) this trade-off entails an iterative process.

2.3.1.3 Single Channel Per Carrier/Frequency Modulation (SCPC/FM) Television

(a) TV signal standards. There are several CCIR defined TV standards in the 525-line and 625-line systems (CCIR Rep. 624–2, 1982). The nominal amplitude of the signal is that of a monochrome video signal and comprises the synchronization pulses and the picture luminance information. This nominal amplitude equals 1 V (0.7 V for the picture luminance signal and 0.3 V for synchronization pulses). The colour (chrominance) and sound information is added by modulation of specific subcarriers.

The signal/noise power ratio at the receiver output, S/N is defined as the ratio expressed in decibels, of the nominal peak-to-peak amplitude of the picture luminance V_{Lpp} signal to the r.m.s. noise voltage V_{Nrms}, in the operating video-frequency band. This corresponds to usual measurement methods: oscilloscope for measuring the picture luminance signal and r.m.s. voltmeter for measuring the noise.

The quality objective for TV corresponds to a minimum signal-to-weighted-noise power ratio that is (CCIR, Rec. 567–1, 1982)

$$S/N \triangleq 10 \log(V_{\mathrm{Lpp}}/V_{\mathrm{Nrms}})^2 \geqslant 53 \text{ dB (for more than 1 per cent of any month)}$$
$$45 \text{ dB (for more than 0.1 per cent of any month)} \tag{31}$$

where V_{Nrms} represents weighted noise.
(b) SCPC/FM transmission of TV signals.

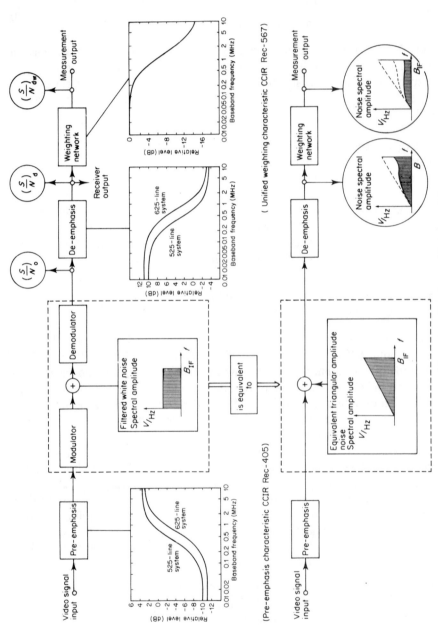

Figure 2.15 SCPC/FM video signal transmission model.

Figure 2.15 illustrates the modulating–demodulating scheme with pre-emphasis of the video signal prior to modulation. The effect of frequency demodulation on white noise introduced on the link is to transform the white noise into a triangular amplitude spectral density noise. This feature is taken into consideration by incorporating in the modulation–demodulation scheme pre-emphasis of the video signal prior to modulation and de-emphasis after demodulation. Recommendation 405–1 of the CCIR provides the pre-emphasis characteristics for various TV standards. At frequency f_r such that pre-emphasis is 0 dB the signal/noise power ratio $(S/N)_0$ at the output of the demodulator is given by:

$$(S/N)_{0,f=fr} = 3 \left[(r\Delta F_{Tpp})^2 / f_m^3 \right] (C/N_0) \tag{32}$$

where ΔF_{Tpp} is the *peak-to-peak frequency deviation* that would be caused by a sinusoidal baseband signal at frequency f_r, and r is the ratio of the peak-to-peak amplitude of the luminance signal to the peak-to-peak amplitude of the video signal including synchronization pulses ($r = 0.7$ for 625/50 systems and $r = 0.714$ for 525/60 systems). At a low frequency, for instance 15 kHz given by a fixed pattern picture luminance signal (half line at peak-white level and half line at black level), the de-emphasis network modifies both the video signal amplitude and the noise power so that the signal/noise power ratio at the output of the de-emphasis network is given by:

$$(S/N)_{d,f=15 \text{ kHz}} = (S/N)_{0,f=fr} \times p \tag{33}$$

where $p = F \times G$ takes into account the effect of pre-emphasis and de-emphasis. F is the result of de-emphasis on triangular noise and G is the result of pre-emphasis on low frequency components of the video signal ($G = -10$ dB for 525-line systems and $G = -11$ dB for 625-line systems.)

Now $(S/N)_d$ represents the actual signal to noise power ratio as measured at the output of the receiver. However, the viewer's perception acts like a low-pass filter (i.e. the human eye is not sensitive to high frequency noise components). So taking this fact into account when measuring noise power, it is common practice to use weighting networks, with the result that the weighted noise power is lower than the total noise power. For most television systems the available weighting networks are designed so that the measurements more clearly represent the subjective impression on monochrome pictures than do unweighted noise power measurements. So finally, the quality objective of service defined by (31) should be met by $(S/N)_{dw,f=15kHz}$ as given by:

$$(S/N)_{dw,f=15kHz} = (S/N)_{\theta,f=fr} \times p \times w \tag{34}$$

where w represents the effect of the weighting network.

Table 2.5 shows values of F, G, p and w for different TV systems. The total improvement on signal/noise power ratio $(S/N)_0$ resulting from de-emphasis and weighting can be considered to be of the order of 13 dB to 18 dB. (CCIR, Rep. 637–2, 1982)

The bandwidth of an FM channel can be obtained from Carson's rule which indicates that the necessary bandwidth is given by

$$B_{\mathrm{IF}} = 2f_\mathrm{m} + \Delta F_{\mathrm{Tpp}}$$

where f_m is the highest video-frequency and ΔF_{Tpp} is the peak-to-peak frequency deviation by video signal (including synchronization pulses).

Experience shows that it is preferable to accept a 10 per cent increase in bandwidth, to accommodate both the sound channels and guard bands. A practical value would then be:

$$B_{\mathrm{IF}} = 1.1(2f_\mathrm{m} + \Delta F_{\mathrm{Tpp}}) \tag{35}$$

Considering for instance, a 625-line system with unified weighting ($f_\mathrm{m} = 5$ MHz), assuming $\Delta F_{\mathrm{Tpp}} = 13.5$ MHz, then $B_{\mathrm{IF}} = 25.85$ MHz and

$$(S/N)_{\mathrm{dw},\,f=15\,\mathrm{kHz}} = 10 \log(V_{\mathrm{Lpp}}/V_{\mathrm{Nrms}})^2 = \left(\frac{C}{N}\right)_{\mathrm{dB}} + 30.6\ \mathrm{dB}$$

where $C/N = C/N_0 B_{\mathrm{IF}}$.

In the calculation p was taken equal to 2 dB and $w = 11.2$ dB so that $pw = 13.2$ dB according to Table 2.5. Assuming $(S/N)_{\mathrm{dw}}$ to be 45 dB minimum, then C/N should be at least equal to 14.4 dB. In any case C/N should be higher than the FM demodulator threshold.

2.3.2 Digital transmission

Digital transmission systems are characterized by:

Table 2.5 IMPROVEMENT ON SIGNAL/NOISE RATIO DUE TO DE-EMPHASIS AND WEIGHTING.

Number of lines	Highest video-frequency f_m (MHz)	Name of system	Effect of de-emphasis network on noise F(dB)	Effect of pre-emphasis network on signal at 15 kHz GdB	Effect of pre-emphasis/deemphasis on SNR at 15 kHz $p = F \times G$	Effect of the weighting network on de-emphasized noise w(dB)	Total improvement on SNR (dB)
525	4.2	M	12.9	− 10	2.9	9.9	12.8
525	5	Unified[*]	13.1	− 10	3.1	11.7	14.8
625	5	B, C, G, H	13	− 11	2	14.3	16.3
625	5.5	I	13	− 11	2	10.9	12.9
625	6	D, K, L	13.3	− 11	2.3	15.8	18.1
625	5	Unified[*]	13	− 11	2	11.2	13.2

[*] SNR is measured within the 0.01–5 MHz band with the unified noise weighting network (CCIR, Rec 568, 1982)

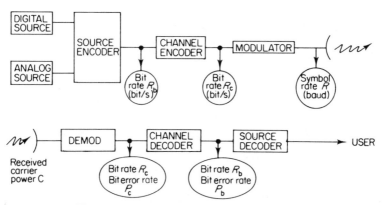

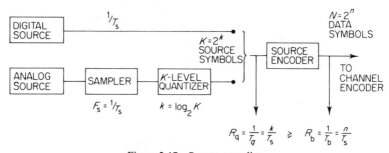

Figure 2.16 Digital communication system model

(1) Bit error rate which is the probability of an erroneous bit at the receiver output.
(2) Bandwidth requirements in the IF amplifier.
(3) Complexity of equipment at transmission and reception.

The principal signalling components of a digital communication system are shown in Figure 2.16. The discrete symbols may come from a discrete data source (e.g. teletype data, computer outputs) or from an analog source (voice, television, facsimile, telemetered data signals) which has been sampled at a periodic rate $F_S = 1/T_S$ with samples being quantized before transmission by means of a K-level quantizer (Figure 2.17).

Each source symbol is applied to the source encoder which converts the source symbols into data symbols. *Source encoding* refers to techniques that alter the information source for efficient transmission over the channel, that is minimize the source bit rate while retaining the desired information. If the number of different data symbols delivered by the source encoder from the $K = 2^k$ source symbols, occurring at rate $1/T_S$ is N, then since N symbols can be represented by $n = \log_2 N$ binary symbols, the input bit rate to the channel encoder is $R_b = 1/T_b = n/T_s$

Channel encoding involves relating n source information bits so as to make efficient use of the communication channel resources—bandwidth and power.

Figure 2.17 Source encoding

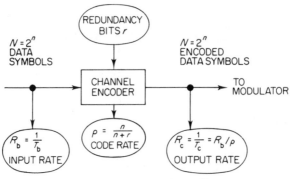

Figure 2.18   Channel encoding

The channel encoder can alter information bits in either a block or a convolutional fashion. Usually the channel encoder inserts redundancy (r redundancy bits for n information bits) for purposes of error control and error correction (Figure 2.18). So the encoded output bit rate R_c is greater than the input bit rate R_b to the channel encoder and the ratio $\rho = R_b/R_c = n/n + r$ is called code rate.

The *modulator* (Figure 2.19) forms M symbols (or signal elements) from m successive encoded bits and translates these symbols into the set of channel signals $s_i(t)$, $i = 1, 2, \ldots, M$ suitable for transmission over the channel. One signal element from the set $s_i(t)$ is transmitted every T seconds so that the symbol rate (signal elements per second) is $R = 1/T$ expressed in *bauds*. The transmitted bit rate is m/T and must be equal to the encoded bit rate R_c. One has consequently:

$$R_c = 1/T_c = m/T = mR = \log_2 M/T = R_b/\rho = n + r/T_s \qquad (36)$$

The function of the modulator is either to provide a one-to-one mapping of the channel symbols into the set of M channel signals, or produce the set of

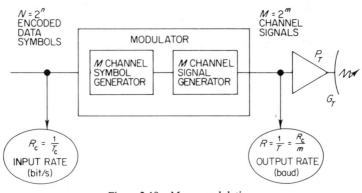

Figure 2.19   *M*-ary modulation

changes in the successive M channel signals. The latter process is called differentially encoded modulation (Lindsey and Simon, 1973, p. 240).

At the receiver side this operation is performed in inverse. The demodulator interprets the received signal as one of the M channel symbols. The channel decoder converts the sequence of bits at the demodulator output into data symbols, which if there is no error, corresponds to the transmitted data symbols. The primary effectiveness measure hence appears to be the probability of a channel symbol error (word error probability) or the probability of a specific bit being in error (bit error probability). The source decoder restores the source symbols from the decoded data symbols. With analogue sources, source regeneration implies the use of a digital to analogue converter and an interpolation filter which compensates for effects of sampling. The output of the interpolation filter has a smooth function with time which should be as similar as possible to the original waveform. The output signal/noise power ratio then depends on the bit error probability and the number K of quantization levels.

As previously stated, in any satellite communication system, the two primary communication resources are the transmitted power and channel bandwidth. In power limited channels, coding schemes would be generally used to save power at the expense of bandwidth. Moreover, satellite communications systems make use of certain types of non-linear devices such as travelling wave tubes (TWT) that call for constant envelope modulation. This is necessary because a non-linearity produces extraneous sidebands when passing a signal with amplitude fluctuations. Such sidebands introduce out of band interference and increase the level of interference in adjacent channels.

2.3.2.1 Digital modulation techniques

Digital modulation entails sending $M = 2^m$ symbols at T second intervals by means of a radio-frequency modulated carrier. The modulated carrier can be written as:

$$S(t) = \sum_i s_i(t)X(t - iT) \tag{37}$$

where $X(t - iT) = 1$ if $0 \leqslant t \leqslant T$ and 0 elsewhere.

$$s_i(t) = A_i \cos(\omega_c t + \theta_i) = (A_i \cos \theta_i)\cos \omega_c t - (A_i \sin \theta_i) \sin \omega_c t$$

where $f_c = \omega_c/2\pi$ is the carrier frequency.

The expressions from A_i and θ_i determine the output modulation according to the following terminology:

(1) *Pulse amplitude modulation* (PAM): A_i is one character from an M-ary alphabet, θ_i is constant.

(2) *Multiple frequency shift keying* (MFSK): A_i is constant and

$$\theta_i = \frac{\pi h}{T} \int_0^T a_i(x)\, \mathrm{d}x$$

where a_i is one character from an M-ary alphabet $(-M/2, \ldots, -1, +1, \ldots, +M/2)$, h being the modulation index.

(3) *Multiple phase shift keying* (MPSK): A_i is constant and θ_i is one character from an M-ary alphabet given by

$$(j-1)2\pi/M, j = 1, 2, \ldots, M$$

If $M = 2$, MFSK and MPSK are usually referred to as *frequency-shift keying* (FSK) and *phase-shift keying* (PSK) respectively.

If the amplitude and phase of a constant frequency carrier are varied discretely one has the hybrid of PAM and MPSK modulation called *combined amplitude phase shift keying* APSK.

In satellite communication systems, reducing envelope fluctuation of the modulated carrier favours the use of BPSK, QPSK, Offset-Keyed QPSK and MSK, to the extent of maintaining low complexity of the modulator and demodulator.

2.3.2.2 Binary PSK (BPSK)

In a binary PSK modulator, input data bits cause discrete $180°$ changes in the phase of the carrier. Two distinct procedures are commonly used:

(1) *Direct encoding*, where there is a one-to-one mapping of the transmitted bit into absolute phase value of the channel signal, as in the following example:

Bit sequence	0	1	1	1	0	1	0	1
Phase sequence	0	π	π	π	0	π	0	π

(2) *Differential encoding*, where there is a one-to-one mapping of the bit into phase shift between two consecutive signals according to the following rule:

Bit	Phase shift
0	0
1	π

so that the previous bit sequence changes to the following phase sequence:

Bit sequence	0	1	1	1	0	1	0	1
Phase sequence	0	π	0	π	π	0	0	π

2.3.2.3. Quaternary PSK (QPSK)

Quaternary modulation relates each pair of bits (channel symbol) of the incoming bit sequence to one of four phases of the carrier, according for example to the following scheme (Gray code)

Channel symbol	00	01	11	10
Carrier phase	0	$\dfrac{\pi}{2}$	π	$\dfrac{3\pi}{2}$

Alternatively, one can map each pair of bits into a change in phase in a manner analogous to differential encoding of BPSK, as for instance:

Channel symbol	00	01	11	10
Phase shift	0	$\dfrac{\pi}{2}$	π	$\dfrac{3\pi}{2}$

The QPSK carrier can be considered as the sum of two BPSK carriers that are in phase quadrature one with another.

2.3.2.4 Offset-keyed QPSK (OK-QPSK)

With QPSK an instantaneous change in both bits of a channel symbol induces a 180° phase shift. Such phase jumps influence the filtered output of a QPSK modulator causing the envelope to go to zero. Envelope fluctuations will then be translated into an unwanted phase modulation as a result of the AM/PM distortion of the non-linear power amplifier. With offset-keyed QPSK envelope nuls are avoided by delaying the inphase bit stream by $T_c/2$ seconds relative to the quadrature bit stream, where $T_c T_c = 1/R_c$ is the duration of an incoming bit giving a 0 or $\pi/2$ phase shift each T_c seconds.

2.3.2.5 Minimum Shift Keying (MSK)

MSK is a special case of continuous phase FSK with modulation index $h = 0.5$, which ensures orthogonal signalling. The carrier waveform phase $\theta_i(t)$ increases by $\pm \Delta\omega T_c$ during each bit duration T_c, with $\Delta\omega = \pi/T_c$, as shown in Figure 2.20. Hence one of two frequencies $f_c = \pm 1/4T_c$ is transmitted during each bit period depending on the input bit and the phase is equal to an integer multiple of $\pi/2$ at the end of each bit period. MSK can also be considered as a special case of QPSK with sinusoidal pulse weighting rather than

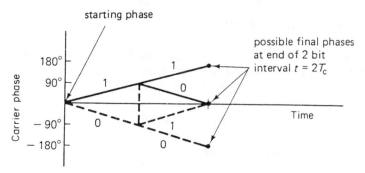

Figure 2.20   Phase of a MSK modulated carrier

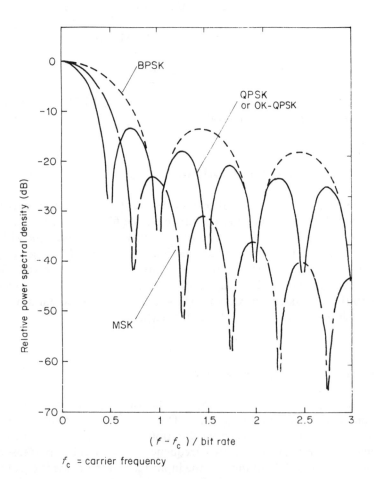

Figure 2.21   Power spectral densities for unfiltered BPSK, QPSK or
OK-QPSK and MSK

rectangular weighting (Gronomeyer and McBride, 1976). Figure 2.21 (Amoroso, 1980) compares power spectra of unfiltered BPSK, QPSK or OK-QPSK, and MSK.

2.3.2.6 Error performance of digital modulation techniques

The commonly used measure of performance for digital communications techniques is the relationship between bit error rate and signal/noise power ratio E_b/N_0, where E_b is the energy per information bit and N_0 the one-sided noise power spectral density. The *energy per information bit* E_b is related to the carrier power, C, and the information bit rate R_b by the following expression:

$$E_b = C/R_b \tag{38}$$

Considering an encoded digital system the *energy per transmitted* bit, E_c, is given by

$$E_c = C/R_c \tag{39}$$

where R_c is the transmitted bit rate on the radio frequency link. Digital communication systems can be compared on the basis of ideal error rate considered as the best performance attainable. In a real digital communication system degradation relative to the ideal error rate is liable to occur, and the various implementation degradations can then be determined by analysis, computer simulation or experiment. The ideal error rate conditions are determined under the assumptions that only additive Gaussian white noise perturbs signals, perfect synchronization is maintained and no error control coding is employed (that is $E_c = E_b$).

Coherent demodulation requires that a reference signal be available at the receiver. This reference can be established by carrier recovery techniques (Lindsey and Simon, 1973; Spilker, 1977), but some phase ambiguity may be encountered from the carrier recovery unit. This phase ambiguity can be avoided by employing differential encoding at the transmitter, which makes possible two different systems:

(1) With coherent demodulation of successive differentially encoded signals we have a *differentially encoded coherent system* (for instance, if MPSK modulation is employed, this system will be DE-MPSK).
(2) If a differential demodulation is employed, we have a *differentially encoded and differentially demodulated* coherent demodulation system (with M-PSK modulation, this would be called D-MPSK). In this case no carrier recovery circuit is needed.

So, finally, three schemes are possible, as indicated below:

Channel symbol to channel signal mapping	Demodulation	Designation
Direct encoding	Coherent	MPSK
Differential encoding	Coherent	DE-MPSK
	Differential coherent	D-MPSK

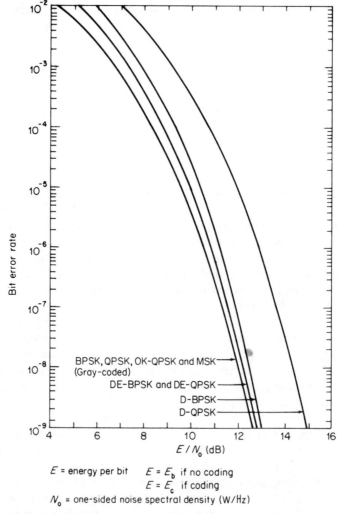

E = energy per bit $E = E_b$ if no coding
$E = E_c$ if coding
N_o = one-sided noise spectral density (W/Hz)

Figure 2.22   Ideal bit error rate performance of various digital modulation schemes

Table 2.6 IDEAL ERROR RATE PERFORMANCE OF VARIOUS DIGITAL MODULATION SCHEMES (E = energy per bit = E_b if no coding or E_c if coding).

Type of modulation technique	Ideal bit error rate performance
MSK BPSK/QPSK (Gray encoded) With coherent detection	$P = 1/2 \operatorname{erfc} \sqrt{\left(\dfrac{E}{N_0}\right)}$
Orthogonal FSK With coherent detection	$P = 1/2 \operatorname{erfc} \sqrt{\left(\dfrac{E}{N_0}\right)}$
Orthogonal FSK With non-coherent detection	$P = 1/2 \exp\left(-\dfrac{E}{2N_0}\right)$
D-BPSK	$P = 1/2 \exp\left(-\dfrac{E}{N_0}\right)$

When considering M-ary modulation schemes the error rate performance is specified in terms of symbol error rate P_S. Assuming encoding of the modulator input bits by a Gray code (Spilker, 1977, p. 55), the bit error rate P approximates to:

$$P \cong P_s/\log_2 M \qquad (40)$$

For QPSK ($M = 4$) a closed form expression for P_S can be given:

$$P_s = \operatorname{erfc} \sqrt{(E_s/2N_0)} - 1/4 \operatorname{erfc}^2 \sqrt{(E_s/2N_0)} \qquad (41)$$

where E_s is the symbol energy, $E_S = E_b \log_2 M = 2E_b$.

With practical values of E_b/N_0, $P_s \cong \operatorname{erfc}\sqrt{E_b/N_0}$ and $P \cong 1/2 \operatorname{erfc}\sqrt{E_b/N_0}$.

Table 2.6 gives the expressions for the ideal bit error rate, P, considering various digital modulation schemes. Numerical values of erfc(x) are given in Appendix 1.

Figure 2.22 shows the ideal binary error rate curves for various digital modulation schemes according to the expressions given in Table 2.6, along with DE-MPSK (Feher, 1983). Although MSK can be viewed as a special case of orthogonal FSK it performs better than orthogonal FSK because bit decisions are made after observing the waveform over two bit periods instead of one with FSK detection (Pasupathy, 1979).

2.3.2.7 Error rate performance with channel encoding

Channel encoding aims at introducing redundancy into the transmitted data for the purpose of achieving error protection, that is to obtain as low error rate as possible given an available transmission bandwidth B_{IF} and an average carrier power C.

The channel encoder accepts binary symbols at a rate of R_b bits/s from the binary data source and generates encoded binary symbols at a rate of R_c binary symbols per second ($R_c = R_b/\rho$, where the code rate ρ is less than 1). Encoding requires that the channel capacity c is not exceeded, as given by (Shannon and Weaver, 1949):

$$c = B_{IF} \log_2(1 + C/N) \tag{42}$$

where $N = N_0 B_{IF}$ is the average noise power on the channel.

Hence, if the binary transmission rate is smaller or equal to c error-free communication is possible. From the above expressions we see that the rate is limited by the signal/noise power ratio. Assuming $R_b = c$:

$$R_b/B_{IF} = \log_2(1 + E_b/N_0 \, R_b/B_{IF}) \tag{43}$$

which may be written as

$$E_b/N_0 = (R_b/B_{IF})^{-1}[\exp(0.69 \, R_b/B_{IF}) - 1] \tag{44}$$

then a minimum value of E_b/N_0 for zero error is -1.6 dB. This occurs when the available bandwidth approaches infinity (R_b/B_{IF} tends to zero).
Two main classes of encoding procedures are used:

(1) *Block encoders* in which r parity check bits formed by linear operations on n data bits are appended to each block of n bits resulting in blocks $k = n + r$ coded symbols long. The encoding operation produces what is known as a (k, n) block code, and there are 2^n code-words contained in the code.
(2) *Convolutional encoders* which generate R_c/R_b encoded binary symbols for each input binary source symbol. These R_c/R_b encoded symbols depend not only on the most recent symbol but also on a specified number of previous source symbols so that a sliding sequence of past data symbols is used to generate several encoded symbols.

It should be noted that if E_b is the received energy per information bit, then the corresponding received energy per coded bit is $E_c = \rho E_b$. Since $E_c/N_0 < E_b/N_0$ more bits will be received in error with a coded system than with a direct, non-coded, system. However, the larger number of channel-induced errors should be more than compensated for by the error correcting capability of the coding/decoding scheme, resulting in a superior overall error performance for the coded system. Figure 2.23 shows the ideal error rate performance obtained from several channel encoding schemes (Bhargava *et al.*, 1981). The effectiveness of communication systems using the channel encoding techniques may be measured by the *coding gain* defined as the difference expressed in decibels in the required E_b/N_0 for a given error rate performance between ideal PSK signalling and the particular coding scheme.

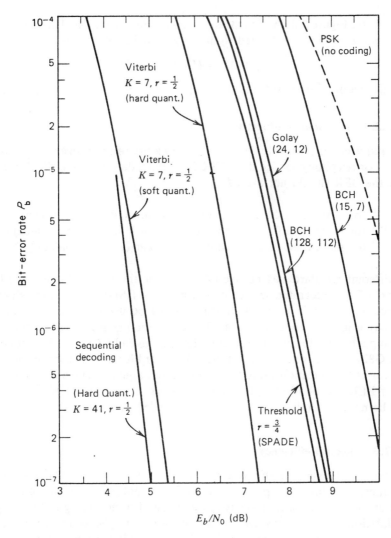

K = constraint length
r = code rate

Figure 2.23 Ideal bit error rate performance of various coding schemes (Bhargava *et al.*, 1981)

Example: Consider a link with bit rate $R_b = 1$ kbit/s. Using BPSK and assuming that the required bandwidth equals $B_{IF} = 1.4 R_b = 1.4$ kHz and considering a bit error rate of 10^{-5} we need $E_b/N_0 = C/N_0 R_b = 9.6$ dB (see Figure 2.22) so that $C/N_0 = 39.6$ dBHz.

Using convolutional encoding with a constraint length $K = 7$, coding rate $\rho = 1/2$ and Viterbi decoding with hard quantization the required E_b/N_0 is

6.5 dB (see Figure 2.23) which means a coding gain of 3.1 dB. This coding gain provides a received modulated signal power saving of 3.1 dB as

$$C/N_0 = (E_c/N_0)R_c = (E_b/N_0)R_b = 35.6 \text{ dBHz},$$

but the required bandwidth is now $B_{IF} = 1.4\,R_c = 1.4(R_b/\rho) = 2.8$ kHz.

2.3.2.8 Performance objectives

The required bit error rate depends on the nature of the transmitted information. For telephony service the CCIR Recommendation 522–1, 1982, requires that the BER should not exceed:

10^{-6} for more than 20 per cent of any month (10 min mean value)
10^{-4} for more than 0.3 per cent of any month (1 min mean value)
10^{-3} for more than 0.01 per cent of any year (1 s mean value)

A different standard has been established for the basic 64 kbit/s 'building block' of the integrated service digital network (ISDN). This quotes performance objectives in terms of average errors occurring in one minute (for speech) or in one second (for data). Table 2.7 summarizes the requirements for a complete link, not only the satellite link, as given by CCITT Recommendation G821. There is no CCITT standard yet for bit rates in excess of 64 kbit/s but for communication using highly processed digitized video it is likely that lower BER will be needed. For data transfers, error rates of 10^{-10} or better will be needed. Coding will probably be needed to attain these figures.

2.3.2.9 Spectral efficiency

Digital transmission with M-level digital modulation makes it possible to transmit more bits per second and yet not increase the bandwidth proportionally. This advantage is more or less clear according to the adopted modulation type, and may demand an increase in the necessary power. To put this in perspective, the performance of different kinds of modulation can be compared in terms of the spectral efficiency Γ defined as the ratio of transmitted bit rate to bandwidth, expressed in bit/s Hz versus E/N_0. The result is shown in Figure 2.24 with bit error probability fixed at 10^{-5}. In this figure coherent detection has been assumed for MPSK and non-coherent detection for MFSK. The spectral efficiency of the modulation schemes is given by Jacob (1967) and Lindsey and Simon (1973):

$$\text{MPSK} \qquad \Gamma = \begin{cases} 2 & \text{for } M = 2 \\ \log_2 M & \text{for } M > 2 \end{cases} \qquad (45)$$

$$\text{MFSK} \qquad \Gamma = \log_2 M/M \qquad (46)$$

Table 2.7 PERFORMANCE OBJECTIVES FOR VOICE AND DATA TRANSFER WITHIN THE INTEGRATED SERVICE DIGITAL NETWORK (ISDN).

For 64 kbit/s channels (CCITT, Rec. G821)

BER in 1 min	Percent of available minutes
more than 10^{-6}	less than 10
less than 10^{-6}	more than 90

BER in 1 second	Percent of available seconds
more than 0	less than 8
zero	more than 92 (Error-free-seconds)

No standard yet for bit rates in excess of 64 kbit/s. Likely, 10 E–9 to 10 E–12

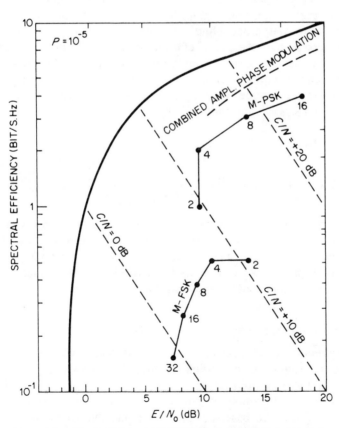

Figure 2.24 Theoretical spectral efficiency of various digital modulation schemes at a given bit error rate (theoretical)

All modulation schemes are well within the curve limit defined by equation (44), as given by the Shannon theorem.

For a fixed bit rate, digital modulation permits a trade-off between bandwidth and power. With phase modulation when increasing M, we have a decreasing bandwidth but this requires a E/N_0 increase and hence, according to (38), more power C. With frequency modulation the same applies when M decreases. Similar trade-offs can be obtained using channel encoding schemes as presented in Section 2.3.2.7.

If the bandwidth is specified, system capacity can be increased by using modulation schemes with high spectral efficiencies, such as phase modulation, or better a combined amplitude and phase modulation. However, the sensitivity of combined phase and amplitude modulation to non-linearities in the transponder on board the satellite must be emphasized. This penalizes these modulations in practice, relative to more 'robust' modulations, such as QPSK. The key to the improvement in satellite telecommunication efficiency lies largely in the use of linear amplifiers, or compensation of channel non-linearities. On the other hand, the availability of more linear satellite transponders favours the use of efficient analogue modulation such as single side band (SSB) amplitude modulation and explains the growing interest towards such techniques, as indicated at the beginning of paragraph 2.3.1. The advantage in transponder capacity is improved when SSB is combined with companding (Szarvas and Suyderhoupd, 1981).

2.4   LINK BETWEEN TWO EARTH STATIONS VIA A SATELLITE

The link between two earth stations via a relay satellite is divided into the earth–satellite, or up-link, and the satellite–earth, or down-link. The results of Sections 2.1 and 2.2 apply to each of these link sections. The ultimate information quality received on Earth, depends on the characteristics of the up-link, the satellite transponder and the down-link.

2.4.1   The transponder

A *conventional* transponder refers to a transponder which simply translates the received signal from up-link frequency to down-link frequency and retransmits it after amplification.

A *regenerative* transponder provides on-board detection of the received up-link signal prior to baseband processing and remodulation for the down-link (Campanella *et al.*, 1977: Koga *et al.*, 1977). Such a transponder is most effective where digital signals are involved.

2.4.2 Conventional transponder: Link budget for a linear channel

2.4.2.1 Signal power to noise power spectral density ratio at ground station receiver input

The up-link is characterized by the ratio of signal power to noise power spectral density at the transponder input $(C/N_0)_U$. Similarly the down-link is characterized by $(C/N_0)_D$.

Considering the complete link from one earth station to another, at the receiver input:

(1) The useful signal is $C = C_U G_S G_T G_R / L$, where G_S, G_T, G_R, are the transponder gain, the satellite transmitting antenna gain and the receiving station antenna gain, respectively, L is the loss on the down-link and C_U is the signal power at the transponder input.

(2) The noise power spectral density is $N_0 = N_{0D} + N_{0U}(G_S G_T G_R)/L$ where N_{0U} is the noise power density at the transponder input and N_{0D} is the noise power density that would be measured at the earth station receiver input, considering down-link only.

It follows therefore that

$$(C/N_0)_T = \frac{C}{N_0} = \frac{C_U}{N_{0U} + N_{0D}(L_D/G_S G_T)G_R}$$

The transponder of bandwidth B, transmits a constant power P_T. Its gain, G_S, is therefore such that:

$$G_S = \frac{P_T}{C_U + N_{0U} \times B}$$

Considering that $C_D = P_T(G_T G_R)/L$
It follows

$$(C/N_0)_T = \frac{(C/N_0)_U \times (C/N_0)_D}{(C/N_0)_U + (C/N_0)_D + B} \tag{44}$$

$(C/N_0)_T$ is expressed in Hz.

In most cases B is much smaller than $(C/N_0)_U$ or $(C/N_0)_D$ so that it is very often considered that:

$$(C/N_0)_T^{-1} = (C/N_0)_U^{-1} + (C/N_0)_D^{-1} \tag{45}$$

Formula (45) is used in Figure 2.25 to represent the number of telephone channels V in terms of $(C/N_0)_D$. The satellite bandwidth is taken equal to 40 MHz. Digital transmission using biphase modulation (BPSK) is assumed and bit error rate is fixed at 10^{-5}, which corresponds to $E_b/N_0 = 9.5$ dB (see Figure

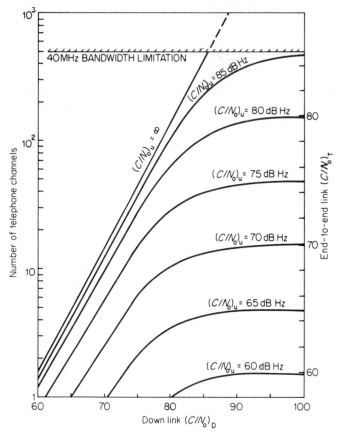

Figure 2.25   Capacity of a 40-MHz bandwidth satellite transponder (single access) according to C/N_0 values.

2.22). As $C/N_0 = (E_b/N_0)R_b$, with $R_b = 67\,200$ V (assuming 64 kbit/s for one voice channel plus 5 per cent signalling) it follows that:

$$10 \log V = (C/N_0)_{T,dBHz} - 57.8$$

The maximum bit rate is fixed by the value of bandwidth. Assuming a spectral efficiency of 0.8 bit/s Hz it follows $V_{max} = 476$.

2.4.2.2   Effect of down-link on link quality

$(C/N_0)_U$ is often much greater than $(C/N_0)_D$; indeed, although N_{0U} is roughly ten times greater than N_{0U}, C_U may be 100 times greater than C_D. Antenna gain in the ground station is not limited, like that of the satellite, by considerations of dimensions or coverage, and there is less limitation on power supply.

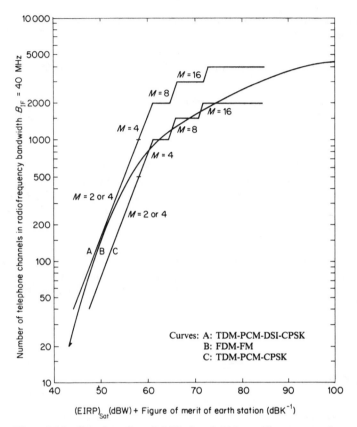

Figure 2.26   Capacity of a 40-MHz bandwidth satellite transponder (single access) according to various multiplexing/modulation schemes.

Hence $(C/N_0)_U$ may be as large as ten times $(C/N_0)_D$. So as the value of $(C/N_0)_T$ is virtually that of $(C/N_0)_D$, the down-link determines link quality.

With a geostationary satellite the value of $(C/N_0)_D$ depends simply on the 'EIRP of the satellite plus figure of merit of the earth station' (see formula (18)). So, the number of telephone channels can be directly expressed in terms of this quantity. Figure 2.26 compares various systems which use single access.

2.4.2.3   Example

Let us assume a Single Channel Per Carrier (SCPC) link between two earth stations via a geostationary satellite. Voice is transmitted in PCM using 8 bits and QPSK modulation.

The satellite ensures total coverage of the Earth. Antennae gain is 20 dB, and the noise temperature at its receiver input is 1000 K. The two earth stations have 2 m diameter antennae, and the receiver noise temperature is 125 K. The same antenna is used for both transmission and reception. The calculation aims at determining the power transmitted by the satellite and the earth station, knowing that the up-link is at 6 GHz and the down-link at 4.6 GHz.

2.4.2.3.1 Characteristics of the baseband and the intermediate frequency bandwidth. When a telephone signal (300–3400 Hz) is sampled at 8 kHz, the bit rate is $R_b = 64$ Kbit/s.

2.4.2.3.2 C/N ratio of carrier power to noise power in the radio frequency bandwidth. Assuming a bit error rate of 10^{-6} and neglecting quantization noise, Figure 2.22 shows that with QPSK, E/N_0 must equal 10.5 dB. Assuming a spectral efficiency of 1.6 bit/s Hz = 2 dB then $C/N = (E/N_0) \times (R/B_{IF}) = 10.5 + 2 = 12.5$ dB.

2.4.2.3.3 Up-link: power transmitted by ground station. Let us assume for simplicity that the transmitter power of the earth station is high enough for the up-link noise to be negligible. Assume then $(E/N_0)_U = 10(E/N_0)_D =$

Table 2.8 UP-LINK BUDGET.

		+	−
Energy per bit/ noise power density	$\left(\dfrac{E}{N_0}\right)_U$	20.5 dB	
Boltzmann constant	k	−228.6 dB WHz^{-1}	
Noise temperature (1000 k)	T	30 dBK	
Bit rate (64 kbit/s)	R_b	48 dB s^{-1}	
Transmitting antenna gain	G_T		40.4 dB^a
Receiving antenna gain	G_R		20 dB
Free space loss	$L_{FS} = \left(\dfrac{\lambda}{4\pi R_0}\right)^2$		−199 dB^b
Additional losses (see eq(7))	L_A		−2.6 dB^c
Subtotal		−130.1	−141.2
Total	Eq. (45)		11.1 dBW (12.9 W)
Correction for zone boundary conditions:			
off axis gain fallout	$\left(\dfrac{R}{R_0}\right)^2$		−3 dB
range variation			−1.3 dB
Total	P_T		15.4 dBW (35 W)

[a] Antenna efficiency $\eta = 0.7$
[b] $R_0 = 35800$ km
[c] $L_{FTX} = 0.5$ dB, $L_{POINT} = 1$ dB, $A_{AG} = 0.4$ dB, $L_{POL} = 0.3$ dB, $L_{FRX} = 0.4$ dB

20.5 dB so that $(C/N_0)_U = 10(C/N_0)_D$. Then $(C/N_0)_T \cong (C/N_0)_D$. From equation (16) written as follows:

$$P_T = \left(\frac{C}{N}\right)_U \frac{kTB_{IF}}{G_T G_R L} = \left(\frac{E}{N_0}\right)_U \frac{kT}{G_T G_R L_{FS} L_A} R_b \qquad (45)$$

it is possible to construct Table 2.8. The required transmitter power is 35 W.

2.4.2.3.4 Down-link: power transmitted by the satellite transmitter. A similar calculation (Table 2.9), based on an earth station noise temperature of 200 K, (where the antenna noise temperature of 75 K adds to the receiver noise temperature of 125 K), and $(E/N_0)_D = 10.5$ dB, leads to a required transmitter power on board the satellite of 1 W. It should be noted that such a low value results from the choice of a narrow bandwidth channel of $(64 \text{ kbit/s})(1/\Gamma) = 40$ kHz. A common transponder bandwidth would be of 36 MHz for instance, and would support many QPSK carriers, sharing the total transmitter power, which would then be of the order of tens of watts.

Variations in power level all along the signal path are shown in Figure 2.27.

Table 2.9 DOWN LINK.

		+	−
Energy per bit/ noise power density	$\left(\dfrac{E}{N_0}\right)_D$	10.5 dB	
Boltzmann constant	k	-228.6 dB WHz^{-1}	
Noise temperature (200 K)	T	23 dBK	
Bit rate (64 kbit/s)	R_b	48 dB s^{-1}	
Transmitting antenna gain	G_T		20 dB
Receiving antenna gain	G_R		36.9 dB^a
Free space loss	$L_{FS} = \left(\dfrac{\lambda}{4\pi r_0}\right)^2$		-195.6 dB^b
Additional losses	L_A		-2.3 dB^c
Subtotal		-147.1	-141
Total Eq. (45)			-6.1 dBW (0.25 W)
Correction for zone boundary conditions:			
off axis gain fallout	$\left(\dfrac{R}{R_0}\right)^2$		-3 dB
range variation			-1.3 dB
Margin			-2 dB
Total P_T			0.2 dBW (1.05 W)

^{a} Antenna efficiency $\eta = 0.7$.
^{b} $R_0 = 35\,800$ km.
^{c} $L_{FTX} = 0.4$ dB, $L_{POINT} = 1$ dB, $A_{AG} = 0.4$ dB, $L_{POL} = 0.3$ dB, $L_{FRX} = 0.2$ dB.

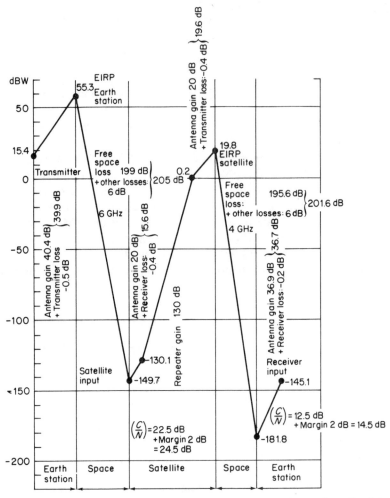

Figure 2.27   Carrier power level variations throughout the end-to-end satellite link.

2.4.3   Regenerative transponder: link budget for a linear channel

2.4.3.1   Overall link bit error rate

To estimate the performance with a regenerative transponder assuming digital signalling, it is necessary to estimate the bit error rate (BER) on the up- and down-links separately and to obtain the overall error rate by combining up- and down-link error rates as follows:

$$BER_T = BER_U(1 - BER_D) + BER_D(1 - BER_U) \simeq BER_U + BER_D$$

2.4.3.2 Performance of a regenerative satellite system

The above formula has the following implications: let us assume, for instance, a required overall bit error rate of 10^{-4} with QPSK modulation and same link performance on both up- and down-links. As $(C/N)_U = (C/N)_D$, $BER_U = BER_D$, the required bit error rate on each link must equal 0.5×10^{-4} which asks for $E/N_0 = 8.6$ dB. Assuming a spectral efficiency Γ of 2 bit/s Hz leads to $C/N = (E/N_0)\Gamma = 11.6$ dB.

Now let us assume that the performance is slightly better on the up-link than on the down-link. Then the total BER_T equals that of the down-link, since the variation of the BER on any link is a steep function of E/N_0 (see Figure 2.22).

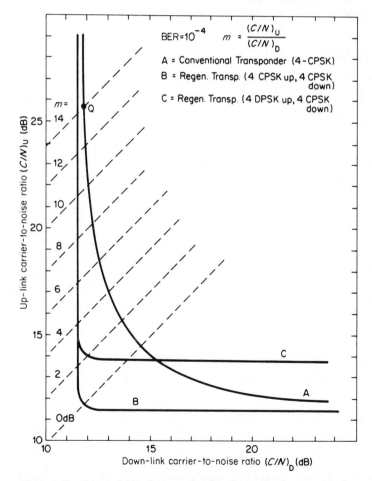

Figure 2.28 Comparison of conventional and regenerative transponder performances (linear channel)

For instance, if the up-link E/N_0 equals 10.6 dB, that is a 0.2 dB improvement, then the up-link bit error rate equals 10^{-6} and $BER_T = BER_D = 10^{-4}$. It could be seen, in the same way, that a slightly better performance on the down-link than on the up-link leads to an overall BER_T equal to that of the up-link. So finally, the overall performance of a regenerative satellite system is most often constrained by that of the worse link. This is clearly illustrated in Figure 2.28 (curves B and C) which gives the relationship between up- and down-link carrier/noise power ratios for a given total BER of 10^{-4}, assuming a linear channel and no degradation in the demodulator performance.

2.4.4 Comparison between conventional and regenerative transponders (Maral and Bousquet, 1984)

2.4.4.1 Linear channel

Figure 2.28 shows a comparison between conventional and regenerative transponders, in terms of up- and down-link carrier/noise power ratios, considering a specified bit error rate of 10^{-4}.

Ideal error rate conditions are assumed, that is no degradation resulting from filtering or non-linear distortions. Notice that the maximum power gain saving is obtained when up- and down-links are the same ($\alpha = 0$ dB). In that case the advantage of a regenerative linear system compared with a conventional one is a saving of 3 dB on both up- and down-link transmitted power. However, when the up-link power is much larger than the down-link power (for instance, $\alpha = 14$ dB) the power saving is much smaller (about 0.5 dB). Unfortunately, this may be a practical situation as the satellite, contrary to earth stations, is power limited. However, the regenerative system presents the advantage of a total BER similar to that of the conventional one with the same down-link power, but with much less up-link power, and this may have some impact on earth station design (see paragraph 3.7.2).

2.4.4.2 Influence of non-linear and filtering distortions

In the previous section a linear channel was assumed with no filtering effects and no interference. In practice, a real channel is non-linear and bandwidth limited.

At the transmitting earth station, a transmit bandpass filter is used to band-limit the spectrum of the digitally modulated carrier. This band-limiting causes envelope modulation (or AM). The AM to PM non-linearity of the earth station high power amplifier (HPA) produces phase modulation of the RF carrier during the individual symbol periods and the sidebands of the modulated carrier that were reduced by the band-limiting filtering, build up again after the HPA so that an earth station transmitter filter and a satellite receiver filter are necessary to avoid adjacent channel interference.

The distorted carrier is then fed to the transponder which again exhibits filtering and non-linear effects. In the conventional satellite system the modulated carrier suffers from cascading both filtering and non-linear distortion caused by the earth station HPA and the satellite power amplifier (TWTA). The resulting overall degradation is higher than that which is produced by the single non-linear distortion caused either by the earth station HPA or the satellite TWTA alone, as in the case of a regenerative satellite system, where the up- and down-links are separated by a demodulation/remodulation process.

Figure 2.29 shows results obtained from computer simulations under various conditions of input back-off for the earth-station HPA, and a 2 dB

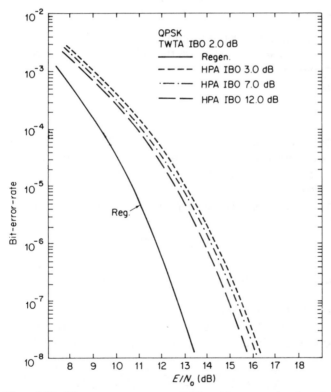

Figure 2.29   Bit error rate versus E/N_0 with conventional and regenerative transponders (non-linear channel, no interference) (Wachira *et al.*, 1981). (TWTA IBO: conventional transponder travelling wave tube input back-off. HPA IBO: earth station high power amplifier input back-off.)

input back-off for the satellite TWTA (Wachira *et al.*, 1981). This figure, and other results (Chiao and Chethik, 1976; Cuccia *et al.*, 1977), show that the regenerative system may offer a 2–5 dB gain over a conventional one even though the up-link power is much larger than the down-link power.

2.5 FREQUENCY BANDS

The use of a specific frequency band depends on operational constraints, propagation conditions and radio regulations.

2.5.1 Operational constraints

Wavelength appears in the fundamental relationship

$$\frac{C}{N_0} = \frac{P_T G_T G_R}{N_0} \left(\frac{\lambda}{4\pi R}\right)^2 \frac{1}{L} \tag{46}$$

where R is the range between transmitting and receiving antennae. The essential operational constraints acting on this relationship are antenna beamwidth θ_{3dB} and diameter D. These parameters are linked to gain by the following expressions:

$$G = \eta \left(\frac{4\pi D}{\lambda}\right)^2 \quad \text{and} \quad G = \eta \frac{\pi^2 70^2}{\theta_{3dB}^2}$$

In most cases the satellite antenna beamwidth $(\theta_{3dB})_{SAT}$ is determined by the required earth coverage. If cost considerations, or operational constraints (for instance the antenna may need to be small enough to fit on a vehicle and to be transported and put into operation quickly) impose a given earth station antenna diameter D_{ES}, then for a given C/N_0 one obtains from the above formulae

$$\frac{C}{N_0} = \frac{P_T}{LN_0} \eta_T \eta_R \frac{\pi^2 70^2}{R^2} \frac{D_{ES}^2}{(\theta_{3dB})_{SAT}^2}$$

It appears, therefore, that frequency choice, at least in the first approximation, does not come into it.

Now under other circumstances, the earth station antenna beamwidth $(\theta_{3dB})_{ES}$ is constrained to a given value, either to avoid or simplify tracking, or to avoid too large interference. In those cases the required C/N_0 is given by

$$\frac{C}{N_0} = \frac{P_T}{LN_0} \eta_T \eta_R \frac{\pi^2 c^2 70^4}{(4R)^2 (\theta_{3dB})_{SAT}^2 (\theta_{3dB})_{ES}^2} \frac{1}{f^2}$$

in which case C/N_0 increases with low frequencies so low frequencies would be preferred.

Finally, it may be that the antenna diameters D_{SAT} and D_{ES} are given. For a satellite the requirement may come from the restricted dimensions of the launcher fairing and from the need to avoid a deployable antenna. In that case the required C/N_0 is given by

$$\frac{C}{N_0} = \frac{P_T}{LN_0} \eta_T \eta_R \left(\frac{\pi D_{\text{SAT}} D_{\text{ES}}}{4Rc}\right)^2 f^2$$

In such a case, high frequencies are preferable.

The above discussion assumes that noise temperature varies very little with antenna characteristics, which is not actually true.

2.5.2 Conditions of propagation

See Section 2.2

Table 2.10 FREQUENCY ALLOCATION FOR THE FIXED SATELLITE SERVICE.
(WARC, 1979 and RARC, 1983).

Frequency band (GHz)	Region 1	Region 2	Region 3	Use	Designation (IEEE Radar Std 521)
3.4–4.2	D	D	D	Commercial	C-band
4.5–4.8	D	D	D		
5.725–5.85	U	—	—		
5.85–7.075	U	U	U		
7.25–7.75	D	D	D	Military	X-band
7.9–8.4	U	U	U		
10.7–11.7	U*/D	D	D	Commercial	Ku-band
11.7–12.2	—	D	—	*Restricted to	
12.5–12.7	U/D	—	D	BSS feeder link	
12.7–12.75	U/D	U	D		
12.75–13.25	U	U	U		
14–14.5	U	U	U		
14.5–14.8	U	U	U	BSS feeder link	
17.3–18.1	U	U	U	BSS feeder link	
18.1–20.2	D	D	D	Commercial	K-band
20.2–21.2	D	D	D	Military	
27–27.5	—	U	U	Commercial	Ka-band
27.5–30	U	U	U		
30–31	U	U	U	Military	

D = Down-link
U = Up-link
BSS = Broadcasting Satellite Service
Region 1 = Europe, Africa, Soviet Union
Region 2 = North and South America
Region 3 = Indonesia, Australia

2.5.3 Radio regulations

The frequency allocation is ruled by the International Telecommunication Union (ITU). The International Radio Consultative Committee (CCIR), a branch of the ITU, formulates recommendations relative to international radiocommunications.

For the Fixed-Satellite Service, the World Administrative Radio Conference (WARC) held in Geneva in 1979 and the Regional Administrative Radio Conference (RARC) held in Geneva in 1982 defined the frequency allocation as given in Table 2.10.

Many of those frequency bands are also used by other services and therefore can be used by satellite systems in the Fixed-Satellite Service on a shared basis only, which implies the possibility of mutual interference.

The CCIR publishes recommendations aimed at reducing interference levels. These recommendations deal with:

(1) The 'co-ordination area' of an earth station, such that a terrestrial microwave station outside this zone does not experience or cause interference higher than the acceptable level. (CCIR, Rep. 382–4, Vol. IV/IX-2, 1982) (Figure 2.30);
(2) Limitation of EIRP and power of terrestrial radio relay transmitters. (Rec. 406–5, Rep. 393–3 and 790–1, CCIR, Vol. IV/IX-2, 1982)
(3) Limitation of power transmitted towards the horizon by earth stations (CCIR Rep. 386–3 Vol. IV/IX-2, 1982); limitations of power flux density produced at the earth's surface by satellite in the Fixed-Satellite Service (Rec. 358–3 and Rep. 387–4, CCIR, Vol. IV/IX-2, 1982) (See Table 2.11):

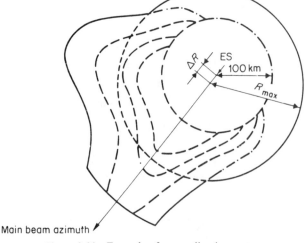

Figure 2.30   Example of a coordination contour.

Table 2.11 LIMITS OF POWER FLUX-DENSITY.

Frequency range (GHz)	Limit of power flux-density dB (W/m^2)			Reference bandwidth
	$\theta \leqslant 5°\,^a$	$5° < \theta \leqslant 25°$	$25° < \theta < 90°$	
1.7–2.5	−154	$-154 + 0.5\ (\theta - 5)$	−144	
2.50–2.69	−152	$-152 + 0.75\ (\theta - 5)$	−137	in any
3–8	−152	$-152 + 0.5\ (\theta - 5)$	−142	4 kHz
8–11.7	−150	$-150 + 0.5\ (\theta - 5)$	−140	band
11.7–15.4	−148	$-148 + 0.5\ (\theta - 5)$	−138	
15.4–23	−115	$-115 + 0.5\ (\theta - 5)$	−105	in any
31.0–40.5	−115	$-115 + 0.5\ (\theta - 5)$	−105	1 MHz band

^{a} θ: the angle of arrival of the wave (degrees above the horizontal).

(4) Station keeping of geostationary satellites (Rec. 484–2) and the required orbital spacing between two satellites (CCIR Rep. 713–1, Vol. IV-1)

(5) The characteristics of earth station antennae radiation pattern (Rec. 465–1, Rec 580, Rep. 391–4, Rep. 390–4 CCIR, Vol. IV-1, 1982) and satellite antennae radiation pattern (CCIR, Rep. 558–2, Vol IV-1, 1982)

Recommendations 27, 28, 29 and Appendix 28 of the Radiocommunications Regulations (1982 Edition), are also relevant.

REFERENCES

AMOROSO, F. (1980) The bandwidth of digital data signals, *IEEE Comm. Magazine*, **18**(6), pp. 13–24.

ARNBAK, J. C. (1983) The systems background for satellite communications antennae, in *Communication Antenna Technology*, North-Holland.

BHARGAVA, V. K., D. HACCOUN, R. MATYAS and P. NUSPL (1981) *Digital Communications by Satellite*, John Wiley, New York.

BROWN, R. J., M. L. GUHA, R. A. HEDINGER and M. L. HOOVER (1983) Companded single sideband (CSSB) implementation on COMSTAR satellites and potential applications to INTELSAT V satellites. *ICC83*, pp. E2.1.1–E2.1.6.

CAMPANELLA, S. J., F. ASSAL and A. BERMAN (1977) On-board regenerative repeater, *ICC77*, paper 6.2.

CHIAO, J. T. and F. CHETHIK (1976) Satellite regenerative repeater study, *Canadian Proc. Comm. Power Conf.*, pp. 222–225.

CUCCIA, C. L., R. S. DAVIS and E. W. MATTEWS (1977) Baseline considerations of beam switched SS/TDMA satellites using baseband matrix switching, *ICC77*, Chicago, pp. 6.3.126–6.3.131.

DUPONTEIL, D. (1981) Binary modulation for digital satellite communications link with differential demodulation, *5th Int. Conf. Digital Satellite Comm.*, Genoa. pp. 89–98.

FEHER, K. (1983) *Digital Communications*, Prentice-Hall.

GRONOMEYER, S. and A. MCBRIDE (1976) MSK and offset QPSK modulation *IEEE Trans. Comm.*, Vol. COM-24, No. 8 (Aug.), pp. 809–820.

JACOB, I. (1967) Comparison of M-ary modulation systems, *The Bell Technical Journal* (May–June), pp. 843–864.

KOBAYASHI, T. (1977) Degradation of cross polarization isolation due to rain, *Journ. Radio Research Laboratories*, **24**, 115.

KOGA, K., T. MURATANI, and A. OGAWA (1977) On-board regenerative repeaters applied to digital communications, *Proc. IEEE*, **65**(3) (Mar.), pp. 401–410.

LABORDE, E. and P. J. FREEDENBERG (1984) Analytical comparison of CSSB and TDMA/DSI satellite transmission and techniques, *Proc. IEEE*, **72**, (11) (Nov.), pp. 1548–1555.

LINDSEY, W. C. and M. K. SIMON (1973) *Telecommunications Systems Engineering*, Prentice-Hall.

MARAL, G. and M. BOUSQUET (1984) Performance of regenerative/conventional satellite systems, *Intern. Journ. of Satellite Communications*, **2** (3), pp. 199–207.

PASUPATHY, S. (1979) Minimum shift keying: a spectrally efficient modulation, *IEEE Comm. Magazine*, **17**(4) (July), pp. 14–22.

SHANNON, C. E. and W. WEAVER (1949) *The Mathematical Theory of Communication*, University of Illinois Press, Urbana.

SPILKER, J. J. (1977) *Digital Communications by Satellite*, Prentice-Hall.

SZARVAS, G. and H. SUYDERHOUND (1981) Enhancement of FDM–FM satellite capacity by use of compandors, *COMSAT Tech. Rev.*, **11**(1) (Spring), pp. 1–58.

TRAFFON, P. J., B. J. TUNSTALL, J. C. ELLIOTT *et al.* (1973) Error protection manual for AFCS, *Final Report AD 759-836, Air Force Communication Service*, Computer Sciences Corporation.

WACHIRA, M., V. ARUNACHALAM, K. FEHER and G. LO (1981) Performance of power and bandwidth efficient modulation techniques in regenerative and conventional satellite systems, *ICC81*, Denver, pp. 37.2.1–37.2.5.

CHAPTER 3   Satellite networks

Chapter 2 considered the case of a single satellite link between one transmitting and another receiving earth station. However, a satellite more often operates with several stations simultaneously. This chapter discusses the various aspects of multiple access by several stations to one satellite considering the simplest on-board non-regenerative transponders and the more complex regenerative transponders with on-board processing.

3.1   ONE-WAY LINK

A one-way link from a transmitter T_X to a receiver R_X on the earth's surface is very simple as shown in Figure 3.1. The satellite has two antennae and one repeater. As reception and transmission are continuous, a frequency conversion is essential to ensure isolation between the repeater input and output, even though the satellite antennae may be equipped with narrow beam antennae, they are not directional enough to ensure proper isolation by themselves. So

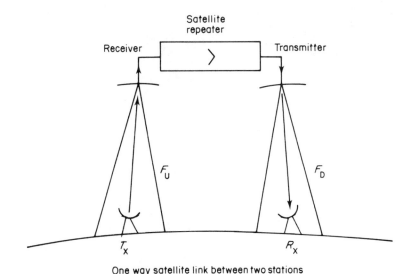

One way satellite link between two stations

Figure 3.1   One-way satellite link between two stations.

97

one carrier frequency, F_U, is allocated to the up-link, and a second, F_D to the down-link.

3.2 BROADCAST NETWORK

When broadcasting, several receiving stations R_{X1}, R_{X2}, R_{X3}, etc. are located within the down-link's zone of coverage and receive the signals broadcast by the satellite (Figure 3.2). If the transmitting station is in this same zone, the two antennae in Figure 3.2. can be replaced by a single one; the separation between the transponder input and output is then achieved by a duplexer and by filters tuned to F_U and F_D.

3.3 TWO-WAY LINKS BETWEEN TWO EARTH STATIONS

A two-way communication link between two earth stations can be achieved by simply doubling the elements in Figure 3.1 to achieve the configuration in Figure 3.3. The system comprises two up-links and two down-links and four frequencies are required: one can have a single receiving antenna and a single transmitting antenna if the earth stations are located in the same coverage area. The first commercial satellite, INTELSAT I used this principle: one transponder received at 6301.02 MHz and transmitted at 4081 MHz. The second received at 6389.97 MHz and transmitted at 4160.75 MHz, with both

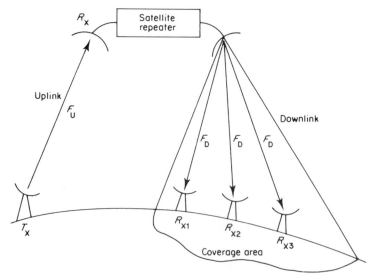

Figure 3.2 One-way satellite link: broadcasting mode

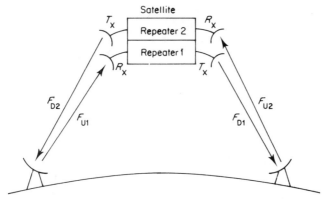

Figure 3.3   Two-way links between two earth stations

antennae ensuring the same earth coverage. Moreover, the two repeaters can be replaced by a single wide-band repeater, and eventually the two antennae by a single one using a duplexer.

3.4   MULTIPLE ACCESS

The trend towards small, inexpensive, direct-to-user terminals means that a satellite needs to interconnect a large number of scattered stations by means of two-way links. A simple case is where the two-way links are not simultaneous and the traffic in both directions is well defined (exchange of data for example) or if waiting time is acceptable. Then communication links can then be allocated on a regular basis (every hour, once daily, etc.), and for a limited duration. This ensures the satellite is in constant use, at the disposal of its associated stations successively, two at a time. The satellite is therefore time-shared between pairs of stations, the time unit being the duration of one communication. Meteorological data can be exchanged on this principle. However such situations are rare and usually many links are needed simultaneously between earth stations.

To achieve multiple two-way links one could simply increase the number of transponders in the system of Figure 3.3. However, as the number of transponders grows too large, due to the increasing numbers of links needed, it becomes advantageous to use a single wide-band transponder with appropriate multiple access techniques. This is the system used by the generation of commercial satellites that began with INTELSAT II.

Actually two problems must be solved as illustrated in Figure 3.4.

(1) Access to an earth station by a community of users.
(2) Access to the satellite by all earth stations.

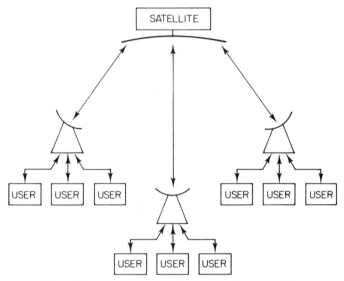

Figure 3.4   Access to an earth station and to a satellite

The former is usually solved in the same way as for terrestrial networks by the well known techniques of multiplexing and demultiplexing. The latter, which is specific to satellite communications, is solved by multiple access techniques.

3.4.1   Multiplexing and demultiplexing

Multiplexing consists of combining individual signals into a single one. Demultiplexing restores the individual signals from the multiplexed one assuming the signals that were multiplexed were not overlapping. There are several types of multiplex:

(1) Frequency Division Multiplex (FDM): signals occupying non-overlapping frequency bands are added and any one of these can be recovered by filtering.
(2) Time Division Multiplex (TDM): signals are compressed into high speed bursts which are placed in non-overlapping time slots within a time frame. Recovery of the original burst is accomplished by selection of the specific time slot in which the burst is positioned. Clearly this procedure requires timing references.
(3) Code Division Multiplex (CDM): signals are given a unique signature by a code transformation before they are combined in the frequency-time plane. Recovery is achieved by cross-correlation techniques which require synchronization of a known reference to the incoming bit stream.

3.4.2 Multiple access techniques

Multiple access is a variant of multiplexing which is specific to satellite communications. It may be defined as any technique by which a number of earth stations form communications links through one or more satellite radio frequency channels. Multiple access is the multiplexing of RF signals in the satellite channel.

Hence, there are three basic forms of multiple access:

(1) Frequency Division Multiple Access (FDMA).
(2) Time Division Multiple Access (TDMA).
(3) Code Division Multiple Access (CDMA).

There are also hybrids of these multiple access techniques.

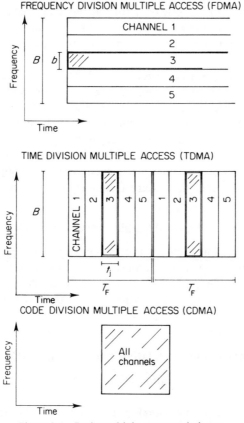

Figure 3.5 Basic multiple access techniques.

Figure 3.5 illustrates the basic types of multiple access using a time–frequency graph. The transponder bandwidth is B.

With FDMA each link requires a frequency band b, and if b is the same for all links, the number of simultaneous links that can be established is $n = B/b$. A particular address, i among the n channels is defined by the band b_i in B. This band is permanently assigned, at least for the duration of the link.

With TDMA, each link is only established during a time slot t_j, every T_F second. Such a procedure requires a burst mode transmission. The total frequency band B, is common to all links, and time is divided into sections of T_F length, called frame period. If t_j is the same for all links, $n = T_F/t_j$ simultaneous links can be established. The address of each link is defined by the position of the time slot t_j within T_F. All forms of TDMA require some form of control and synchronization of each station in the satellite network. If these are not present simpler systems termed time random multiple access (TRMA) result.

With CDMA, transmissions from each earth station are spread over the time–frequency plane by a code transformation. These techniques are referred to as spread spectrum systems. In addition to their multiple access capabilities, they are useful in combating jamming and are for this reason principally used in military systems.

Note that the multiple access mode only defines the address technique, and does not characterize the communication system completely. To fully characterise a satellite communication link the type of modulations and the multiplex technique of the baseband information should be specified in addition to the multiple access technique used.

For instance, the term FDMA/FM/FDM states that frequency division multiple access (FDMA) is used, that each carrier is frequency modulated (FM), and that baseband signals are frequency division multiplexed (FDM).

3.4.3   Frequency Division Multiple Access (FDMA)

With frequency division multiple access the available repeater bandwidth is divided into a number of smaller bands. An earth station transmits on one or more of these bands. Guard bands between subbands allow for imperfect filters and oscillators. Receiving earth stations select their desired carriers by RF frequency tuning and IF filtering.

3.4.3.1   Multiplexing and modulation schemes with FDMA

Three cases are worthy of mention (Figure 3.6):

(1) Multichannel per carrier analog transmission: the transmitting earth stations frequency division multiplex (FDM) several single sideband sup-

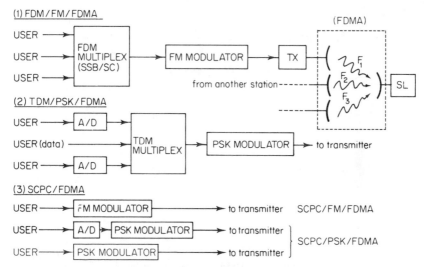

Figure 3.6   Multiplexing and modulation schemes with FDMA.

pressed carrier (SSB/SC) signals into one baseband assembly which frequency modulates (FM) a carrier. All carriers transmitted by different earth stations access the satellite at different frequencies. This is called FDM/FM/FDMA.

(2) Multichannel per carrier digital transmission: several pulse code modulated (PCM) baseband signals are time multiplexed TDM and used to phase shift key (PSK) a radiofrequency carrier. As above, all carriers, transmitted by different earth stations, have different frequencies. This is called TDM/PSK/FDMA.

(3) Single channel per carrier transmission: each baseband signal independently modulates a separate radiofrequency carrier. This is called SCPC/FDMA and can be accomplished by transmitting either a carrier modulated by digital information (SCPC/PSK/FDMA) or by analog information (SCPC/FM/FDMA). The digital information is original data or a quantized form of analog signals. The number of users per earth station may vary from one earth station to another. All carriers and hence users, access the satellite at different frequencies.

3.4.3.2   Network organization with FDM/FM/FDMA

Considering an FDMA network with N stations, there are three ways to allow each station to communicate simultaneously with the other $N-1$ stations.

Case (a) Using one carrier per link (Figure 3.7). In this case each station transmits as many carriers as there are receiving stations (that is $N-1$

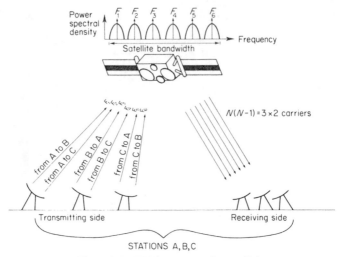

Figure 3.7   FDMA: one carrier per link.

carriers). As there are N transmitting stations, there are $N(N-1)$ carriers received and retransmitted by the satellite repeater and each receiving station only receives the $N-1$ carriers the station is concerned with. Considering a network with 10 earth stations, this implies that the satellite transponder handles 90 carriers.

Case (b) Transmission of a single carrier per station (Figure 3.8). Contrary to the above scheme, this scheme requires multiplexing of the baseband signals

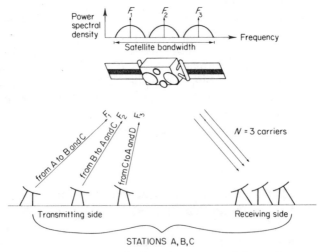

Figure 3.8   FDMA: single carrier per transmitting station.

(a) TRANSMITTED CARRIERS

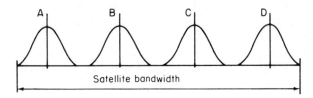

(b) BASEBAND SIGNAL MULTIPLEX (FDM or TDM)

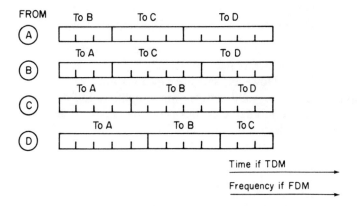

(c) EARTH STATION A EQUIPMENT BLOCK DIAGRAM

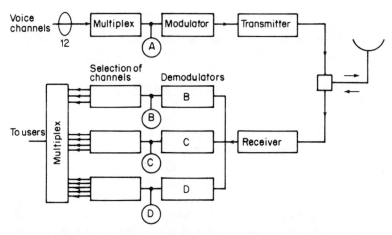

Figure 3.9 Example of a FDMA system with four earth stations and one carrier per station. *Reproduced by permission of Centre National d'Etudes Spatiales from Payet (1977).*

prior to transmission. The $N - 1$ baseband signals that a station must transmit are multiplexed and this multiplex modulates a single carrier. As one carrier is associated to one transmitting station, N carriers are received and retransmitted by the transponder. These are demodulated in the receiving stations which retain from the multiplex only the signals the station is concerned with (Figure 3.9). Considering a network with 10 earth stations, the satellite transponder handles only 10 carriers.

Case (c) Hybrid methods. Each station transmits several carriers (two or three) but each carrier is intended for several stations.

3.4.3.3 Intermodulation products (Shaft, 1965; Sevy, 1966; Spoor 1967)

The satellite repeater amplifies several frequency separated signals simultaneously which remain completely separated at output if the repeater is perfectly linear. However, satellite repeaters are equipped with a high power amplifier (TWT or solid state power amplifier, SSPA) which exhibits non-linear input output characteristics (see Chapter 6). If several carriers are fed to a non-linear device one finds at the output of the non-linear device the input carriers plus a number of signals which are linear combinations of the input frequencies:

$$F_{IM} = M_1 F_1 + M_2 F_2 + M_3 F_3 + \ldots + M_n F_n$$

where F_{IM} is the frequency of one of these signals called 'intermodulation products'.

The order of any intermodulation product is defined as:

$$M = |M_1| + |M_2| + |M_3| + \ldots + |M_n|$$

If the input frequencies are contained within a small bandwidth relative to the central frequency, then only odd order intermodulation products with $|\sum M_i| = 1$ fall within the bandwidth.

In most cases only third order and fifth order intermodulation products are significant.

Intermodulation products have two indesirable effects:

(1) A fraction of the available power is wasted.
(2) The odd intermodulation products fall within the transponder bandwidth and interfere with the wanted signals.

In cases where the n carriers have the same level at the transponder input, the gain is the same for all the carriers, but if one of the carriers has an input level higher than the others, it benefits from greater gain (capture effect). If two signals have differences in level of 10 dB at a repeater input, their output levels differ by 16 dB if the repeater has a hard limiter, and 14.5 dB with TWT limiting (Sevy, 1966).

Reducing intermodulation products is achieved either by using a linearizer or by driving the high power amplifier (HPA) with a signal power level lower than that required to operate at saturation (input back-off). The HPA then acts more linear. The corresponding relative reduction of the output power is called 'output back-off'.

The spectral density of intermodulation products generated by modulated carriers appears to be nearly constant within the bandwidth, so that intermodulation products can be considered as filtered white noise with constant power density $(N_0)_{IM}$.

For systems calculations, one introduces the ratio $(C/N_0)_{IM}$ which expresses the ratio of the carrier power to the intermodulation equivalent noise power spectral density. Figure 3.10 shows an example of variation of this ratio in terms of the input back-off. Note that intermodulation noise increases, and subsequently $(C/N_0)_{IM}$ decreases with the number of carriers.

The ratio $(C/N_0)_T$ for the total link is expressed according to the following formula, which assumes the repeater bandwidth to be negligible compared with $(C/N_0)_U$ and $(C/N_0)_D$ and introduces the effect of the intermodulation products (Dicks *et al.*, 1972)

$$(C/N_0)_T^{-1} = (C/N_0)_U^{-1} + (C/N_0)_D^{-1} + (C/N_0)_{IM}^{-1} \qquad (1)$$

$(C/N_0)_U$ represents the ratio of carrier power to the noise power spectral density for the up-link; $(C/N_0)_D$ represents the ratio of carrier power to the noise power spectral density for the down-link; and $(C/N_0)_T$ represents the ratio of carrier power to the noise power spectral density for the total link (measured at the receiving earth station).

Figure 3.11 shows $(C/N_0)_T$ versus input back-off. It can be seen that $(C/N_0)_T$ is maximum for a given back-off value, which would be the optimum value to adopt in practice.

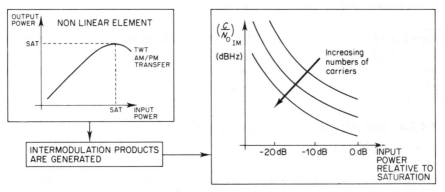

Figure 3.10   Signal-to-intermodulation noise ratio

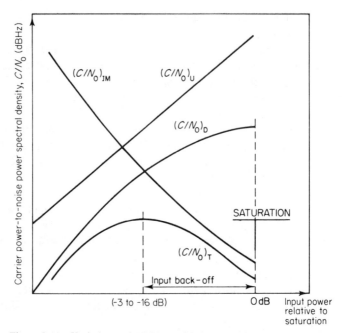

Figure 3.11   Variations of $(C/N_0)_T$ with input back-off. *Reproduced with permission from Dicks et al (1972).*

As the number of carriers at the input of the transponder increases, several effects combine together and a reduction in transponder capacity results:

(1) The total power at the output of the transponder is reduced due to back-off operation.
(2) Useful power for carriers is reduced as a fraction of the total power turns into intermodulation noise power.
(3) The intermodulation noise contribute to increase the overall noise power density at the earth station receiver input.

Figure 3.12 shows the capacity of a 36 MHz, global coverage repeater of an INTELSAT IV or IVA satellite according to the number of carriers. The number of baseband channels for a single access (one carrier) is around 1000, and drops to nearly 400 for multiple access with 14 carriers.

3.4.3.4   Summary and conclusions on FDMA

FDMA has the following problems:

(1) Need for linear transponder: use of a linearizer, or back-off operation (Figure 3.11);

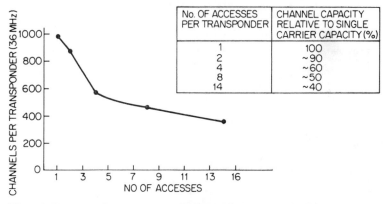

No. OF ACCESSES PER TRANSPONDER	CHANNEL CAPACITY RELATIVE TO SINGLE CARRIER CAPACITY (%)
1	100
2	~90
4	~60
8	~50
14	~40

Figure 3.12 Channel capacity of an INTELSAT 36-MHz bandwidth transponder versus the number of accesses.

(2) Loss in transponder capacity relative to a single access with back-off operation (Figure 3.12);

(3) Lack of flexibility due to complicated procedures when changing frequencies allocated to different stations.

(4) Necessity for transmitted power control at the earth stations, so that all carriers access the transponder with a specified equal power.

The main advantage of FDMA is that it involves well mastered techniques, and uses equipment which for years has been employed in the field of terrestrial microwave networks. In addition, no synchronization is needed between the different stations in the network, unlike TDMA.

3.4.4 Time Random Multiple Access (TRMA)

With time multiple access schemes, signals are transmitted by earth stations in a burst mode. If no scheduling is provided between the transmitting earth stations, this type of access is called time random multiple access. The simplest method is for the earth stations to transmit bursts without regard for other stations (Figure 3.13). A burst is defined as a finite sequence of bits. Owing to the lack of co-ordination among the distributed earth stations, packets from different stations may reach the satellite at the same time and collide, thereby destroying the information content (Figure 3.14). Therefore a subsequent retransmission of the packet is required. In this simplest form, this protocol is known as the ALOHA protocol (Abramson, 1977; Hayes, 1981).

A more elaborate version of the ALOHA protocol, namely the slotted ALOHA (or S-ALOHA), reduces the probability of collisons by rudimentary synchronization among the terminals which are now requested to transmit bursts according to a common clock. Only messages arriving at the satellite in

SATELLITE NETWORKS

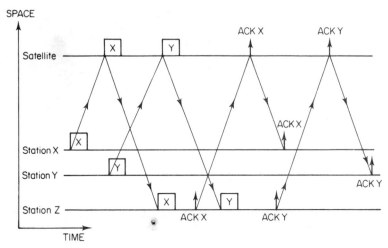

Figure 3.13   Space/time graph of a time random multiple access.

the same time interval interfere with one another (see Figure 3.15). This method reduces the rate of collisions by half (Abramson, 1977; Derosa *et al.*, 1979; Bhargava *et al.*, 1981).

Performance measures of such protocols are *throughput* and *delay*. The throughput is defined as the ratio of the information bits delivered to the user to the total number of bits transmitted and measures the ratio of successful bursts to the total that were sent. The delay is the time between initial transmission of information and successful delivery to the user.

Figure 3.16 shows that maximum throughput is 18 per cent with ALOHA

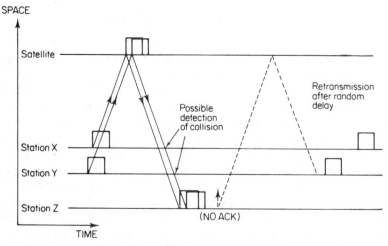

Figure 3.14   Packet collision with a time random multiple access system

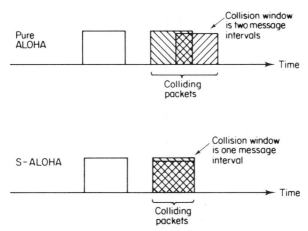

Figure 3.15   Comparison between pure-ALOHA and slotted-ALOHA collision schemes.

protocol and 36 per cent by using the S. ALOHA protocol (Tobagi, 1980). Figure 3.17 illustrates the variation of delay versus throughput. (Tobagi *et al.*, 1984).

The low bandwidth utilization of the ALOHA and the S-ALOHA protocols has led to many proposals for increasing utilization by means of *slot reservation* schemes. The object of slot reservation schemes is to reserve a particular time slot for a given station. This ensures that no collision takes place. This increase in channel utilization efficiency is obtained at some overhead cost, either in terms of allocation of part of the bandwidth for reservation purposes

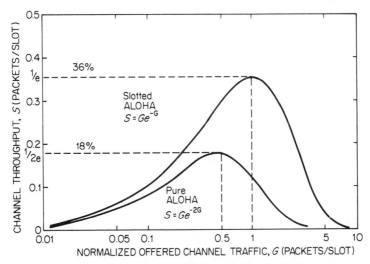

Figure 3.16   Throughput of pure-ALOHA and slotted-ALOHA schemes.

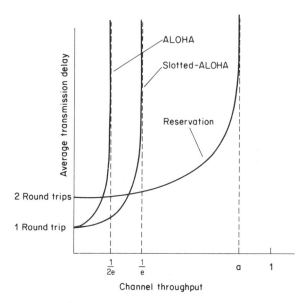

a = fraction of bandwidth not used for reservation
(typ. 0.7 to 0.9)

Figure 3.17   Mean transmission delay versus channel throughput.

and/or increased complexity of the control mechanisms in transmitting stations. All reservation methods use some form of framing approach and the reservation scheme can be either *implicit* or *explicit* (Retnadas, 1980). The *implicit* method involves *reservation* by use—so that whenever a station successfully transmits in a slot, all the stations internally assign that slot in subsequent frames for exclusive use by the successful station. This is called R-ALOHA (Crowther *et al*, 1973; Roberts, 1973). In this scheme, there is no way to prevent a station capturing most or all of the slots in a frame for an indefinite time. *Explicit reservation* is a distinct and unique assignment of slots to a user by the network scheduler. Two examples are R-TDMA and C-PODA (Contention based Priority Oriented Demand Assignment) (Jacobs *et al.*, 1978; Bhargava *et al.*, 1981).

The R-TDMA protocol establishes a permanent association between slots and stations. The slots not claimed by the original owner may be reassigned on a 'round robin' basis to the stations that have traffic to send ('round robin' means that each station gets access in an ordered, sequential, but not necessarily predetermined manner).

The C-PODA is a satellite channel access scheme which allocates slots to users based on prior reservations. Channel time is divided into an information subframe and a control subframe. The control subframe is used exclusively to send reservations which cannot be sent in the information subframe by piggybacking them in the header of ongoing traffic. Random access is used to access to the reservation slots of the control subframe.

Figure 3.17 shows that with reservation schemes a higher throughput is achieved at the expense of a larger transmission delay, specially at low load.

3.4.5 Time Division Multiple Access

In TDMA each earth station is required to transmit bursts in short non-overlapping timing intervals (Figure 3.18). Therefore a TDMA scheme requires some form of *frame structure* and a global *timing mechanism* to achieve non-overlapping transmission. For this reason a TDMA system is complex as it is necessary to establish co-ordination among the earth stations. In the simplest *fixed assignment* TDMA scheme, called F-TDMA, the frame time is divided into slots of fixed duration with the slots equally divided among the stations (Gerla *et al.*, 1977).

3.4.5.1 Frame structure

Figure 3.19 shows a typical TDMA frame structure. A frame is the time interval over which a signal format is established and repeated. A frame is subdivided into time slots and a burst consists of an exact number of slots and occupies a precise position in the frame. The set of bursts, one from many of the different earth stations, builds up into the frame. The frame duration T_F is typically of the order of a few milliseconds ($T_F = 2$ ms with INTELSAT V

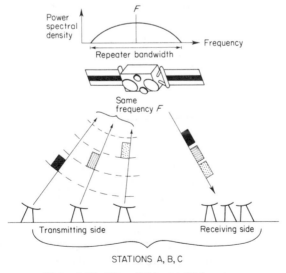

Figure 3.18   Time division multiple access.

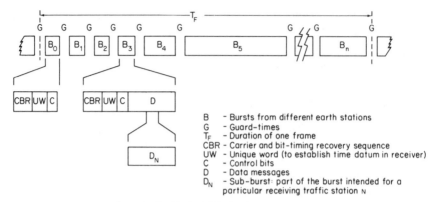

Figure 3.19   Typical TDMA frame structure.

TDMA system, $T_F = 20$ ms with TELECOM 1 TDMA system). A small time gap called guard time (G) is inserted between bursts to ensure that adjacent bursts do not interfere. The first burst in the frame (B_0 in Figure 3.19) called *reference burst* contains no traffic but serves to synchronize and identify the frame. It is transmitted by the controlling earth station, or reference station. Other bursts ($B_1, B_2, \ldots$) are transmitted by traffic stations and have two parts:

(1) A *header*, or preamble, which consists of a part for carrier and bit timing recovery (CBR), a burst codeword or *unique word* (UW) for burst synchronization, and *control bits* (C) for station identification and housekeeping functions including signalling bits and orderwire for voice or data.
(2) A *data message* (D) which consists of as many sub-bursts as destinations of the burst and supports the traffic from one station to the other.

3.4.5.2   Synchronization

Clearly TDMA bursts transmitted by earth stations must not interfere one with another. Therefore each earth station must be capable of first locating and then controlling its transmit burst time phase. Each burst must arrive at the satellite transponder at a prescribed time τ_N relative to the reference burst transmitted by the reference station. This ensures that no two bursts overlap and that the guard time between any two bursts is small to guarantee a high transmission efficiency (see Figure 3.20).

Synchronization is the process of providing timing information at all stations and controlling the TDMA bursts so that they remain within their prescribed slots. All this must operate even though each earth station is at a different distance from the geostationary satellite, and considering the motion

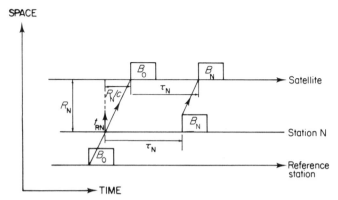

Figure 3.20 Space/time graph illustrating TDMA synchronization.

of the satellite with respect to the Earth: Geostationary satellites are located at a nominal longitude and typically specified to remain within a 'window' with sides of $0.1°$ as seen from the centre of the Earth. Moreover, the satellite altitude varies as a result of a residual orbit eccentricity of about 0.001. The satellite can thus be anywhere within a volume of space which is typically 75 km × 75 km × 85 km (Figure 3.21). The tidal movement of the satellite causes an altitude variation of about 85 km, resulting in a round trip delay variation of about 500 microseconds and a frequency change of signals due to the Doppler effect (a typical maximum radial velocity would be 20 km/hour).

It is assumed that all stations are able to receive the reference burst. Each station uses this reference burst to establish its own local TDMA frame, defined as a fictive frame beginning at time t_{RN} (Figure 3.20). Due to transmission delay the local TDMA frame should be aligned with the actual frame at

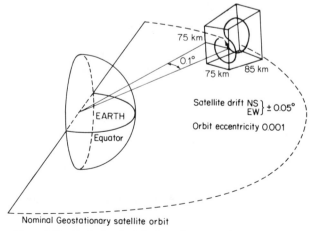

Figure 3.21 Domain of evolution of a geostationary satellite.

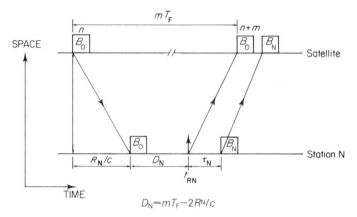

Figure 3.22   Space/time graph illustrating the synchronization of an
earth station.

the satellite transponder input. Synchronization is achieved with the
knowledge of t_{RN}, which is obtained by introducing a delay D_N relative to the
reception of the reference burst (see Figure 3.22). Detection of the unique
word of the reference burst establishes the reference time and is performed
using the autocorrelation properties of the unique word sequence. Figure 3.22
shows that D_N is equal to the difference between the round trip propagation
time and mT_F, where m is an integer. Thus:

$$D_N = mT_F - 2R_N/c \qquad (2)$$

where R_N is the distance between the satellite and the earth station, and c the
velocity of light.

m is selected so that the delay D_N is the smallest, but still positive, when R_N
is the distance between the satellite and the furthest earth station. For instance,
with Telecom I (Bousquet, 1981), where $T_F = 20$ ms and $2R_N/c \simeq 280$ ms, m
is larger or equal to 14.

Two main classes of TDMA synchronization methods can be identified
according to whether D_N is determined by the earth station directly from
monitoring its own transmission ('closed loop control' method), or transmitted bursts are not received by transmitting stations, and D_N must be determined by other means ('open-loop control' method) (Nuspl *et al.*, 1977).

Figure 3.23 shows an example of the *closed loop synchronization* method.
The station observes the position of its own burst relative to the reference burst
by measuring time interval between the reference burst unique word and the
traffic burst unique word, computes the error, and determines a new value for
the delay D_N. If $\varepsilon_N(j)$ is the observed error then the corrected delay $D_N(j+1)$
is given by

$$D_N(j+1) = D_N(j) - \varepsilon_N(j) \qquad (3)$$

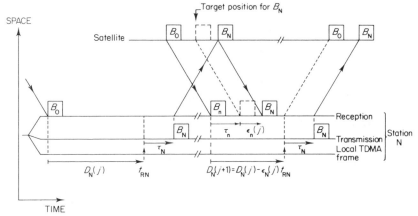

Figure 3.23   Space/time graph illustrating TDMA closed-loop synchronization.

Open loop synchronization refers to control of traffic burst position based on the distance d_N between the satellite and the earth station (Campanella and Hodson, 1979). R_N is calculated from accurate measurements of satellite position and precise knowledge of earth station locations. This method does not require that earth station should receive their own transmitted bursts. The delay D_N relative to the reception time of the reference burst is calculated using (2).

Open-loop synchronization for a TDMA network does not offer the accuracy and high precision of closed-loop methods, but have the advantage of reduced TDMA terminal costs.

3.4.5.3   Frame acquisition methods

Prior to synchronization, *acquisition* must be performed. Frame acquisition is the process whereby a station is brought into synchronism with an operating TDMA network.

Closed-loop acquisition is achieved by transmitting either a low power burst, or a pseudo-noise sequence modulated burst, observing its position, correcting it to an assigned position and then transmitting the traffic burst at full power. A pseudo-noise (*PN*) random sequence is a binary sequence which has random properties, including high length before repetition of the same pattern and specific auto-correlation properties. By using wideband signals at low power level and taking advantage of the auto-correlation properties, precise synchronization is possible with minimal interference to other signals.

Open-loop acquisition is initiated by informing a station entering the network of the time shift D_N relative to the reference burst at which the station should transmit a short burst consisting of a preamble and a unique word. Due

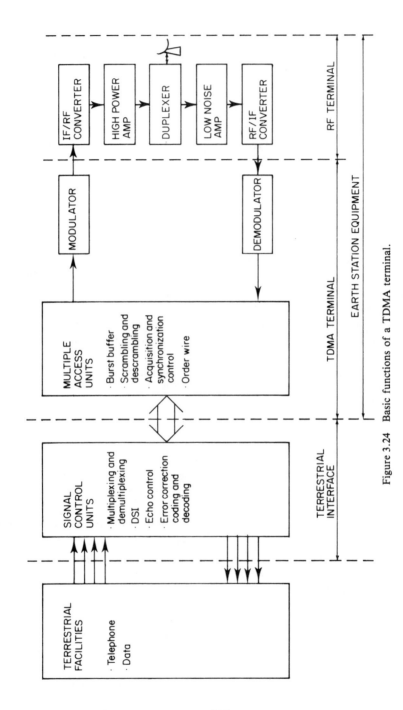

Figure 3.24 Basic functions of a TDMA terminal.

118

to uncertainty in determining the range between the satellite and the station, the delay should be calculated so that the burst falls near the centre of the assigned epoch of the TDMA frame.

3.4.5.4 TDMA earth terminals

Figure 3.24 shows the basic functions of a TDMA terminal which handles voice and data traffic. Signal control units interface terrestrial facilities with the earth station equipment. The unit which processes speech ready for transmission is separate from that which prepares data. Data requires an error-correcting code whereas speech may not.

Echo control consists in annuling the return signal of a telephone speaker due to partial reflection at the listener's side. Echo results from delay transmission and impedance mismatches at the four-wire to two-wire hybrid junction of a long distance telephone circuit (Demytko and English, 1977).

Digital speech interpolation (DSI) is used to condense the telephone traffic. This is obtained by accommodating calls in the idle time between calls and conversation pauses during calls. DSI provides a bit rate reduction with a factor 1.5 to 2. The theoretical limit for the DSI advantage is 3 to 4 depending on the voice activity factor assumed (Campanella, 1976) for the speech. DSI is a technique which makes better use of transponder capacity by allowing for transmission of more bit/s per unit of bandwidth.

Bit streams from both the speech and data access units are sent to the multiple access facility which buffers and scrambles this traffic before transmitting it in appropriately timed high bit rate bursts. The size of the buffers is conditioned by the frame duration. *Scrambling* consists of module-2 adding of a pseudo-noise (PN) random sequence to bit streams in order to avoid long sequences of 1's and 0's and distribute the transmitted frequency spectrum more evenly across the satellite transponder. A header, or preamble, is generated and added to each burst prior to modulation.

The received signals are demodulated. Unique word from each received burst preamble is detected and bursts intended for the given earth station are processed into a terrestrial signal format. The multiple access unit extracts various control criteria from the received signal, particularly for the correction of D_N, and provides a clock signal for the different units of the system. For further interest Chapter 8 of Feher (1983) by S. J. Campanella and D. Schaeffer offers a detailed description of TDMA earth terminals implementation and signal processing.

3.4.5.5 TDMA frame efficiency and system capacity

A practical measure of the efficiency of a TDMA system is the ratio of the time devoted to transmission of information bits in the frame to the total frame length. This is called *frame efficiency* and is expressed as:

$$\eta = 1 - \sum t_i / T_F \tag{4}$$

where $\sum$ is the sum of all the guard times and preambles including the reference burst. With operational systems the requirement for high availability may imply that two redundant reference stations be implemented, each one transmitting a reference burst. There are then two reference bursts to be considered in the frame.

Expression (4) then becomes:

$$\eta = 1 - \frac{(n+2)P}{RT_F} \tag{5}$$

where R is the satellite transmission link bit rate, n is the number of traffic bursts in a frame considering two reference bursts per frame, P is the number of bits in the preamble plus guard time, and T_F is the frame period.

Frame efficiencies are currently of the order of 0.9. For efficient systems $\sum t_i = (n+2)P/R$ should be small and T_F should be large. Guard times are governed by synchronization accuracy and preambles are governed by demodulation and signalling requirements. The limiting constraint on frame length is that the size of the buffers used for compression and expansion of the data to form the TDMA bursts increases with the frame length and so increases the costs of the earth station equipment. Moreover, transmission delay increases with frame length.

As an example, for the INTELSAT BG 42.65 TDMA system, the number of telephone channels that can be supported by a TDMA system is

$$v = \frac{R}{r\eta}$$

where r is the bit rate of a voice channel. Assuming $n = 12$ transmitting stations, $P = 680$, $T_F = 2$ ms, $r = 64$ kbit/s, $R = 120.832$ Mbit/s, hence efficiency is 96 per cent and $v = 1813$ voice channels.

The capacity can be further increased by use of DSI which is a facility incorporated in the TDMA/DSI BG 42.65 Intelsat system.

3.4.5.6 Summary and conclusions on TDMA

Compared with FDMA, TDMA has many advantages:

(1) No intermodulation products are generated within the transponder, and the satellite transmitter operates at saturated output power. However, the non-linearity effect remains and results in intersymbol interference which introduces some degradation in the demodulation process and hence in the bit error rate.

(2) No precise adjustment of carrier power transmitted by earth stations is necessary.
(3) All stations transmit and receive on a single frequency, whatever the destination of the bursts is. This simplifies RF tuning.
(4) Facilities provided by digital techniques can be used, such as: storage, coding, DSI.

The main drawbacks are the requirements for network synchronization and the increased complexity of ground station equipment, but costs tend to reduce with decreasing costs of digital circuitry.

3.4.6 Code Division Multiple Access (Simon *et al*, 1985; Dixon, 1976)

With code division multiple access all users simultaneously operate within the same frequency band and each user occupies all the time the entire transponder bandwidth. Each user combines the signal to be transmitted with a signature sequence which displays two main correlation properties: (1) each sequence can easily be distinguished from a time shifted version of itself (2) each sequence can be easily distinguished from every other one in the set. Using these properties the receiver is able to separate the received signals even though they occupy the same bandwidth at the same time.

Two most widely used CDMA techniques are *direct sequence* (DS) and *frequency hopping* (FH). CDMA which results in an occupied bandwidth in excess of the minimum necessary to send the information is often called Spread Spectrum Multiple Access (SSMA).

3.4.6.1 Direct sequence

Figure 3.25 shows the principle of Direct Sequence operation. The binary baseband message at rate $1/T_b$ is multiplied by a pseudo-noise (PN) random sequence at rate $1/T_c$, where the chip duration T_c is much smaller than T_b. The composite signal is then used to PSK modulate a carrier with the same frequency for all stations. The transmitted signal expresses as:

$$s(t) = m(t)p(t)\cos \omega_c t$$

where $m(t)$ represents the baseband mesage, $m(t) = \pm 1$, and $p(t)$ represents the PN sequence, $p(t) = \pm 1$.

At the receiver side, the carrier is demodulated by multiplying the incoming modulated carrier with a coherent replica of the carrier. Neglecting thermal noise, this results in the following:

$$r(t) = m(t)p(t)\cos \omega_c t \times 2 \cos \omega_c t = \underset{\text{low frequency component}}{m(t)p(t)} + m(t)p(t)\cos 2\omega_c t \qquad (7)$$

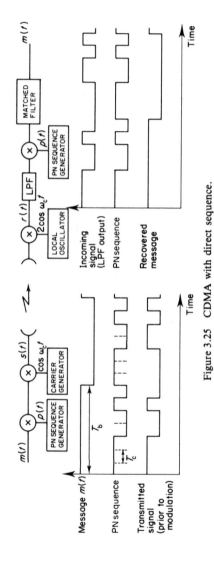

Figure 3.25   CDMA with direct sequence.

A low-pass filter retains the low frequency component. This component is then multiplied with a local PN sequence, the same as that of the transmitter. If the local sequence is in phase with the received signal sequence, then the signal $x(t)$ at the matched filter output is the desired message:

$$x(t) = m(t)p(t) \times p(t) = m(t)p^2(t) = m(t) \tag{8}$$

The power density spectrum of the radio-frequency signal is given by:

$$S(f) = \frac{P}{R_c} \left[\frac{\sin \pi (f - f_c)/R_c}{\pi (f - f_c)/R_c} \right]^2 \tag{9}$$

where P is the carrier power $f_c = \omega_c/2\pi$ is the carrier frequency.

It can be noted that the bandwidth corresponding to the first nulls of this function is $2 R_c$, and is (R_c/R_b) times that of the PSK modulated carrier by the baseband message alone. This is the result of the spreading action of the PN sequence.

3.4.6.2 Frequency hopping:

Figure 3.26 shows the principle of frequency hopping operation. The binary baseband message at rate $1/T_b$ is used to PSK modulate a carrier the frequency of which is hopped over the total available bandwidth under the control of a PN random sequence with chip rate $1/T_c$, larger than the information bit rate

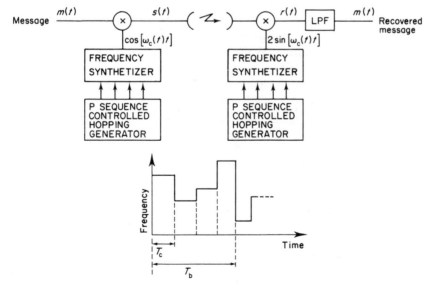

Figure 3.26 CDMA with frequency hopping.

$1/T_b$. The transmitted signal expresses as:

$$s(t) = \cos[\omega_c(t)t + \phi(t)] \qquad (10)$$

where $f_c = \omega_c/2\pi$ is the instantaneous carrier frequency as given by the frequency synthetizer controlled by the PN sequence, and $\phi(t)$ is the $m(t)$ message related phase (for instance with BPSK $\phi(t) = m(t)\pi/2$ and $m(t) = \pm 1$)

At the receiver side, the carrier is demodulated by multiplying the incoming modulated carrier with a in-phase replica of the carrier synthetized within the receiver under the control of the same PN sequence. Neglecting thermal noise, this results in the following:

$$r(t) = \cos[\omega_c(t)t + \phi(t)] \, x2 \sin[\omega_c(t)t] = \underset{\substack{\text{low frequency} \\ \text{component}}}{\sin \phi(t)} + \sin[2\omega_c(t)t + \phi(t)] \qquad (11)$$

A lowpass filter retains the low frequency component. Hence, if the local PN sequence is in phase with the received signal hopping sequence, then the signal at the output is the desired message. For instance, with BPSK:

$$x(t) = \sin \phi(t) = \sin m(t)\frac{\pi}{2} = m(t) \qquad (12)$$

3.4.6.3 CDMA performance

The above description has shown how the receiver is able to recover the message transmitted by a specific user. The CDMA operation is based on the fact that multiplication of two unrelated signals produces a signal whose spectrum is the convolution of the spectra of the two components signals. Thus, if the baseband message is narrow band compared with the spreading PN sequence, the product signal will have nearly the wider spectrum of the spreading sequence. This is illustrated on Figure 3.27 where the radio frequency signal bandwidth B is much larger than the baseband message bandwidth W (which for an information bit rate of R_b is approximately $1/R_b$). At the demodulator, where the received signal is multiplied by an in-phase replica of the same spreading sequence than that used by the transmitter of the desired signal, the original message is restored. If there is any undesired signal at the receiver, the local sequence spreads it just as it did the original signal at the transmitter. Thus any undesired interfering signal in the band of interest will have a bandwidth of at least B. If its power is $\dot{I}$ watt, its average density assuming uniformity is;

$$N_0 = \dot{I}/B \qquad (13)$$

If we assume that this noise density is larger than that of the thermal noise,

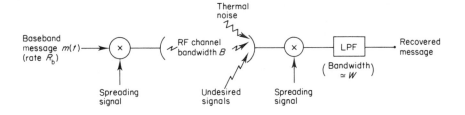

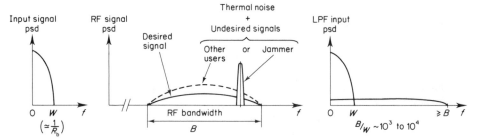

Figure 3.27   Spectrum spreading with CDMA.

the energy-per-bit to noise power density ratio expresses as:

$$E_b/N_0 = PB/IR_b \qquad (14)$$

where P is received carrier power.

Hence the undesired to desired signal power ratio is:

$$I/P = (B/R_b)/(E_b/N_0) \qquad (15)$$

B/R_b appears as the ratio of two signal-to-noise ratios. For a given bit error rate, E_b/N_0 is specified, and the larger B/R_b the higher the level of acceptable interference is. B/R_b is often called the *processing gain*. Practical values range from 1000 to 10 000.

The interfering signal may be an intentional jammer, as in the case of military satellite communications where the hostile interferer transmits for instance a high power narrow bandwidth radiofrequency signal, or the other users of a commercial satellite communications system who share the same radiofrequency bandwidth. In the first case the ratio I/P, often called the *jamming margin*, expresses the capability of the system to perform in an hostile environment. In the second case the effect of the other users can be considered to consist of additive broadband noise. If one considers N users with equal power, the interfering power I expresses as:

$$I = (N-1)P \qquad (16)$$

and using expressions (15) and (16) the maximum number of active users that

Table 3.1 MAXIMUM NUMBER OF USERS IN A CDMA SYSTEM WITH A
36 MHz BANDWIDTH TRANSPONDER AND A PROCESSING GAIN OF
10^3 USING BPSK OR QPSK MODULATION.

Required BER (BPSK or QPSK)	E_b/N_0	Maximum number of users
10^{-4}	8.4 dB	145
10^{-5}	9.6 dB	110
10^{-6}	10.5 dB	90

can share the total bandwidth for a given bit error rate is given by:

$$N_{\max} = 1 + \frac{B/R_b}{E_b/N_0} \qquad (17)$$

The above formula is the result of a simplified analysis which assumes that
thermal noise power density is negligible compared with interference noise
density, that interfering noise densities are additive and that they express as
given by (13). In a practical situation the number of users would be less.

As an example, considering a satellite transponder bandwidth of
$B = 36$ MHz and assuming a practical processing gain of 1000, the message

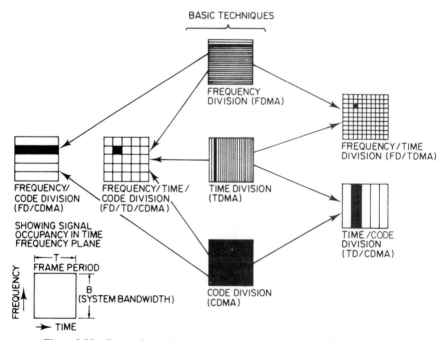

Figure 3.28 Generation of hybrid multiple access tecniques (Wittman, 1967).

information bit rate is $R_b = 36\ 10^6/1000 = 36$ kbit/s. Using formula (17) the maximum number of users depending on the required BER is given in Table 3.1.

3.4.7 Hybrid multiple access techniques (Wittmann, 1967)

The three basic techniques previously discussed can be combined, as graphically suggested in Figure 3.28 to generate four distinct hybrids:

(1) Combined frequency division and time division (FD/TDMA).
(2) Combined frequency division and code division (FD/CDMA).
(3) Combined time division and code division (TD/CDMA).
(4) Combined frequency division, time division and code division (FD/TD/CDMA).

3.5 FIXED AND DEMAND ASSIGNMENT

In the previous sections multiple access techniques have been discussed as a function of the division of the time and frequency domain. In this section the question of how the channels are assigned to the users is discussed. There are two basic alternatives:

(1) *Fixed assignment* of the channels.
(2) *Dynamic (demand) assignment* of the channels (DAMA—Demand Assigned Multiple Access).

Needs are shared between the two following extreme cases:

(1) High capacity links between a few international transit centres.
(2) Low capacity links between a large number of small, low traffic stations, sometimes only necessitating a single channel.

3.5.1 FDMA fixed assignment

Fixed assignment (preassigned channels) is best applied to high capacity commercial systems. It can be implemented using FDM/FM/FDMA with capacities in the order of 14 000 circuits (INTELSAT V) with each circuit having a high time occupancy. This approach is not appropriate for low traffic links such as are encountered in many national systems where isolated villages are distributed over vast areas with little infrastructure. So links are more often preassigned on the basis of a small number of voice channels per carrier, or even only one channel per carrier (SCPC: single channel per carrier). Each carrier uses a very narrow band, and the spacing between two carriers can be as small as a few tens of kHz, compared with frequencies of 4, 6, 11 and 14 GHz. This poses a problem with the stability of the received frequency. To compensate for drift in the local oscillator of the transponder and the effect of movements of the satellite, a very stable pilot frequency may be transmitted by one of the stations, and monitored at reception. Moreover, the use of a large number of carriers leads to a reduction in the total transponder capacity (see Figure 3.12).

3.5.2 FDMA demand assignment

The problem with fixed assignment SCPC is that an unused channel cannot be used to establish any other link. The less each circuit is used, the greater the penalty. *Demand assignment* offers a solution where channels are pooled, and so one channel can be assigned to a link between two stations just for the duration of the communication.

3.5.2.1 Example: the SPADE system

The SPADE system (Single Channel per Carrier PCM Multiple Access Demand Assigned Equipment) (Puente *et al.*, 1971; Edelson and Werth, 1972), is an example of demand FDMA system where the 36 MHz bandwidth of an INTELSAT transponder is accessed by 800 discrete carriers (400 circuits) establishing a common frequency pool available on demand for any station in the network (Figure 3.29).

Each voice channel modulates a particular carrier in the allocated band. Words are transmitted digitally using pulse code modulation (PCM) and transmission is by quadrature phase shift keying (QPSK). The channel carrier is voice activated, i.e. the carrier is on only if voice is detected. Since the average talker speaks only 40 per cent of the time, only 40 per cent of the total carriers are on. This means that for an 800-channel system, only 320, on the average are supported by the transponder.

A common signalling channel (CSC) operating in a time division multiple access mode allows for a distributed control of the system among all stations. Table 3.2 summarizes the main characteristics of the SPADE system.

3.5.2.2 Comparison of SPADE with preassignment operations

Table 3.3. shows a capacity comparison of a 36 MHz bandwidth transponder using either preassignment of carriers (FDM/FM/FDMA) or the SPADE system (Puente *et al.*, 1971). The advantages of SPADE must be weighted against the higher cost of ground equipment needed for signal processing.

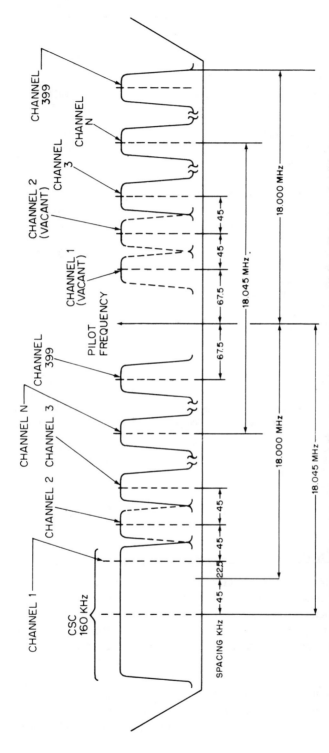

Figure 3.29 SPADE transponder frequency plan. *Reproduced by permission of Communications Satellite Corporation from Edelson and Werth (1972).*

Table 3.2　CHARACTERISTICS OF THE SPADE SYSTEM.

Communications channel characteristics

Channel encoding	PCM
Modulation	4-Phase PSK (coherent)
Bit rate	64 kbps
Bandwidth per channel	38 kHz
Channel spacing	45 kHz
Stability requirement	± 2 kHz (with AFC)
Bit-error rate at threshold	10^{-4}

Common signalling channel characteristics

Access type	TDMA
Bit rate	128 kbps
Modulation	. 2-Phase PSK
Frame length	50 ms
Burst length	1 ms
Number of accesses	50*
Bit-error rate at threshold	10^{-7}

*49 stations plus 1 reference.

Table 3.3　COMPARISON OF PREASSIGNMENT AND DEMAND ASSIGNMENT OPERATION (36 MHz bandwidth transponder).

Frequency division multiple access type	RF bandwidth per carrier	Voice channels per carrier	Total accesses per transponder	Total voice channels per transponder
Preassignment				
FDM/FM	2.5 MHz	24	14	336
FDM/FM	5 MHz	60	7	420
Demand assignment				
SPADE	0.045 MHz	1	800	800

3.5.3　TDMA demand assignment

The simplest fixed assignment TDMA scheme (discussed in Section 3.4.5), has a time frame divided into slots of fixed duration with the slots permanently allocated to each of the stations. However, in fixed assignment TDMA, when there are many lightly loaded earth stations, these slots are wasted when, as often happens, these stations do not use their dedicated slots.

An efficient TDMA scheme requires demand assignment which permits the earth stations to continually vary their demand for channels. This scheme should allow stations to be allocated variable duration slots or more slots per frame than others. With demand assignment TDMA, several consecutive frames such as described in Figure 3.19 are combined to form a *superframe*.

The signal format does not vary during the duration of a superframe, so an earth station transmits the same burst length and starts its burst at the same time after the synchronization burst in every frame of the superframe. Periodically, say every second or so, the allocated slot durations are adjusted from one superframe to another so as to allow stations to vary their burst length and hence meet the changing demands.

3.5.3.1 Example: the TELECOM 1 demand assignment TDMA system:

The 14/12 GHz TELECOM 1 satellite payload supports the first French public multiservice digital network (Bousquet, 1981; Lombard and Rancy, 1981; Guénin *et al.*, 1983). This network is a fully demand assigned system in which TDMA links are established on a call-by-call basis, according to the requests of terrestrial switching centres. Figure 3.30 shows the different channels used for both data transfer and demand assignment operation by a traffic station.

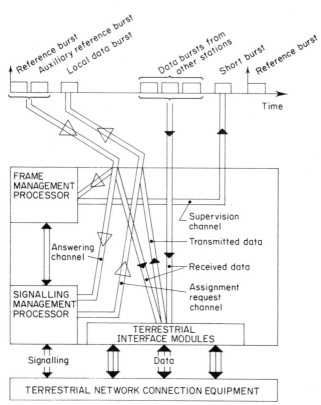

Figure 3.30   Signal flow in the demand assignment TELECOM 1 system.

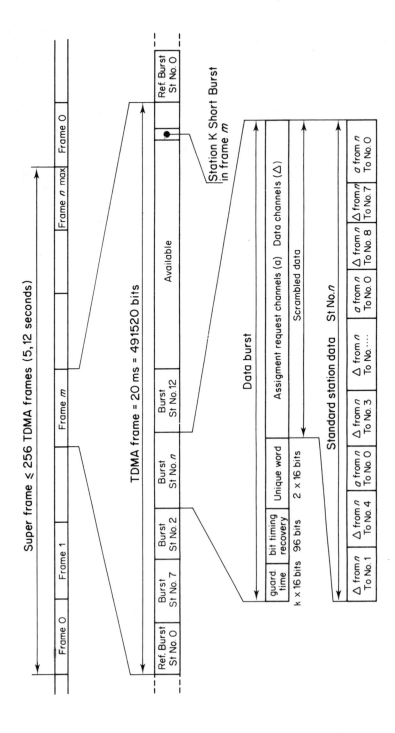

TDMA FRAME STRUCTURE

Figure 3.31   TDMA frame structure in the demand-assignment TELECOM 1 system.

Demand assignment is performed by means of assignment request and answering channels. The traffic station requests a capacity change (variation of burst length) using specific time slots located in data bursts (see Figure 3.31 (Lombard *et al.* 1983) which gives more insight of the frame structure). The reference station transmits each new burst time plan using specific time slots within the reference burst. Table 3.4 gives the main characteristics of the TELECOM 1 demand assignment TDMA system.

Table 3.4 CHARACTERISTICS OF THE TELECOM I DEMAND ASSIGNMENT TDMA SYSTEM, (BOUSQUET, 1981; LOMBARD AND RANCY, 1981; GUÉNIN, 1983).

Modulation	2-PSK with differential encoding
Frame duration	20 ms
Superframe duration	5.12 (256 frames)
Bit rate	24.576 Mbit/s
Bit-error rate	10^{-6} (99% of time)
Transponder bandwidth	36 MHz
User bit rate	2.4 kbit/s, 4.8 kbit/s, 9.6 kbit/s (data on reservation)
	32 kbit/s (DPCM telephony)
	48 kbit/s, $n \times 64$ kbit/s
	($n = 1, 2, 4, 8, 16, 24, 30$)(data on reservation or call by call)
	2.048 Mbit/s (data on reservation only)
Demand assignment operation	Centralized control of assignment
	Frame management messages are sent to stations in the reference burst
	Request for capacity by traffic stations by means of specific time slots, located in data (CCITT No. 7 signalling channel) with capacity from 32 kbit/s to 12.8 kbit/s, using a 4/5 error correcting code
	Answering channel (CCITT No. 7 signalling channel) in reference burst with 64 kbit/s capacity using a 4/5 error correction
	Control station can set up 10 calls every second. A 64 kbit/s point-to-point link is set up in less than 6 seconds, in more than 95% of the calls.
Traffic station	Antenna diameter = 3.5 m
	255 bursts transmit/receive (i.e. 255 links both ways)
	Maximum number of stations:
	62 stations/transponder × 5 transponders = 310

3.5.3.2 Example: the SBS demand assignment TDMA system:

The SBS system is designed to provide all-digital, fully integrated voice, data, and image transmission capability with demand assignment of the satellite capacity to the various traffic stations (Schnipper, 1980). Digital data rates range from 2.4 kbit/s to 6.4 Mbit/s. Bit error rate is 10^{-4} or better 99.5 per cent of the time. The user may request an improved bit error rate performance of 10^{-7} or better 99.5 per cent of the time using forward error correction. The frame duration is 15 ms with a transmitted bit rate of 48 Mbit/s. Traffic stations at the customers premises are equipped with 5.5 or 7.6 m diameter antennas.

3.5.4   Demand assignment control

An important consideration in the design of a demand assignment system is whether the assignment control should be either distributed or centralized.

In *centralized control demand assignment*, such as the TELECOM 1 system cited above, requests for channels are sent over a control channel to the central controller, which arbitrates requests and returns orderwire information to the requesting terminals.

Table 3.5 DEMAND ASSIGNMENT CONSIDERATIONS: CENTRALIZED CONTROL VERSUS DISTRIBUTED CONTROL

Description	Advantages	Drawbacks
Centralized control demand assignment		
Requests for channels are sent over a control channel to the central controller. The central controller arbitrates requests and returns orderwire information to the requesting terminals	Individual terminals do not need to perform the function of channel assignment → low terminal cost Less overheads as control data can be compact and status data for the whole network need not be transmitted	As the network relies on the controller → low reliability need for spare, stand-by controller
Distributed control demand assignment		
Each terminal sends its request over a common channel, and keeps a data base of the status of the channels. Each terminal selects a channel as needs arise, if a channel is available.	No unique controller → high reliability	Equipment for maintaining a network data base is complex → high terminal cost Lack of coordination may result in a poor use of satellite capacity

In *distributed control demand assignment*, such as the SPADE system cited previously, each terminal keeps a data base of the status of the channels and selects, if available, a channel as needs arise.

Table 3.5 lists the main advantages and drawbacks of the two concepts.

3.6 MULTIBEAM SATELLITE NETWORKS

The necessity of sharing the electromagnetic spectrum between all Radiocommunication Services results in the allocation of a limited number of frequency bands to each Service. On the other hand, the available power on-board a satellite is limited. The communication *capacity* of a satellite link is therefore constrained either by the *limited EIRP* or by the *bandwidth limitation*.

Assuming a given satellite transmitted power output, the satellite EIRP is greater with spot beam antennae than with global earth coverage antennae. Hence the link budget is improved, resulting in increased capacity. Therefore, *shaped beams* which embrace the shapes of continents or regions where earth stations are concentrated, have been developed. If the satellite is equipped with antennae delivering several beams, it is called a *multibeam satellite*. When the beams are sufficiently separated, so as to avoid a too high level of interference, the same allocated frequency bands can be reused in different beams. This is known as *frequency reuse* by spatial separation (as distinct from frequency reuse by orthogonal polarization) (Foldes and Dienemann, 1980).

Multibeam satellite networks with frequency reuse display higher capacities; this is why most satellites, such as INTELSAT V and VI, ECS, etc. are now multibeam satellites (Lombard, 1980).

However, multibeam satellite networks are faced with two types of problems; *beam-to-beam interference* and the *need to interconnect* stations from different beams.

3.6.1 Beam-to-beam interference

Beam-to-beam interference occurs when beam isolation is not perfect. A beam can be defined as a solid angle illuminated by the antenna, characterized by the covered geographical area and the wave polarization. Frequency reuse implies beams with same frequency, either with distinct coverage and same polarization (*frequency reuse by spatial separation*) or same coverage and distinct polarization (*frequency reuse by orthogonal polarization*). If isolation between beams is not perfect, interfering signals from other channels are introduced into a given channel. The origin of the interfering signals defines its type, so if the interference originates from channels within the same frequency band it is called '*co-channel interference*' (CCI); and if it originates from channels with different frequency bands it is termed '*adjacent channel interference*' (ACI).

Since in general interfering signals do not display a flat spectrum distribution, interference cannot be dealt with in the same manner as thermal noise. However, it can be transformed in an equivalent noise with constant power spectral density $(N_0)_I$ which serves to define a *signal power to interference noise power density ratio* $(C/N_0)_I$ at the input of the receiver of the receiving ground station, given by:

$$(C/N_0)_I^{-1} = (C/N_0)_{IU}^{-1} + (C/N_0)_{ID}^{-1} \qquad (18)$$

where $(C/N_0)_{IU}$ is the equivalent C/N_0 due to interference in the up-link and $(C/N_0)_{ID}$ is the equivalent C/N_0 due to interference in the down-link.

Since individual noise sources can be regarded not to be correlated, the overall signal power to noise power spectral density ratio at the receiving ground station, considering a non-linear satellite channel, is given by an extension of formula (1):

$$(C/N_0)_T^{-1} = (C/N_0)_U^{-1} + (C/N_0)_D^{-1} + (C/N_0)_{IM}^{-1} + (C/N_0)_I^{-1} \qquad (19)$$

where $(C/N_0)_{IM}$ is the carrier power to intermodulation noise power density ratio.

3.6.2 FDMA with multibeam satellites

Basic FDMA operational aspects, as described in Section 3.4.3, remain the same. If interconnectivity among all stations must be retained, then the entire frequency spectrum must be divided into sub-bands by means of filters. The required interconnectivity is obtained by arranging the connections from these filters to the output beams. This approach requires a number of channelizing filters and transponders equal to the square of the number of beams. Moreover, this arrangement must consider possible traffic pattern changes during the satellite life, and so flexible interconnectivity must be thought of when designing the satellite. This can be achieved by means of switching connections between input filters and output beams, as implemented, for example, on the INTELSAT V satellite (Fuenzalida *et al.*, 1977; Dicks and Brown, 1978).

3.6.3 TDMA with Multibeam Satellites

3.6.3.1 Beam switching with transponder hopping

As previously with FDMA the satellite contains as many transponders as the square of the number of beams, and these transponders are connected between the various beams to provide necessary coverage connectivity. A particular transponder can be selectively accessed by a unique combination of frequency and beam. Each transponder occupies a part of the allocated bandwidth.

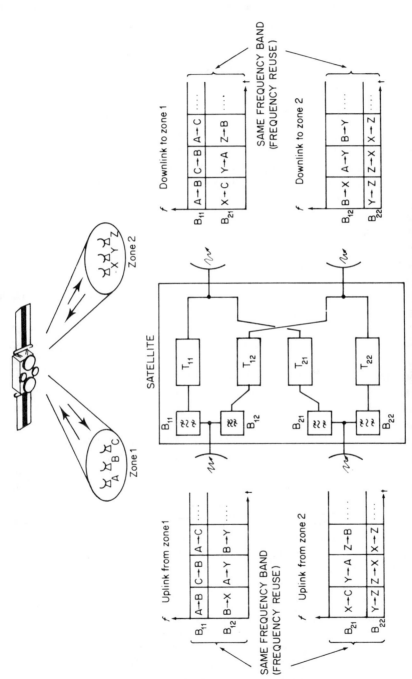

Figure 3.32   Transponder hopping concept and implementation.

Switching is achieved by each station hopping from one frequency to another according to the destination of the transmitted burst. Each earth station selects its transmission time on the corresponding frequency so as to transmit its burst within the correct time slot. Figure 3.32 shows an example of *transponder hopping* concept applied to a two beam satellite system.

3.6.3.2 On-board switching

Transponder hopping is achievable as long as the number of beams is small. The number of transponders increases as the square of the number of beams and the available bandwidth for each transponder reduces as much. One must face on-board switching called SS/TDMA. In a *satellite switched time division multiple access* (SS/TDMA) system the satellite embodies several spot beam antennas and a programmable fast acting switch matrix to route TDMA bursts arriving on different up-link beams to different down-link beams. This concept is illustrated in Figure 3.33. Hence the SS/TDMA system restores the traffic routing flexibility which is characteristic of the global beam system but lost in the multibeam system.

The on-board Distribution Control Unit (DCU) programmes the switch matrix to execute a cyclic set of *switch states*, each consisting of a set of connections between the up-link and down-link beams, so that the traffic from various regions is routed to designated regions without conflict. A *switch state sequence* (see Figure 3.34) is a succession of switch states during a frame period regardless of the duration of each switch state.

The *switching mode allocation* describes both the succession and the duration of each switch state so as to route the desired amount of traffic among the beams. Each *switching mode*, related to a given switch state, comprises as many *submodes* or 'windows' as beams; a submode indicates the period of time when a connection is established between an up-link beam and the corresponding down-link beam. The switching mode allocation must be stored in the on-board switch controller memory and may be reprogrammed by the ground station command as network traffic patterns change.

Figure 3.35 illustrates a typical SS/TDMA frame which consists of a synchronization field and a traffic field. The switching modes dedicated to the *synchronization field* are used for the purpose of network synchronization (Campanella and Inukai, 1981). Network synchronization entails earth station synchronization as discussed hereunder, but also synchronization of reference stations to the on-board switching sequence. A review of solutions is presented in Muratani (1977) and Carter (1980). The *traffic field* consists of a number of switching modes used to route the traffic and a growth space.

The time slot assignment problem is to schedule the network burst traffic so as to maximize the satellite transponder utilization. The satellite transponder utilization is maximum when the submodes are fully occupied by the traffic

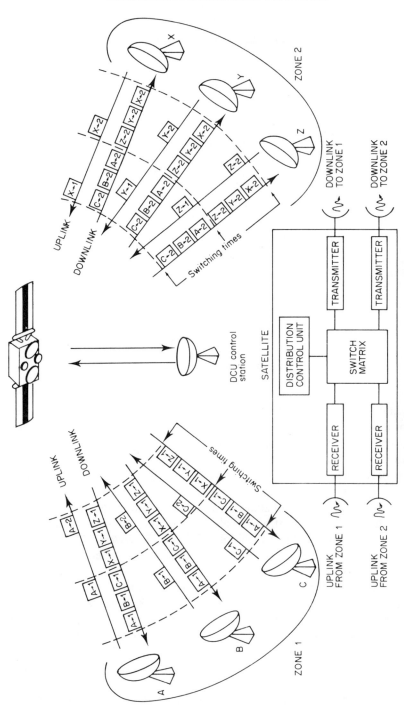

Figure 3.33   SS/TDMA concept and transponder implementation.

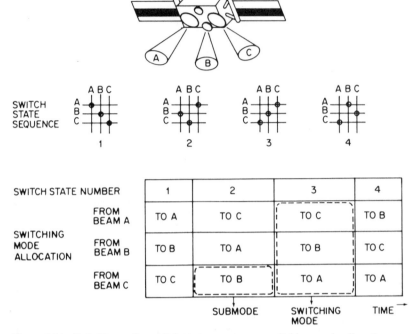

Figure 3.34   Definitions of a switch state sequence, a switching mode allocation, a switching mode and a submode.

bursts and the growth space is zero, but owing to unbalanced traffic between pairs of up-link and down-link beams some submodes are not fully occupied. Moreover, some growth space must be reserved to cope with traffic pattern changes.

It has been shown (Ito *et al.*, 1977) that it is impossible to communicate all the traffic in a period shorter than the required transmitting or receiving time duration T for the spot beam zones with the maximum traffic. This maximum traffic is equal to the sum of all the elements of the line with maximum line sum (low or column), called critical line of the *traffic matrix* which represents the amount of traffic to be routed from one beam to another (Inukai, 1979). The time slot assignment problem has often been solved using algorithms which yield a switching mode allocation with a duration equal to T and a number of switching modes bounded by $n^2 - 2n + 2$, where n is the number of spot beams (Bongiovanni *et al.*, 1981a; 1981b; Camerini *et al.*, 1981; Gopall *et al.*, 1982; Maral *et al.*, 1983). The resulting time slot assignment then comprises for each switching mode one fully occupied submode, called *critical submode*. For instance, in Figure 3.35 critical submodes are those routing traffic to beam B.

The input to these algorithms is the traffic matrix at a given time. However,

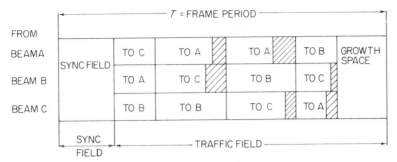

Figure 3.35 Typical SS/TDMA frame.

as the traffic pattern changes the calculated time slot assignment is no longer optimal and the duration of the switching mode allocation may not remain minimal. As the traffic grows, the idle time slots in non-critical submodes represented by the shaded areas in Figure 3.35 are allocated. Once a non-critical submode is fully occupied the corresponding switching mode must be expanded as traffic keeps growing. The duration of this switching mode then exceeds the critical submode duration, and the growth space is reduced as long as the switch state sequence remains the same. As the growth space reduces to zero, one must change the switch state sequence by running an algorithm according to the traffic matrix which reflects the new traffic pattern.

The *efficiency* of a SS/TDMA frame can be defined as the ratio of the actually used capacity to the total capacity available from all transponders. The efficiency is the product of frame efficiency, filling efficiency and assignment efficiency. *Frame efficiency*, as with conventional TDMA, takes into account guard-times and burst preambles and increases if the number of bursts is small and the frame duration is high. *Filling efficiency* is the ratio of the total used capacity for all submodes to the total capacity available from these submodes, and decreases as the traffic between beams becomes more and more unbalanced. *Assignment efficiency* is the ratio of the minimum time to the actual time required for transmission of the traffic given by the critical line of the traffic matrix.

3.6.3.3 Earth station synchronization

With a multiple beam TDMA system, transmitting stations in one beam communicating to stations located in a different beam receive neither the frame transmitted by their reference station, nor their data burst after retransmission by the satellite (Figure 3.36) and because of this, cannot control their synchronization.

One solution is to install a reference station in each beam which observes synchronization errors of stations in other beams and sends them the necessary

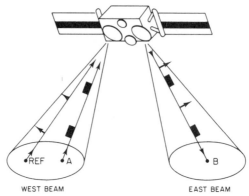

Figure 3.36   Multiple beam TDMA system: the synchronization problem.

corrections (see Figure 3.37). Corrections can only be made at the end of a time period equal to a double hop from Earth to Earth. This is a closed loop process, similar to global coverage synchronization, called *co-operative feedback* synchronization.

Open loop synchronization can also be considered, based on knowing the exact position of the satellite obtained by accurate ranging (Campanella and Colby, 1981).

3.7   MULTIBEAM REGENERATIVE SATELLITE SYSTEMS

As stated in section 3.5, a satellite communications system may operate with various types of earth stations. At the extremes, one may have:

(1) High capacity stations providing the communication needs of zones with a large concentration of activities. These stations, called *'trunking stations'*, will have large high gain antennas and will serve a terrestrial network with a high number of users; they require a high data rate transmission.

(2) Low capacity stations providing the communication needs of an individual user or a small group of users. These stations will be widely dispersed and located near the users (*customer's premises services terminals*). Such stations are more economical if they transmit at low data rate.

Moreover, multibeam satellites are faced with the following constraints:

(1) Beam-to-beam interference: in addition to the adjacent channel inter-ference which is encountered in global beam satellite systems and results

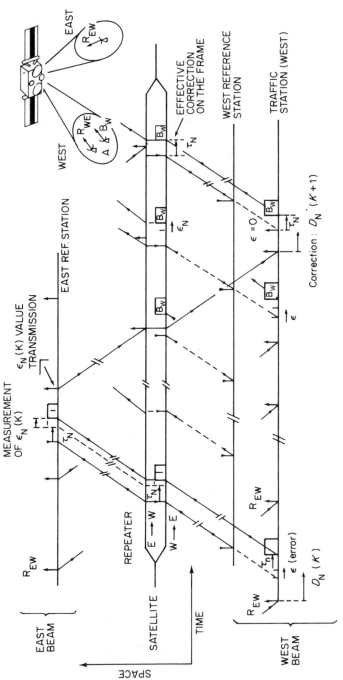

Figure 3.37 Co-operative feedback synchronization with a multiple beam TDMA system. *Reproduced by permission of Communications Satellite Corporation from Campanella and Hodson (1979).*

from satellite channelization, multibeam satellite systems suffer from co-channel interference, caused by antenna beam side-lobes and leakage between microwave switch matrix input–output ports. Davies (1981) has shown that nearly 40 per cent of the total system noise may come from sources other than satellite transponder and earth station receiver noise. Performance of future satellite systems may be limited by interference rather than power or bandwidth.

(2) Loss of link capacity owing to unbalanced traffic.
(3) Same rate and modulation technique are requested for up- and down-links.
(4) High power amplifier non-linearities and filtering effects are cascaded.
(5) A microwave switch matrix is to be used when a large number of beams are to be interconnected.
(6) Synchronization is required between the transmission of the TDMA burst from each earth station and the on-board switching.

Some of the above constraints may be combated with *multibeam regenerative satellite*, the salient features of which are outlined below (Maral and Bousquet, 1984).

3.7.1 Link budget comparison (non-linear channel with interference)

In has been seen in Section 2.4.2. that isolation between the up and down-link avoids effects resulting from cascading the earth station HPA and the transponder TWTA non-linearities. Another advantage which is gained from isolation between up-link and down-link is the reduction of the carrier-to-interference power ratio which can be tolerated.

In a non-regenerative repeater, the total $(C/N)_T$ depends on $(C/N)_T$ without interference and $(C/N)_I$ due to interference on the down-link and on the up-link, i.e.

$$(C/N)_T^{-1} = (C/N)_{T\ \text{without I}}^{-1} + (C/N)_I^{-1} \qquad (20)$$

where $(C/N)_{T\ \text{without I}}^{-1} = (C/N)_U^{-1} + (C/N)_D^{-1}$ and $(C/N)_I^{-1} = (C/N)_{IU}^{-1} + (C/N)_{ID}^{-1}$ as derived from (18).

Figure 3.38 shows the tolerable level of interference $(C/N)_I$ as a function of the overall carrier-to-noise power ratio without interference $(C/N)_{T\ \text{without I}}$ resulting from thermal noise only, to obtain a given carrier-to-noise-plus-interference power ratio $(C/N)_T$.

Suppose we wish to obtain a BER of 10^{-4} using QPSK modulation. Then from curves given in Figure 2.29 for a practical situation $E_b/N_0 = 11$ dB (E_b: energy per bit, N_0: noise power spectral density) and taking a spectral efficiency $\Gamma = R/B_{IF}$ equal to $1.6 = 2$ dB (theoretical value for QPSK as given by Figure 2.24 would be 2) then $(C/N)_T = E/N_0\ R/B_{IF} = 13$ dB. The upper

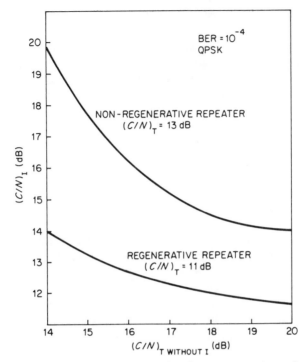

Figure 3.38 Tolerable level of interference: comparison of non-regenerative and regenerative satellite systems.

curve of Figure 3.38 shows the trade-off for $(C/N)_I$ and $(C/N)_{T \text{ without I}}$ according to (20) in that case.

For a regenerative transponder assuming that the up-link SNR is higher than the down-link SNR, $(BER)_T = (BER)_D$ and a BER of 10^{-4} requires $E/N_0 = 9$ dB (Figure 2.29.) Then $(C/N)_T = E/N_0 \, R/B_{IF} = 11$ dB. The lower curve of Figure 3.38 shows the trade-off for $(C/N)_I$ and $(C/N)_{T \text{ without I}}$ in that case.

Comparing the two curves indicates that a higher level of interference can be tolerated with a regenerative satellite system than with a conventional one for a given BER performance.

3.7.2 Influence on earth station design

3.7.2.1 Earth station transmitting power

In such situations where the up-link power is greater than the down-link power the regenerative satellite system has the advantage of a total BER similar to the conventional satellite system with the same down-link power but with much less up-link power. For instance, in the linear case of Figure 2.28 considering a down-link carrier to noise ratio of 12 dB, a BER of 10^{-4} requires an up-link carrier-to-noise ratio of 26 dB (point A) with a conventional transponder. With a regenerative transponder (curve C) only 15 dB are needed on the up-link yielding a possible reduction of 11 dB in the station transmitted power (Reudink, *et al.*, 1981), thereby reducing interference into other transponders or satellites.

3.7.2.2 Phase coherency of down-link bursts

In a conventional satellite system the down-link bursts originate from different earth stations and the ground receiving station must provide a fast carrier recovery for each burst in the frame. Conversely, on-board remodulation permits the carrier to be derived from a common source for all down-link bursts and eliminates station to station and Doppler differences encountered on the up-link (Chiao and Chethik, 1976; Koga *et al.*, 1977; Wachira *et al.*, 1981). With regeneration, data are recovered on-board and after retiming are modulated on a common carrier irrespective of their origin. Therefore, a signal modulated on a continuous carrier is received at the ground and fast carrier recovery circuits are no longer needed.

3.7.2.3 Different up-link and down-link multiplexing/multiple access mode

At each ground terminal the traffic from the active users must be multiplexed together. If FDM is employed using several digital modulated carriers, it becomes necessary to operate the terminal's HPA with back-off to avoid intermodulation products. With time division multiplexing (TDM), the terminal's HPA can be operated near saturation. This is a significant advantage as the same transmitted power is achieved with a cost effective HPA with lower saturated power.

Now consider the problem of providing up-link multiple access to the satellite for all the ground terminals in the network. If the ground terminals are all time division multiplexed together to form a TDMA up-link, as it is desirable with a conventional satellite to allow its TWTA to operate at full power, then each terminal has to transmit at a high burst data rate in a direct proportion to the number of terminals in the network. Such an approach places a burden upon the earth stations. First, burst modems operating at very high transmission rate may not be available. Secondly, if one wants to maintain the same energy per bit and bit error rate performance as in the case where continuous transmission would occur for each station (as in an FDMA up-link), then each ground station must be equipped with a transmitter capable of providing perhaps 20 dB more power although at a duty cycle of 1/100.

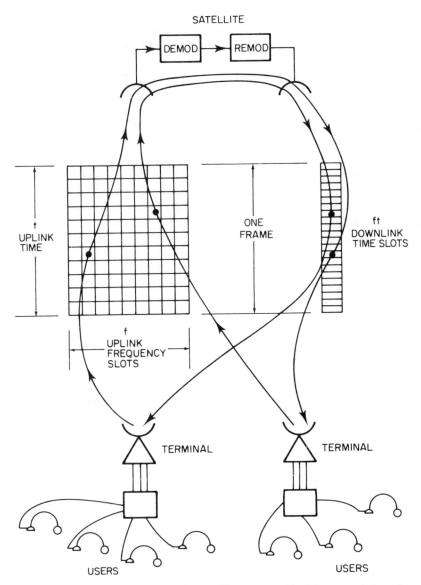

Figure 3.39 Example of a regenerative satellite system with different up-link and down-link multiple access mode (Knick *et al.*, 1981).

To overcome this problem one can conceive of a system where each terminal radiates on a different non-interfering carrier. Hence the up-link employs a hybrid TDM/FDMA scheme for multiplexing/multiple access.

Baseband signal regeneration and remodulation permits this favourable up-link access while maintaining the highly desirable time division multiplex for the down-link, which allows the satellites TWTA to operate at maximum saturated power.

One possible example of up-link frequency channelization considers t uplink slots in a frame and f available frequencies for each time slot (Knick *et al.*, 1981) (Figure 3.39). Each terminal can transmit on one frequency in each up-link time slot. As can be seen this solution requires that the terminal has some frequency hopping capability, from one time slot to another. Calls may be blocked because there are no available channels at the satellite (that is, all of the f frequencies of a given time slot are in use), or because the called or calling terminal is using all of its t available up-link time slots. If terminals are designed to access only a single channel, that is transmit always on only one frequency, which make ground stations more simple and economic, then the access is reduced and the blocking probability is higher. Similar schemes have also been considered in the context of a multibeam switched satellite (Reudink *et al.*, 1981; Holmes, 1983).

3.7.3 On board processing

On-board regeneration allows for *baseband processing facilities* which are not available from a non-regenerative repeater. Among these facilities are modulation conversion and forward-error-correction accordingly to the specific needs of each link, on-board storage which permits rate conversion and baseband switching.

3.7.3.1 Modulation conversion

Carrier recovery is necessary to attain the optimum BER performance provided by coherent demodulation. However, differential demodulation of differentially encoded signals avoids the need for carrier recovery, and this is of interest for reliable implementation of an on-board regenerative receiver. But the counterpart to this advantage is some BER degradation: 1 dB using BPSK, 2.5 dB using QPSK at BER $= 10^{-4}$ (theoretical values). Hence coherent demodulation will be retained at the receiving ground station together with on-board direct encoding, whereas the up-link scheme will be differential demodulation with differential encoding. The resulting up-link degradation has no impact on the total link degradation if the up-link C/N is about 3 dB more than that of the down-link, as can be inferred from Figure 2.28.

3.7.3.2 Coding

Multibeam satellite systems suffer from severe interference, which results in a higher channel bit error rate. Eventually the required information bit error rate can be restored using *forward error correction*.

For satellite communication systems operating at 30/20 GHz, as much as 10–20 dB degradation due to the effect of rain is to be expected (Fiorica, 1978). This degradation can also be combated by the use of forward error correction, either on the up-link or the down-link according to where the rain loss occurs (Eaves and Seay, 1980; Holmes, 1983). For example a 10 dB coding improvement is achievable with a 1/3 coding format; the 10 dB actually including a 5 dB increase in channel E/N_0 due to a 1/3 reduction of the information bit rate without changing the signalling rate, and 5 dB coding gain (Holmes, 1980).

3.7.3.3 On-board storage

On-board storage in a store-and-forward mode has two purposes: *rate conversion* and *baseband switching*.

Rate conversion allows merging traffic from trunking terminals and customer premises service terminals (Schmidt, 1974; Holmes, 1983).

With a multibeam SS/TDMA satellite system, if no on-board storage is provided, a burst from a particular up-link beam must be immediately routed to a particular down-link beam. Because of this immediate flow, scheduling must take into account both the burst origin and destination zones simultaneously. Where on-board storage is provided the switch scheduling process is effectively replaced by a remultiplexing of received bits stored in the satellite buffers prior to transmission on the various down-links.

On-board storage also allows for temporary storage of data in the case of *beam hopping*. In a beam hopping system, the beam is steered so that all parts of the service area can be covered at different times. This works perfectly with a time division multiple access configuration. To achieve total service, a conventional satellite requires both receive and transmit scanning beams with co-ordinated movements. A regenerative satellite with on-board storage ability allows for scanning with a unique beam used for simultaneous reception and transmission (Schmidt, 1974). It should be remarked that on-board storage is more adapted to data transmissions than voice transmission considering the transmission delay requirements.

REFERENCES

ABRAMSON, N. (1977) The throughput of packet broadcasting channels, *IEEE Trans. Comm.*, Vol. COM-25, No. 1 (Jan.), pp. 117–128.

BHARGAVA, V. K., D. HACCOUN, R. MATYAS and P. NUSPL (1981) *Digital Communications by Satellite*, John Wiley, New York.

BONGIOVANNI, G., D. COPPERSMITH and C. K. WONG (1981a) An optimum time slot assignment algorithm in SS/TDMA system with variable number of transponders, *IEEE Trans. Comm.*, Vol. COM-29, No. 5 (May), pp. 721–726.

BONGIOVANNI, G., D. T. TANG, and C. K. WONG (1981b) A general multibeam satellite switching algorithm, *IEEE Trans. Comm.*, Vol. COM-29, No. 7 (July), pp. 1025–1036.

BOUSQUET, J. C. (1981) Time division multiple access system with demand assignment for intracompany network using the satellite Telecom 1, *5th Int. Conf. Digital Satellite Communications*, Genoa.

CAMERINI, P., F. MAFFIOLI and G. TARTARA (1981) Some scheduling algorithms for SS/TDMA systems, *5th Int. Conf. Digital Satellite Communications*, Genoa, pp. 405–409.

CAMPANELLA, S. J. (1976) Digital speech interpolation, *COMSAT Tech. Rev.*, **6**, pp. 127–158 (Spring).

CAMPANELLA, S. J. and K. HODSON (1979) Open loop TDMA frame acquisition and synchronization, *Comsat Tech. Rev.*, **9**(2A), pp. 341–385.

CAMPANELLA, S. J. and R. J. COLBY (1981) Network control for TDMA and SS/TDMA in multiple beam satellite systems, *5th Int. Conf. Digital Satellite Communications.*, Genoa, pp. 335–343.

CAMPANELLA, S. J. and T. INUKAI (1981) SS/TDMA frame synchronization, *ICC 81*, Denver, pp. 5.5.1–5.5.7.

CARTER, C. R. (1980) Survey of synchronization techniques for a TDMA satellite switched system, *IEEE Trans. Comm.*, Vol COM-28, No. 8 (August), pp. 1291–1301.

CHIAO, J. T. and F. CHETHIK (1976) Satellite regenerative repeater study, *Canadian Proc. Comm. & Power Conf.*, pp. 222–225.

CROWTHER, W., R. RETTBERG, and D. WALDEN (1973) A system for broadcast communication: reservation ALOHA, *Proc. 6th Internat. Syst. Science Conf., Hawaii*, pp 371–374.

DAVIES R. S. (1981) Optimization of SS/TDMA communication satellite payload, *5th Int. Conf. Digital Satellite Communications*, Genoa, pp. 435–439.

DEMYTKO, N. and K. S. ENGLISH (1977) Echo cancellation on time variant circuits, *Proc IEEE*, **65**(3) (March), pp. 444–453.

DEROSA, J. K., L. H. OZAROW and L. N. WEINER (1979) Efficient packet satellite communications, *IEEE Trans. Comm.*, Vol COM-27, No. 10 (Oct.), pp. 1416–1422.

DICKS, J. L. and M. P. BROWN (1978) Intelsat V satellite transmission design, *ICC 78*, pp. 2.2.1–2.2.5.

DICKS. J. L., P. H. SCHULTZE and C. H. SCHMITT (1972) The Intelsat IV communications systems: systems planning, *Comsat Tech. Rev.*, **2**(2), pp. 439–476.

DIXON, R. C. (1976) *Spread Spectrum Systems*, Wiley.

EAVES, R. E. and T. S. SEAY (1980) Adaptive TDM satellite links to counter rain attenuation, *ICC 80*, Seattle, pp. 9.6.1–9.6.6.

EDELSON, B. I. and A. M. WERTH (1972) Spade system progress and application, *Comsat Tech. Rev.*, **2**(1), pp. 221–242.

FEHER, K. "Digital Communications" Prentice-Hall (1983)

FIORICA, F. (1978) Use of regenerative repeaters in digital communications satellite, *AIAA, 7th CSSC*, San Diego, pp. 524–532.

FOLDES, P. and M. W. DIENEMANN (1980) Large multibeam antennas for space, *J. Spacecraft*, **17**(4) (July/Aug.), pp. 363–371.

FUENZALIDA, J. C., P. RIVALAN, and H. J. WEISS (1977) Summary of the Intelsat V communications performance specifications, *Comsat Tech. Rev.,* **7**(1), (Spring) pp. 311–326.

GERLA, M., L. NELSON and L. KLEINROCK (1977) Packet satellite multiple access: models and measurements, *NTC 77,* pp. 12.2.1–12.2.8.

GOPAL, I. S., G. BONGIOVANNI M. A. BONUCELLI, D. T. TANG and C. K. WONG (1982) An optimal switching algorithm for multibeam satellite systems with variable bandwidth beams, *IEEE Trans. Comm.,* Vol. COM-30, No. 11 (Nov), pp. 2475–2481.

GUÉNIN, J. P., J. C. BERNARD-DENDE, Y. CHOI and A. HOANG-VAN (1983) The Telecom 1 satellite system: architecture of the common signalling network, *6th Int. Conf. on Digital Satellite Communications,* Phoenix, pp. V.17–V.21.

HAYES, J. F. (1981) Local distribution in computer communications, *IEEE Communications Mag.,* March, pp. 6–14.

HOLMES, W. M. (1980) Multigigabit satellite on board signal processing, *AIAA, 8th CSSC,* Orlando, pp. 623–626.

HOLMES, W. M. (1983) The ACTS multibeam communications package cost-effective advanced communications technology, *NTC 83,* San Francisco, pp. 300–304.

INUKAI, T. (1979) Efficient SS/TDMA time slot assignment algorithm, *IEEE Trans. Comm.* Vol. COM-27, No. 10 (October), pp. 1449–1455.

ITO, Y., Y. URANO, T. MURATANI and M. YAMAGUCHI (1977) Analysis of a switch matrix for an SS/TDMA system. *Proc. IEEE,* **65** (3) (March) pp. 411–419.

JACOBS, I. M., R. BINDER and E. V. HOVERSTEN (1978) General purpose packet satellite networks, *Proc. IEEE,* **66**(11) (Nov.), pp. 1448–1467.

KNICK, E. B., G. KOWALSKI and R. SINGH (1981) User capacity of a demand assigned satellite communication system with a hybrid TD/FDMA uplink and a TDM downlink, *ICC 81,* Denver, pp. 5.6.1–5.6.6.

KOGA, K., T. MURATANI and A. OGAWA (1981) On board regenerative repeaters applied to digital satellite communications, *Proc. IEEE,* **65**(3), pp. 401–410.

LOMBARD, D. (1980) Les satellites de télécommunications de deuxième generation, *L'echo des recherches* (May).

LOMBARD, D. and F. RANCY (1981) TDMA demand assignment operation in TELECOM 1 business service network, *NTC 81,* New Orleans, pp. 6.2.2.1–6.2.2.5.

LOMBARD, D., F. RANCY and D. ROUFFET (1983) Telecom 1—A national satellite communication for domestic and business services, *SCC 83,* Ottawa, pp. 17.4.1–17.5.2.

MARAL, G. and M. BOUSQUET (1984) Performance of regenerative/conventional satellite systems, *Int. Journal of Satellite Communications,* **2** (3), pp. 199–207.

MARAL, G., M. BOUSQUET and P. WATTIER (1983) SS/TDMA time slot assignment algorithm with reduced number of connection time slots, *1st Satellite Communications Conference SCC 83,* Ottawa, pp. 24.5.1–24.5.4.

MURATANI, T. (1977) Satellite switched time division multiple access, *Eascon 74,* pp. 189–196.

NUSPL, P. L., K. E. BROWN, W. STEENAART and B. GHICOPOULOS (1977) Synchronization methods for TDMA, *Proc. IEEE,* **65**(3), pp. 434–444.

PUENTE, J. G., W. G. SCHMIDT and A. M. WERTH (1971) Multiple access techniques for commercial satellites, *Proc. IEEE,* **59**(2), pp. 218–219.

RETNADAS, G. (1980) Satellite multiple access protocols, *IEEE Comm. Magazine,* **18**(5), pp. 16–2.

REUDINK, R. O., A. S. ACAMPORA and Y. S. YEH (1981) Digital satellites with time and frequency divided channels, *NTC 81,* New Orleans, pp. B5.5.1–B5.5.8.

ROBERTS, L. G. (1973) Dynamic allocation of satellite capacity through packet reservation, *NCC,* pp. 711–716.

SABOURIN, D. J. and R. J. JIRBERG (1982) Baseband processor development for SS/TDMA communication system, *ITC 82*, pp. 521–529.

SCHMIDT, W. G. (1974) Satellite switched TDMA: transponder switched or beam switched?, *AIAA*, 5th *CSSC*, Los Angeles, pp. 1–7.

SCHNIPPER, H. (1980) Market aspects of satellite business services, *Eascon 80*, pp. 92–94

SEVY, J. L. (1966) The effect of multiple CW and FM signals passed through a hard limiter or TWT, *IEEE Trans. Comm.*, Vol. COM-14, No. 5, pp. 568–578.

SHAFT, P. D. (1965) Limiting of several signals and its effect on communication system performance, *IEEE Trans Comm.*, Vol. COM-13, No. 4, pp. 504–512.

SIMON, M. K., J. K. OMURA, R. A. SCHOLTZ and B. K. LEVITT (1985) *Spread Spectrum Communications,* Computer Science Press.

SPOOR, J. H. (1967) Intermodulation noise of FDM/FM communications through a hard limiter, *IEEE Trans Comm.*, Vol. COM-15, No. 4, pp. 557–565.

TOBAGI, F. A. (1980) Multiaccess protocols in packet communication systems, *IEEE Trans. Comm.*, Vol COM-28, No.4 (April), pp. 468–488.

TOBAGI, F. A., R. BINDER and B. LEINER (1984) Packet radio and satellite networks, *IEEE Comm. Magazine,* Vol. 22, No 11, (Nov.), pp 24–40.

WACHIRA, M., V. ARUNACHALAM, K. FEHER and G. LO (1981) Performance of power and bandwidth efficient modulation techniques in regenerative and conventional satellite systems, *ICC 81,* Denver, pp 37.2.1–37.2.5.

WITTMANN, J. H. (1967) Categorization of multiple access/random modulation techniques, *IEEE Trans. Comm.*, Vol. COM-15, No. 5, pp. 724–725.

CHAPTER 4 Geometric considerations for a geostationary satellite communication system

In this chapter we shall examine various topics related to the geometry of the system composed of orbiting bodies such as the satellite, the Earth, the Moon and the Sun. This will lead us to give several important practical results such as the determination of the orbit of the satellite, the distance between satellite and earth stations, coverage areas, earth station antenna pointing angles, eclipses and solar interference.

4.1 UNDISTURBED KEPLERIAN MOVEMENT

Satellites orbit the Earth in accordance with Kepler's laws with the following hypotheses:

(1) The satellite mass m is small compared with the mass M of Earth (assumed to be homogeneous and spherical).
(2) Movement is in the vacuum of space, the only bodies present are the satellite and Earth. In actual fact the true movement must take into account that the Earth is neither spherical nor homogeneous and that both Sun and Moon are present.

4.1.1 Newton's law

Two bodies of mass m and M are attracted by a force (F) proportional to their mass, and inversely proportional to the square of the distance r between them.

So
$$F = GM \frac{m}{r^2}$$

where G is the gravitational constant, and is:

$$G = 6.672 \ 10^{-11} \ \text{m}^3 \ \text{kg}^{-1} \ \text{s}^{-2}.$$

With Earth of mass $M = 5974 \ 10^{24}$ kg, $\mu = GM = 3.986 \ 10^{14} \ \text{m}^3/\text{s}^2$

4.1.2 Kepler's laws

Kepler's laws govern the movement of planets around the Sun and state that:

(1) The orbit of each planet describes an ellipse with the Sun at one focus.
(2) The line joining a planet to the Sun sweeps over equal areas in equal intervals of time.
(3) The square of the orbital period of a planet is proportional to the cube of its semi-major axis a.

4.1.3 Orbital parameters (Figure 4.1)

The orbit of a satellite in its orbital plane is described in polar coordinates by the equation:

$$r = \frac{a(1 - e^2)}{1 + e \cos v} \tag{1}$$

where v is the true anomaly, a is the semi-major axis and e is the eccentricity. v specifies the position of the satellite on the orbit. a and e define the shape of the ellipse.

The true anomaly v is the positive angle oriented according to the velocity vector of the spacecraft, between the Earth centre to satellite axis and the Earth centre to perigee axis. v specifies the position of the satellite on the orbit.

The semi-major axis a is linked to the orbital period by the formula:

$$T = 2\pi \sqrt{\left(\frac{a^3}{\mu}\right)} \tag{2}$$

Satellite velocity V_s at its orbital point S (distance to the centre of the Earth r) is given by

$$V_S^2 = \frac{2\mu}{r} - \frac{\mu}{a} \tag{3}$$

The eccentricity e varies from 0 to 1. With a circular orbit ($e = 0$) the velocity is constant:

$$V_S^2 = \frac{\mu}{a} \tag{4}$$

The argument of the perigee ω, specifies the orientation of the ellipse in the orbital plane and is defined as the positive angle oriented according to the satellite motion between the Earth centre to the ascending node axis and the

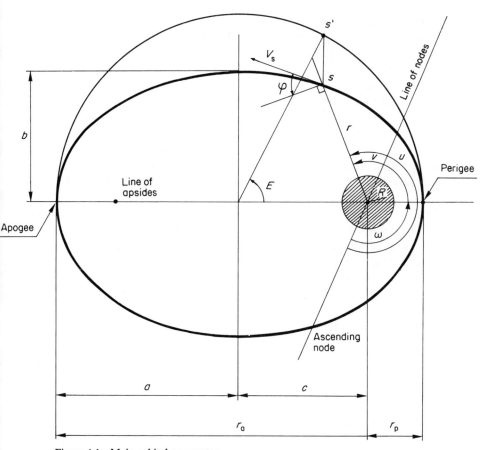

Figure 4.1   Main orbital parameters.

$$b = a\sqrt{1 - e^2} \qquad e = \frac{c}{a} \qquad r = \frac{a(1 - e^2)}{1 + a(1 - e^2)\cos v}$$

$$c = \sqrt{a^2 - b^2}$$

$$r_a = a + c = a(1 + e) \qquad\qquad M = \frac{2\pi t}{T} = E - e\sin E$$

$$r_p = a - c = a(-e)$$

$$r = \frac{a(1 - e^2)}{1 + e\cos v} \qquad\qquad \cos v = \frac{\cos E - e}{1 - e\cos E}$$

Earth centre to perigee axis. The *ascending node* is the point of the orbit located in the equatorial plane the satellite passes from below to above the equatorial plane. ω varies from $0°$ to $360°$.

The satellite's orbit in its orbital plane is then defined by three parameters: a, e, ω.

In addition to the above parameters two further quantities are required so that the orientation of the plane of the orbit in three-dimensional space is fully defined. These quantities are usually specified as the *inclination i* of the orbital plane and the *longitude of the ascending node* Ω (Figure 4.2).

The *inclination of* the orbital plane (*i*) is the angle of the orbital plane to the equatorial plane, this inclination *i* can take any value from 0 to 180°.

The *longitude of the ascending node* (Ω) is the counter-clockwise angle specified from 0 to 360° in the equatorial plane from the direction of the *vernal equinox* (γ_{50}) to the direction of the ascending node.

The direction of the vernal equinox γ_{50} is the direction of the Earth–Sun line when it lays in the Earth's equatorial plane at the spring equinox of the year 1950.

The five parameters (*i, a, e, Ω, ω*) completely define the satellite orbit in space. The motion of the satellite on its orbit requires a sixth parameter, either the true anomaly *v*, or the angular nodal elongation, or the mean anomaly, or the eccentric anomaly, or the time of perigee passage.

The *angular nodal elongation u* is the positive angle oriented according to the satellite motion between Earth centre to ascending node axis and the Earth centre to satellite axis ($u = \omega + v$). This is a useful parameter for a circular orbit where the perigee cannot be determined.

The *mean anomaly M* is the true anomaly the satellite would have if it proceeded along a circular orbit of the same period *T*.

$$M = 2\pi \frac{t}{T} \tag{5}$$

where $2\pi/T$ is the satellite mean motion.

The *eccentric anomaly* (*E*) is the argument of the satellite image S' (Figure 4.1) in the operation which transforms the ellipse into its auxiliary circle, i.e. the circle whose diameter is the major axis. S' is obtained from S by drawing

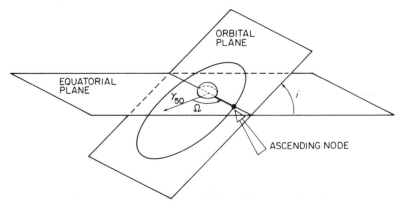

Figure 4.2   Orbital plane positioning

a line parallel to the semi-minor axis. It is linked to the mean anomaly M by:

$$M = E - e \sin E \qquad (6)$$

and to the true anomaly by:

$$\cos v = (\cos E - e)/(1 - e \cos E) \qquad (7)$$

4.2 GEOSYNCHRONOUS SATELLITES

4.2.1 Definition

A geosynchronous satellite has a period of revolution equal to that of the rotation period of Earth: $T = 23$ h 56 min 4.1 s = 86164.1 s. The inclination i of the satellite's orbital plane to the equator and its eccentricity e may have any value.

4.2.2 Ground track of geosynchronous satellites

The track of a satellite is the curve drawn on the Earth's surface by intersection

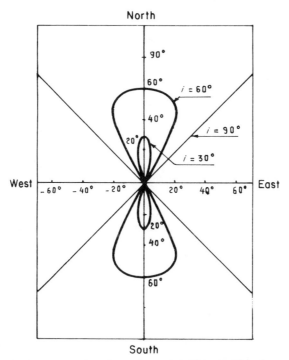

Figure 4.3 Track of a geosynchronous satellite for different orbit inclination angles: $i = 30°$; $i = 60°$; $i = 90°$.

of the axis joining the satellite to the centre of the Earth, with Earth's surface. The various orbital parameters determine the ground track. The following discussion applies to non-retrograde orbits, that is when the satellite orbits according to the direction of rotation of the Earth.

4.2.2.1 Synchronous satellites with circular orbits ($e = 0$)

Figure 4.3 gives the track of synchronous satellites for three inclination values: $i = 30°$, $60°$, $90°$. In any case, latitude displacement is greater than the longitude displacement.

4.2.2.2 Synchronous satellites with equatorial orbits ($i = 0$)

The satellite track remains on the Earth's equator, according to a periodic motion of the satellite (period T) centered around a point corresponding to the sub-satellite point at the perigee.

The longitudinal displacement amplitude of the sub-satellite point is dependent on the orbital eccentricity. Figure 4.4 gives the amplitude (longitude L_{max}), and the time t needed to reach it from the perigee, as a function of

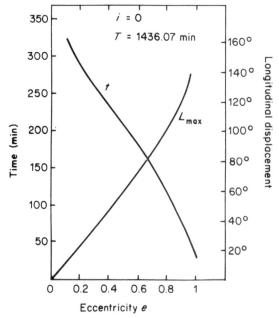

Figure 4.4 Longitudinal drift and related drift time of a geosynchronous equational satellite versus orbit ecentricity.

eccentricity e. For eccentricity e lower than 0.4, the following formula holds:

$$L_{max}(\text{degrees}) = 114e \qquad (8)$$

For eccentricity of a few 10^{-3} the maximum is reached after 360 min (6 h).

4.2.2.3 General case

The shape of the track of geosynchronous satellites depends on three parameters: the eccentricity e, the inclination i, and the argument of the perigee ω.

4.2.2.4 Geosynchronous satellites with orbits of low inclination and small eccentricity

This is the case of actual geostationary satellites as the orbit is never perfectly circular (residual eccentricity of about 0.001) and the orbital plane displays a residual inclination of about $0.1°$.

(1) The inclination i gives the satellite an apparent 'figure of eight' motion: the maximum latitude is equal to $\pm i$, the maximum longitude L_{max} and the related latitude l expressed in degrees are given, with less than 5 per cent error for values of i less than $30°$, by:

$$\begin{aligned} L_{max} &= 4.36\ 10^{-3}\ i^2 \\ l &= 0.707i \end{aligned} \qquad (9)$$

From the above formulae, one sees that the maximum latitude shift is much greater than the maximum longitude shift.

The maximum longitude max is reached at the end of the time t such that

$$\frac{2\pi t}{T} = \frac{1}{\sqrt{(2)}\cos\left(\dfrac{i}{2}\right)} \qquad (10)$$

(2) The longitudinal motion resulting from the effect of e must be added to the previous motion: the shift resulting from e reaches its maximum from the perigee sub-satellite point after 6 h. If the perigee is located at one of the nodes of the orbit at that time the latitude shift due to inclination is maximum. Then the peak of the 'figure of eight' is slightly shifted in longitude (with $e < 10^{-3}$, the longitude shift is less than $0.114°$).

4.3 GEOSTATIONARY SATELLITES

4.3.1 Definition

A geostationary satellite is a geosynchronous satellite of non-retrograde circular orbit in the equatorial plane ($e = i = 0$).

4.3.2 Nominal orbit

If the satellite follows a Keplerian orbit, that is, without perturbations, it has the features shown in Table 4.1. Altitude is determined from orbit radius value, assuming an average radius of Earth equal to $R_e = 6378.1$ km (equatorial radius).

Table 4.1 GEOSTATIONARY SATELLITE ORBIT ($e = i = 0$).

Orbital radius	42 164.2 km
Altitude	35 786.1 km
Orbital period (sideral day)	23 h 56 min 4.1 s
Satellite velocity	3075 m/s

Figure 4.5   Earth as seen from the geostationary satellite METEOSAT. *Reproduced by permission of the European Space Agency.*

4.4 SATELLITE TO EARTH STATION RANGE

The coordinates of the earth station in the reference system of Figure 4.6 are:

$$x = R_e \cos l \sin L$$
$$y = R_0 + R_e(1 - \cos l \cos L) \qquad (11)$$
$$z = R_e \sin l$$

where R_0 is the satellite altitude ($R_0 = 35\,786$ km), and R_e is the average radius of the Earth ($R_e = 6378$ km). l is the earth station latitude and L the earth station's relative longitude with respect to the satellite's longitude.

The range R from satellite to earth station is given by:

$$R^2 = R_0^2 + 2R_e(R_0 + R_e)(1 - \cos l \cos L) \qquad (12)$$

and as $R_e/R_0 = 0.178$, it finally becomes:

$$(R/R_0)^2 = 1 + 0.42(1 - \cos l \cos L) \qquad (13)$$

Variations of $(R/R_0)^2$ versus l, for various values of L, are given by the curves in Figure 4.7.

The maximum value of $(R/R_0)^2$ is 1.356. Approximating R^2 by R_0^2 leads to a maximum error of 1.3 dB in the power link budget.

4.5 PROPAGATION DELAY

The distance between two earth stations via satellite varies between $2R_{max}(L = 0°, l = 81.3°) = 83\,357.6$ km and $2R_0 = 71\,572$ km. This causes a single hop propagation delay of more than 0.238 s and can reach 0.278 s. To this delay should be added the delay caused by terrestrial extensions, the

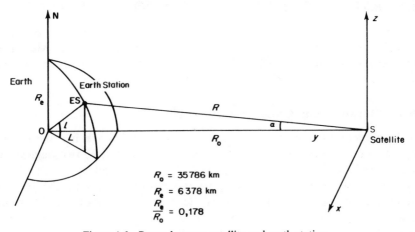

Figure 4.6   Range between satellite and earth station.

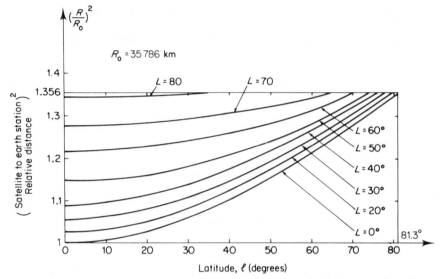

Figure 4.7   Variations of range R from satellite to earth station with latittude l and relative longitude L.

minimum being of the order of 10 ms, the maximum of the order of 50 ms (CCIR, Rep. 383–4, 1982). This delay is not particularly troublesome to a telephone conversation as long as echo suppressors or echo cancellers are used (Suyderhoud *et al.*, 1976; Demytko and English, 1977).

A double hop employing two satellites would have a propagation time exceeding the 0.4 s limit fixed by the CCITT Recommendation G.114 for telephone signals.

4.6   EARTH STATION ANTENNA POINTING ANGLES

In order to receive and/or transmit signals from/to a satellite the antenna of an earth station must be correctly pointed toward the satellite. Moreover whenever the radiowaves are linearly polarized the polarization of the antenna must be matched to the polarization plane of the radiowave.

4.6.1   Elevation angle and azimuth angle

The direction on earth station antenna boreshight should have, in order for the antenna to point correctly at a specific satellite, is defined by two angles: the azimuth angle A and the elevation angle E. figure 4.8(a) allows for the determination of these angles according to the latitude l and to the relative longitude L of the earth station (L is the absolute value of the difference between the geostationary satellite longitude and that of the earth station) (Smith, 1972).

	SL EAST OF ES	SL WEST OF ES
NORTH HEMISPHERE	$A = 180 - a$	$A = 180 + a$
SOUTH HEMISPHERE	$A = a$	$A = 360 - a$

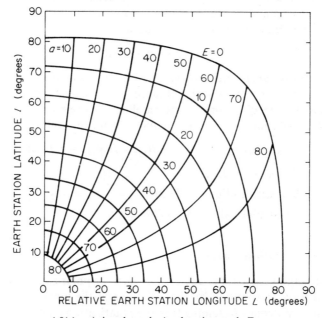

4.8(a): Azimuth angle A, elevation angle E.

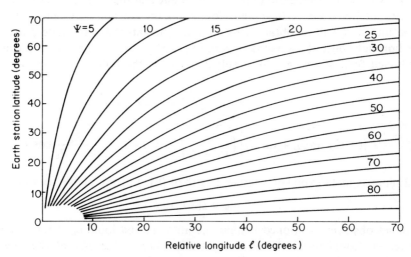

4.8(b): Polarization angle Ψ.

Figure 4.8   Earth station antenna pointing angles.

The *azimuth angle* is the angle by which the antenna, pointing at the horizon, must be rotated clockwise around its vertical axis, from the geographical north, to bring the antenna boreshight into the vertical plane containing the satellite direction. The azimuth angle A is between $0°$ and $360°$. It value is obtained from figure 4.8(a), by determining the value of angle a from the chart and deriving A from angle a using the insert. Angle a may also be calculated from the following formula:

$$a = \text{arc tg}(\text{tg}L / \sin l)$$

The *elevation angle* is the angle by which the antenna boreshight must be rotated in the vertical plane containing the satellite direction from the horizontal to the direction of the satellite. The elevation angle E can be obtained from figure 4.8(a), or calculated from the following formula:

$$E = \text{arc tg}[(\cos \phi - R_e/(R_e + R_0))/(1 - \cos^2 \phi)^{\frac{1}{2}}] \qquad (14)$$

where: $\cos \phi = \cos l \cos L$

R_e = earth radius = 6378 km

R_0 = satellite altitude = 35786 km

4.6.2 Polarization angle

The polarization angle is the angle ψ between the polarization plane of a linear polarized wave transmitted by the satellite and the polarization plane of the earth station antenna. Assuming that the polarization plane of the transmitted radio wave is perpendicular to the orbital plane and that the polarization plane of the antenna is the vertical plane containing the antenna boresight, polarization matching will be obtained by rotating the antenna with angle ψ around its boresight once the antenna is pointing at the satellite. The angle ψ can be obtained form Figure 4.8(b) or calculated from the following formula:

$$\cos \psi = \frac{\sin l \left(1 - \dfrac{R_e}{R_e + R_0} \cos L \cos l\right)}{\sqrt{(\cos^2 l \sin^2 L + \sin^2 l)\left[\left(\dfrac{R_e}{R_e + R_0}\right)^2 \cos^2 l + 1 - 2\dfrac{R_e}{R_e + R_0}\cos L \cos l\right]}}$$

An approximate expression (error less than $0.1°$) of the above formula is:

$$\text{tg } \psi = \sin L / \text{tg } l$$

For an observer positioned behind the antenna and looking at the satellite, rotation should be clockwise if the station lies to the east of the satellite meridian, and anti-clockwise if the station lies to the west of the satellite meridian.

4.7 EARTH COVERAGE AND ANTENNA BEAMWIDTH

Two situations should be examined depending on whether the earth coverage required is the maximum possible, or limited to a relatively small zone (such as one country).

4.7.1 Global beam

Coverage should be as great as possible so as to allow far apart stations to communicate by means of a single beam satellite.

4.7.1.1 Geometrical coverage

Maximum geometrical coverage is given by the portion of the Earth within a cone with the satellite at its apex and tangent to the Earth's surface. The apex angle 2α of this cone is

$$2 \text{ Arc sin } \frac{R_e}{R_0 + R_e} = 17.4°.$$

4.7.1.2 Effective radio frequency coverage

As earth stations at the outer limit of the geometrical coverage area would need to point their antenna horizontally, the link performance (signal-to-noise power ratio) would degradate, as a result of the increase in antenna temperature, and attenuation of propagation due to the increased path length through the atmosphere. The limit of the radio frequency coverage zone is actually defined by the curve along which the earth station antenna elevation angle has a minimum specified value. The radio-frequency coverage for different minimum elevation angles is shown in Figure 4.9 for a geostationary satellite the antenna of which points towards the centre of the Earth. (CCIR, Rep. 206–4, 1982)

Figure 4.10 shows the elevation angle E and the nadir angle α, i.e. the angle between the satellite to earth station axis and the satellite to centre of earth axis, as a function of the angle ϕ given by

$$\cos \phi = \cos l \cos L \qquad (15)$$

The required satellite antenna beamwidth for a global coverage corresponding to a minimum earth station antenna elevation angle E_{min} is determined by the value 2α, calculated from $E = E_{min}$ using the following formula:

$$2\alpha = 2 \text{ Arc sin} \left(\frac{R_e}{R_0 + R_e} \cos E_{min} \right) = 2 \text{ Arc sin}(0.15 \cos E_{min}) \qquad (16)$$

The maximum latitude coverage is then given by:

$$l_{max} = 90 - (\alpha + E_{min}) \qquad (17)$$

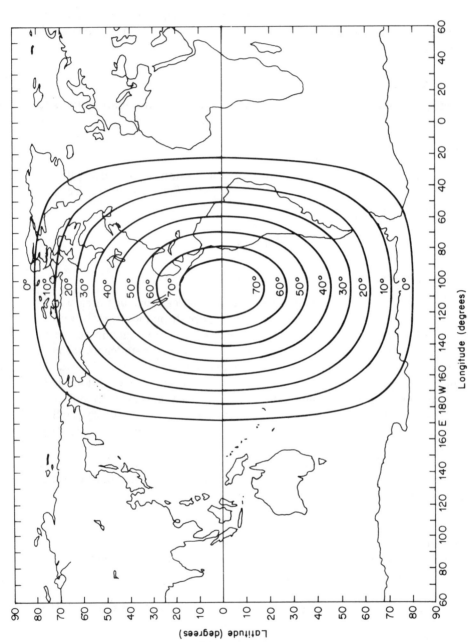

Figure 4.9 Global beam: coverage areas from a geostationary satellite for a given minimum elevation angle *E* of the earth station antenna.

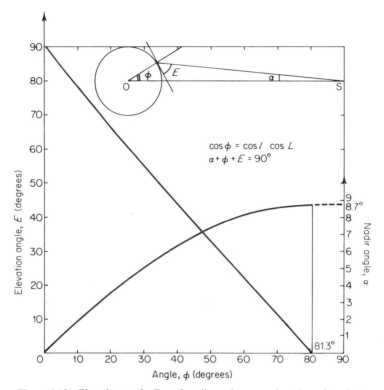

Figure 4.10 Elevation angle E, and nadir angle α as a function of angle ϕ.

4.7.1.3 Influence of satellite antenna depointing

If the satellite antenna depointing relative to the centre of the earth is ε, the satellite antenna beamdwidth must be $2\alpha + 2\varepsilon$. If, for example, $\varepsilon = 1°$, the antenna beamwidth for a global coverage assuming a minimum elevation angle $E_{\min}$ equal to $10°$ should be $2\text{Arc sin}(0.15 \cos 10°) + 2\varepsilon = 19°$.

4.7.1.4 World coverage

Figure 4.11 shows that where E is greater than $5°$, and with four satellites placed at equidistant points over the equator, world coverage is ensured for all regions except those above $70°$ latitude. However, as double hop is not recommended for commercial telephone systems, the maximum distance between two communication stations is limited to approximately $17\,000$ km.

Whenever propagation delay is not an inconvenience, one can accept several hops by the use of two (or possibly three) relay satellites (ATS experiments). Another solution is the use of intersatellite links (Figure 4.12).

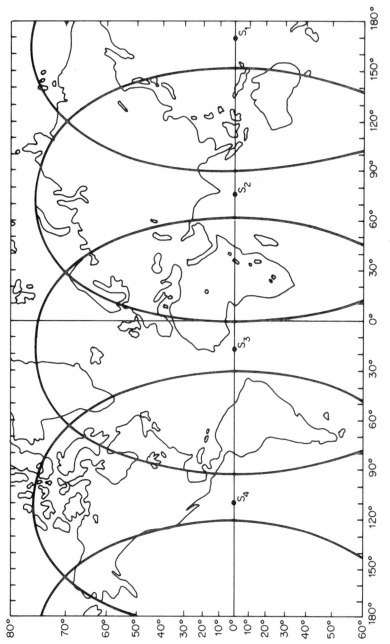

Figure 4.11   Worldwide coverage for latitude up to 70 ° using for satellites.

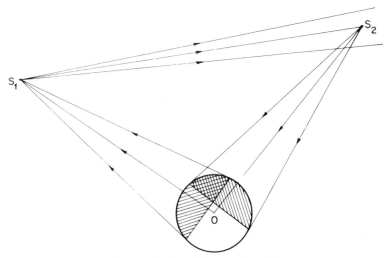

Figure 4.12   Use of intersatellite links.

4.7.1.5   Intelsat system

The Intelsat system ensures almost world coverage using three satellites. Only the polar regions and the centre of the United States are not covered, the system being intended for intercontinental links (see Figure 1.4).

4.7.2   Use of shaped and spot beams

As two-thirds of the Earth's surface is covered by water, global beam coverage is often not optimum for what is essentially land communications. A coverage area limited to the continents can be thought of, using narrow angle beams, possibly shaped according to the contours of these continents.

Narrower beams, often called *spot beam*, can also be used, either to ensure short distance links (e.g. within a national network) using a single spot beam satellite, or to ensure links between zones of small size located at different parts of the Earth using a multiple spot beams satellite. Of course, these zones must be located within the largest coverage of Figure 4.9.

INTELSAT V, for example, as well as providing global coverage, ensures hemispheric and zone coverage at frequencies of 6/4 GHz (Figure 4.13(a), (b), (c)) and routes trunk traffic at 14/11 GHz by means of spot beams (Fuenzalida *et al.*, 1977; Rusch and Dwyre, 1978). Another example is TELECOM 1 which ensures hemispherical coverage for the French overseas territories and France (6/4 GHz), as well as spot beam coverage of France (14/12 GHz) for domestic intercompany business traffic (Figure 1.14). The future will probably see more and more spot beam use. Figure 4.14 shows a possible example of this trend (Stamminger, and Stein, 1980).

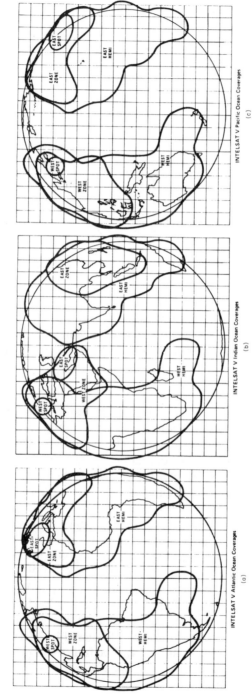

Figure 4.13 Intelsat V coverage areas. *Reprinted with permission from Rusch and Dwyre (1978). Copyright (1978) Pergamon Press Ltd.*

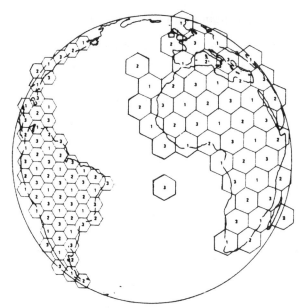

Figure 4.14   Possible Earth coverage by multiple spot beams.

4.7.2.1   Calculation of a spot-beam antenna beamwidth

Figure 4.15 shows an example of spot beam. It is assumed that:

(1) The beam is very narrow.
(2) The coverage area is an ellipse defined by points M_1 and M_2 of latitude l_1 and l_2, and by points N_1 and N_2 of maximum relative longitudes $-L_1$ and $+L_1$.
(3) The satellite is in the meridian plane of centre M of the coverage area.

4.7.2.1.1   *Calculation of the beamwidth in the meridian plane*

At points M_1 and M_2 (Figure 4.15(b)) the distances R_1 and R_2 are given by curve $L = 0$ in Figure 4.7. When the coverage area is narrow enough in latitude ($l_2 - l_1$ small) the three distances R, R_1, and R_2 can be confused: indeed, using extreme latitudes, in the worst case (Figure 4.7), a one degree variation in latitude brings a $(R/R_0)^2$ variation of less than 1 per cent. The antenna beamwidth in the meridian plane $\theta_1 = \theta_1 + \theta_2$ results from the calculation of $\alpha_2 - \alpha_1$, angles α_2 and α_1 being read on Figure 4.10 in terms of l_1 and l_2, taking $L = 0$.

4.7.2.1.2   *Calculation of longitudinal beamwidth*

Assuming points N_1, N_2 and M have the same latitude (Figure 4.15(c)) the longitudinal antenna half-beamwidth θ_3 is given by:

$$\sin \theta_3 = R/R_3 \cos l \sin L$$

where l and L are the latitude and longitude relative to S of N_1. Distance R_3 is obtained from Figure 4.7.

The antenna longitudinal beamwidth is therefore $\theta_L = 2\theta_3$.

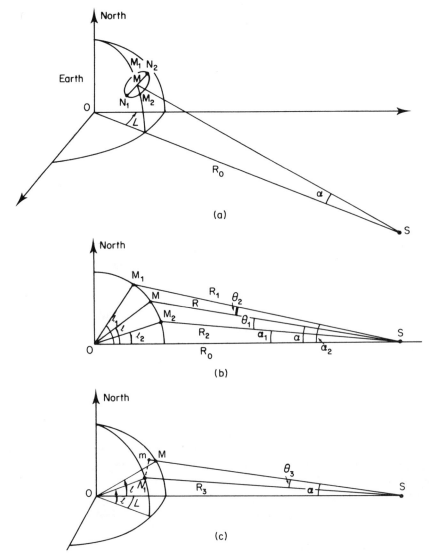

Figure 4.15 Beamwidth of a spot beam antenna: (a) overall view; (b) beamwidth in the meridian plane; (c) longitudinal beamwidth.

4.7.2.2 Example: coverage of Metropolitan France

The territory of Metropolitan France is represented within a 520 km radius circle (Figure 4.16). Table 4.2 gives the coordinates of some characteristic points.

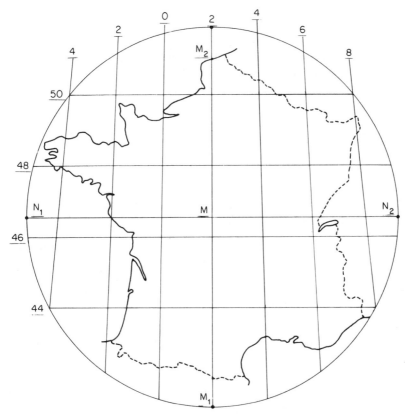

Figure 4.16 Metropolitan France territory coverage (except Corsica).

Table 4.2 CHARACTERISTIC POINTS OF METROPOLITAN FRANCE COVERAGE.

Point	Latitude	Longitude	Relative longitude
M	$46.5°$	$2°$	0
M_1	$42.5°$	$2°$	0
M_2	$51°$	$2°$	0
N_1	$46.5°$	$-5.5°$	$-7.5°$
N_2	$46.5°$	$+9.5°$	$+7.5°$

Beamwidth in the meridian plane. Figure 4.10 gives:

$$\alpha = 7° \qquad \text{for } l = 46.5°$$
$$\alpha_1 = 6.55° \qquad \text{for } l_1 = 42.5°$$
$$\alpha_2 = 7.4° \qquad \text{for } l_2 = 51°$$

The beamwidth in the meridian plan is therefore:

$$\theta_1 = \alpha_2 - \alpha_1 = 0.85°$$

Longitudinal beamwidth. For $l = 46.5°$ and $L = 7.5°$, Figure 4.7 gives $(R_3/R_0)^2 = 1.14$, hence $R_3 = 1.07 R_0$.

$$\sin \theta_3 = R/R_0 \; 1/1.07 \cos l \cos L = 0.0149$$
$$\theta_3 = 0.85°$$

The longitudinal beamwidth is:

$$\theta_L = 2\theta_3 = 1.7°$$

4.7.2.3 Influence of a residual inclination of the orbit on a satellite spotbeam antenna coverage

A residual inclination i of a geostationary satellite orbit manifests itself by an apparent daily movement of the satellite according to the 'figure of eight' described above. The inclination i represents the maximum shift in satellite latitude during the course of this movement. The residual inclination has two effects:

(1) It causes a *pointing error* (Figure 4.17): as the satellite passes from S to

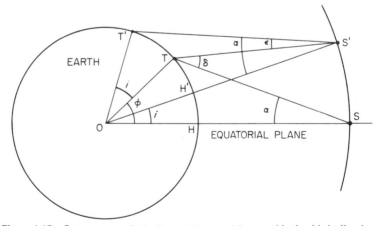

Figure 4.17 Coverage area latitudinal drift caused by a residual orbit inclination.

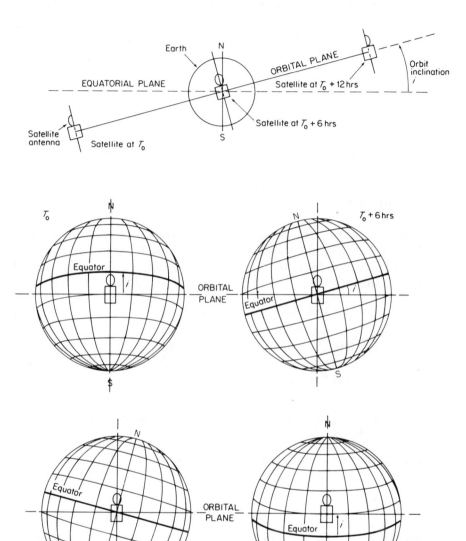

Figure 4.18   Depolarization due to a residual orbit inclination.

S', the direction of the vertical axis becomes $S'H'$, and the direction of the antenna axis, initially ST, becomes $S'T'$. So the spot-beam coverage undergoes a periodic displacement in latitude. The pointing errors of the antenna of the earth station and the satellite antenna are respectively δ and ε.

$$\delta \simeq i$$

$$\varepsilon = \frac{R_e \cdot i}{R} \cos(\alpha + \phi) \tag{18}$$

where R is the distance between satellite and earth station.

Assuming $\phi = 45°$, then $R = 38\,000$ km and $\alpha = 6.7°$ with $R_e = 6380$ km, $\varepsilon = 0.1i$ and if $i = 3°$, the pointing error is $0.3°$.

(2) It causes a *relative rotation* of the earth station antennae and of the satellite around the axis which joins them (Figure 4.18) with a maximum value equal to the residual inclination i. Where links are established using linear cross-polarized waves, a wave component of orthogonal linear polarization is added to the nominal linear polarization wave transmitted. The maximum relative level in power of the interference is $I = \sin^2 i$, that is -25.6 dB for $i = 3°$.

Such a level of interference would place too high a constraint on frequency reuse by orthogonal linear polarizations. Frequency reuse by orthogonal linear polarizations, which allows capacity to be doubled within a given frequency band, implies that the satellite orbit is kept at a sufficiently small inclination.

4.7.2.4 Influence of the satellite's movements around its reference axes

Movements of the satellite around its axis bring about a displacement of the coverage of the satellite antenna beam. With spot beams, it is mandatory to provide precise attitude maintenance (see Chapter 6).

4.8   ECLIPSE

Knowledge of the duration and periodicity of eclipses is important for satellites as they use solar cells as a primary power source. In addition, eclipses give rise to thermal shocks, which must be taken into account when designing the satellite. Eclipses are caused both by the Moon and the Earth.

4.8.1   Eclipse due to the Earth

Figure 4.19 represents the movement of the Earth while orbiting around the Sun. Figure 4.20 shows the apparent movement of the Sun relative to Earth where the satellite has an orbit perpendicular to the plane of the paper. Figure

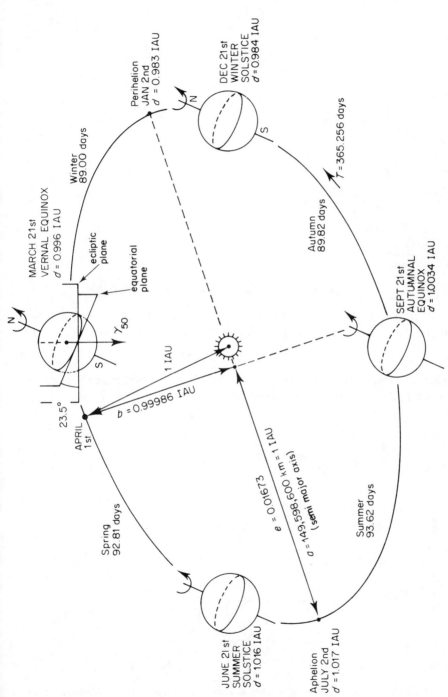

Figure 4.19 Orbit of the Earth around the Sun. *d* is the relative Sun to Earth distance in astronomical units.

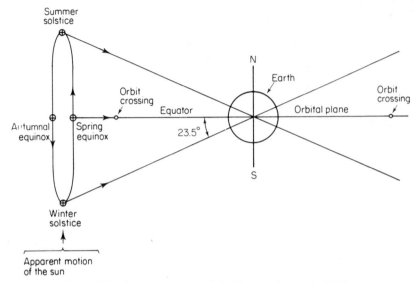

Figure 4.20   Apparent motion of the Sun relative to the Earth.

4.21 the direction of the Sun relative to the equatorial plane, which makes an angle δ, called declination angle, with an annual sinusoidal variation given by

$$\delta = 23.5 \, \sin(2\pi t/t) \tag{19}$$

where δ is expressed in degrees, t in days and $T = 365$ days. Near the equinoxes the formula (19) can be approximated by:

$$\delta = 23.5° \, 2\pi t/T \tag{20}$$

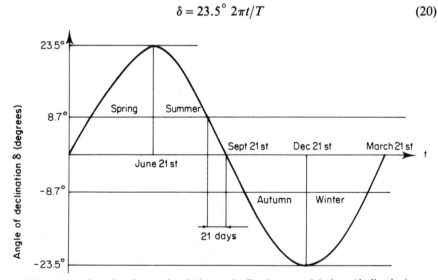

Figure 4.21   Sun elevation angle relative to the Earth equatorial plane (declination).

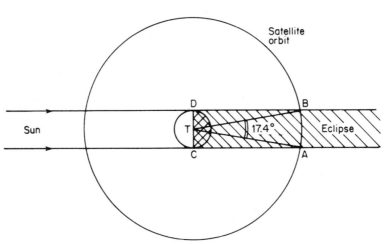

Figure 4.22   Solar eclipse due to the Earth at equinoxes.

At the solstices the satellite is always illuminated. On the other hand, it undergoes the maximum duration of eclipse at the equinoxes, lasting $(17.4°/360°) \times (23\ h \times 60\ min + 56\ min) = 69.4$ minutes (Figure 4.22). The first eclipse day before the equinox corresponds to the relative position of the Sun, such that the ray tangential to Earth passes through the satellite orbit (Figure 4.23). The last day corresponds to a symmetrical figure relative to the equatorial plane. On the first eclipse day before the equinox the inclination of the Sun on the equator is half of $17.4°$ that is $8.7°$; from Figure 4.21, one sees that 21 days are needed to reach the equinox. Figure 4.24 gives the eclipses' daily duration. At half this daily duration, the satellite crosses the plane formed by the Sun and the Earth's axis. Then local time at the satellite's longitude is midnight. If the satellite has no battery back-up, services are interrupted at night.

4.8.2   Eclipse due to the Moon

In addition to the eclipses caused by the Earth, sunlight is blocked totally or partially for a geostationary satellite when the Moon passes in front of the

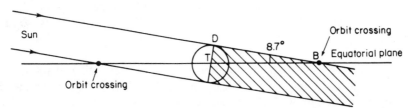

Figure 4.23   Sun–Earth configuration at the first day of eclipse for a geostationary satellite.

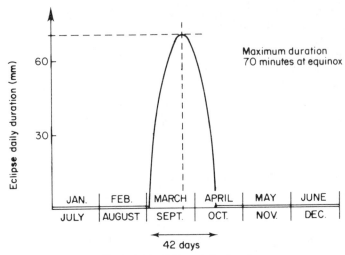

Figure 4.24 Eclipse daily duration.

Sun. Compared with Earth eclipses those caused by the Moon occur irregularly in time of duration and depth (Siocos, 1981). The number of Moon solar eclipses occurrences per orbital location per year ranges from zero to four with an average of two per year. Eclipses can occur twice within a 24 h period. The duration of eclipses range from a few minutes to over two hours with an average duration of about 40 minutes. Special problems in connection with battery recharging and spacecraft thermal reliability could arise when Moon solar eclipses of long duration and appreciable depth occur just before or soon after Earth solar eclipses.

4.8.3 Operation during eclipse

If the satellite uses solar energy as a primary power source, an energy reserve allowing normal functioning during eclipses must be provided if the satellite is to ensure continuous service. This can be achieved with on-board batteries. Another solution is to provide a spare satellite. This spare satellite requires a longitudinal separation great enough so that one of the two satellites is always lit when the other is in shade. This requires the distance between the two satellites to be greater than $17.4°$ (Figure 4.22). However, this solution presents two main drawbacks.

(1) The shift from one satellite to another requires repointing of the ground antennae and therefore a break in service; unless two antennae are employed, or one with electronic beam shaping.
(2) The service area reduces to that which is covered by both satellites simultaneously.

Moreover, the time required to bring the first satellite up to temperature after the eclipse must be considered. Assuming that this asks 30 minutes, then the separation between the satellites should be

$$100 \times 360/24 \times 60 = 25°$$

For certain types of satellites, direct broadcast satellites for example, it is acceptable not to ensure service during the Earth solar eclipses as they always occur at night. The further west the satellite is from the service area the later at night the eclipse occurs (Sawitz, 1980).

4.8.4 Solar interference

Solar interference occurs when the Sun is within the earth station antenna beam. This implies that the declination angle δ is equal to the angle that an earth station antenna axis makes with the equatorial plane. Any earth station antenna pointing at the satellite makes a maximum angle relative to the equatorial plane of $8.7°$. Therefore, solar interference occurs at the periods of the year neighbouring the equinoxes:

(1) Before the vernal (spring) equinox and after the autumnal equinox for a northern hemisphere station;
(2) After the vernal equinox and before the autumnal equinox for a southern hemisphere station.

Feeding $\delta = 8.7°$ in formula (20), results in the following:

(1) Solar interference takes place at the earliest t_{max} days before the equinox and t_{max} days after such that

$$t_{max} = 8.7/23.5 \ T/2\pi = 21 \text{ days}$$

(2) For an antenna of major lobe beamwidth of θ, solar interference takes place during Δt days such that

$$\Delta t = \theta/23.5° \times T/2\pi = 2.5\theta \text{ days}$$

if $\theta = 2°$, then $\Delta t = 5$ days. Each solar interference lasts $\Delta t'$ with:

$$\Delta t'_{(s)} = 23.5° \times 3600/360° \times \theta_{(degrees)} = 240\,\theta_{(degrees)}$$

Taking $\theta = 2°$: $\Delta t' = 480 \text{ s} = 8 \text{ min}$.

During solar interference the antenna noise temperature is severely increased. It reaches around 10^4 K for an antenna of beamwidth equal to or less than the collimation of the Sun $(0.5°)$. Noise temperature is less if the antenna beamwidth is greater (around 10^3 K for a $1°$ antenna beamwidth).

REFERENCES

DEMYTKO, N. and K. ENGLISH (1977) Echo cancellation on time variant circuits, *Proc. IEEE*, **55**(3), pp. 444–453.

FUENZALIDA, J. C., P. RIVALAN and H. J. WEISS (1977) Summary of the Intelsat V communications performance specifications, *Comsat Tech. Rev.*, 7(1), 311–326.

RUSCH, R. J. and D. G. DWYRE (1978) Intelsat V spacecraft design, *Acta astronautica*, Vol. 5, No. 3–4, Pergamon Press, pp. 173–188.

SAWITZ, P. (1980) The effects of geography on domestic fixed and broadcasting satellite systems in ITU region 2, *AIAA, 8th CSSC*, Orlando, Paper 80–0509, 7pp.

SIOCOS, C. A. (1981) Broadcasting Satellites power blackouts from solar eclipses due to the moon, *IEEE Trans. on Broadcasting*, Vol. BC-27, No. 2, pp. 25–28.

SMITH, F. L. (1972) A nomogram for look angles to geostationary satellites, *IEEE Trans. Aerosp. and Electron Sys.*, Vol. AES, No. 5 (May), p. 394.

STAMMINGER, R. and J. A. STEIN (1980) Business satellite developments, *Eascon '80 Record*, pp. 95–101.

SUYDERHOUD, H. G., M. ONUFRY and S. J. CAMPANELLA (1976) Echo control in telephone communications, *National Telecom. Conf.*, Vol 1, pp. 8.1.1–8.1.5.

CHAPTER 5 Space Environment and its effects

This chapter aims at a description of the main environmental factors which condition the design and operation of the satellite on its orbit during its lifetime. The effects of the satellite environment are:

(1) Mechanical, as a result of perturbing forces and torques which modify the satellite attitude and orbit.
(2) Thermal, as a result of heat absorbed by the satellite from the Sun or the Earth, and radiated from the satellite to deep cold space.
(3) Degrading, as a result of the action of high energy radiations and particles on materials and surface finish.

First a description of the space environment will be given, and then the effects of this environment on the satellite itself.

5.1 DESCRIPTION OF SPACE ENVIRONMENT

The paramount factors in space on spacecraft missions are:

(1) Gravitational fields.
(2) Space vacuum.
(3) Space radiation.
(4) Earth's magnetic field.
(5) Meteoroids and space debris.

5.1.1 Earth's gravitational field

The gravitational field originating from its terrestrial host is the source of the main external force affecting an earth orbiting satellite.

The Earth's gravitational field is not uniformly spherical as a result of the non-homogeneous distribution of the mass of the Earth. The related potential function may be expressed as:

$$U = \frac{\mu}{r}\left[1 - \sum_{h=2}^{\infty}\left(\frac{R_e}{r}\right)^n J_n P_n(\sin l) + \sum_{n=2}^{\infty}\sum_{q=1}^{\infty}\left(\frac{R_e}{r}\right)^n J_{nq} P_{nq}(\sin l)\cos q(L - L_{nq})\right]$$

(1)

where:

$\mu = 3.986\,10^{14}\,\mathrm{m}^3/\mathrm{s}^2$, Earth's gravitational constant,

r = distance of the satellite from the Earth's centre,

L = longitude of the satellite (positive towards the east), measured from the Greenwich meridian,

l = latitude of the satellite (positive towards the north),

$R_e = 6378$ km, Earth's equatorial radius,

$$P_n(x) = \frac{1}{2^n!}\frac{\mathrm{d}^n}{\mathrm{d}x^n}(x^2 - 1)^n$$

$$P_{nq}(x) = (1 - x^2)^{q/2}\frac{\mathrm{d}^q}{\mathrm{d}x^q}P_n(x)$$

J_n, J_{nq}, are constants characterizing the Earth's mass distribution. J_n are zonal harmonics related to the Earth's oblateness and J_{nm} are tesseral harmonics related to the ellipticity of the Equator which displays a 150 m difference between the Earth's major and minor axis. The term J_2 is of the order 10^{-3} and is 10^3 times larger than all the J_n terms ($n > 2$).

Value for J_{22} is 1.86×10^{-6} and J_{nq} ($n > 2$) is about 10^{-6}.

As a geostationary satellite orbits the Equator (latitude $0°$ so sin $l = 1$), and R/r is small, we can rewrite (1) ignoring higher order terms above 2:

$$U = \frac{\mu}{r}\left[1 + \left(\frac{R_e}{r}\right)^2\left\{-J_2 + 3J_{22}\cos 2(L - L_{22})\right\}\right] \qquad (2)$$

with $L_{22} = -15°$, which is $15°$ longitude west.

The perturbing potential can be expressed as

$$U_p = U - \frac{\mu}{r} \qquad (3)$$

5.1.2 Gravitation field of the Moon and the Sun

In addition to the Earth, the Moon and the Sun also have gravitation fields which can be expressed as follows:

$$U_p' = \mu_p\left(\frac{1}{|\mathbf{r}_p - \mathbf{r}|} - \frac{\mathbf{r}_p \cdot \mathbf{r}}{|\mathbf{r}_p|^3}\right) \qquad (4)$$

where μ_p is the constant of the attraction of the perturbing body (Moon or Sun), $\mathbf{r}_p$ is the vector from the centre of the Earth to the perturbing body, and $\mathbf{r}$ is the vector from the centre of the Earth to the satellite.

5.1.3 Vacuum

Vacuum is a dominant feature of the space environment. As altitude increases, the density of the particles (atoms, ions, electrons) diminishes very rapidly.

The effect of the aerodynamic drag can be considered negligible if the satellite's altitude is high enough.

5.1.4 Thermal radiation

Recall that a body radiates energy depending on its temperature T and its emittance ϵ. The Stefan Boltzmaan law expresses the radiant exitance M, i.e. the radiat flux leaving an element of surface of the body (W/m^2), as:

$$M = \epsilon\sigma T^4$$

where $\sigma \cong 5.67\,10^{-8}\,W\,m^{-2}\,K^{-4}$.

For a black body, $\epsilon = 1$, so that the emittance appears as the ratio of the exitance of any body to that of the blackbody at the same temperature.

The Planck radiation law expresses the spectral radiance $L_\lambda(W/m^3\,sr)$ of a blackbody, i.e. the power emitted per unit wavelength and per unit solid angle in a given direction of an element of surface of the blackbody divided by the area of the orthogonally projected area of the element on a plane perpendicular to the given direction, as:

$$L_\lambda = C_{1L}\lambda^{-5}(C_2/\lambda T) - 1)^{-1}$$

where $C_{1L} = 1.19\,10^{-16}\,W\,m^2\,sr^{-1}$
$C_2 = 1.439\,10^{-2}\,m\,K$

A functional form of the Planck law, known as the Wien law, relates the wavelength at which maximum spectral radiance L_m of a blackbody occurs to its temperature T:

$$\lambda_m T = b$$
$$L_m/T^5 = b'$$

where $b\ = 2.9\,10^{-3}\,m\,K$
$b' = 4.1\,10^{-6}\,W\,m^{-3}\,K^{-5}\,sr^{-1}$

Space behaves like a blackbody at approximately 5 K; it acts as a heat sink, with an absorbancy equal to 1, so all thermal energy radiated is completely absorbed. The radiation received by an earth orbiting satellite originates mainly from the Sun and the Earth.

5.1.4.1 Solar radiation

Solar radiation flux on any part of the surface of a geostationary satellite facing the Sun is equal to $W = 1353\,W/m^2$ when the satellite is at a distance from the Sun equal to 1 international astronomic unit (IAU), that is eleven days after the equinoxes. Figure 5.1 shows the variations of this solar radiation flux during the year as related to the Sun–Earth geometry displayed on Figure 4.19, taking into account both the Earth to Sun distance variations (from 0.983 IAU to 1.017 IAU) and the variations of the declination of the Sun (from 0° to 23.5°).

Figure 5.2 which shows the spectal irradiance, i.e. the spectral radiant flux

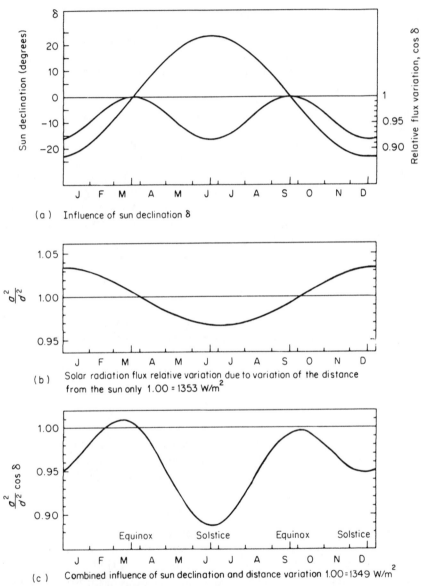

(a)   Influence of sun declination δ

(b)   Solar radiation flux relative variation due to variation of the distance from the sun only $1.00 = 1353$ W/m^2

(c)   Combined influence of sun declination and distance variation $1.00 = 1349$ W/m^2

Figure 5.1   Solar radiation flux variations on a north–south Sun-facing surface of a geostationary satellite.
$a = 1$ IAU (semi major axis of the earth orbit), $d =$ sun to earth distance

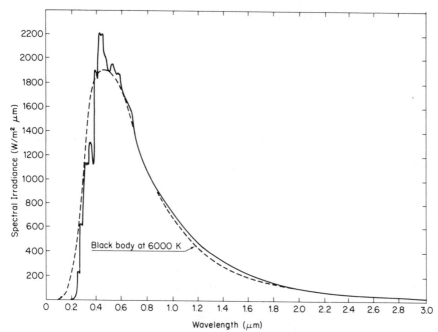

Figure 5.2 Sun irradiance versus wavelength.

incident on the surface of a geostationary satellite facing the Sun, indicates how the above radiation flux W is distributed versus wavelength. The Sun acts as a blackbody at the temperature of 6000 K. 90 per cent of the radiated energy is located in the range 0.3 to 2.5 microns wavelength, with a maximum near 0.5 micron.

The Sun is seen from the Earth with an angle equal to $0.5°$.

5.1.4.2 Terrestrial radiation

Terrestrial radiation results from the reflected solar radiation (albedo) and from its own radiation. The latter corresponds with that of the blackbody at the temperature of 255 K (Infra-red $10-12\ \mu$).

For a geostationary satellite the total terrestrial radiation intensity is less than 40 W/m^2, therefore it is negligible in comparison with the intensity of the Sun (1353 W/m^2).

5.1.4.3 Cosmic particles

Cosmic particles consist mainly of electrons and protons originating from the outer space and from the Sun (King, 1974; Stassinopoulos, 1980). The flux and the energy of these particles depend mainly on the altitude and the solar activity.

Galactic cosmic rays consist mainly of protons (90 per cent) and a few alpha particles. The energies are in the GeV range but the fluxes are extremely low, e.g. 2.5 particles/cm^2 s.

The *solar wind* consists mainly of protons and electrons with less energy. The average density of protons during quiet sun periods is about 5 protons/cm^3 moving away from the Sun at about 400 km/s. The corresponding average flux at the earth orbit is 2×10^8 protons/cm^2 s with an average energy of the order of a few keV. Depending on the solar activity, flux value may fluctuate by a factor of 20. During active sun periods solar flares occur more frequently and they comprise mainly protons of energy between a few MeV to a few hundreds of MeV. Occasionally once every few years proton energy may reach 1 GeV.

As these particles are charged, some are trapped by the Earth's magnetic field in the so-called *Van Allen belts*. Energetic Van Allen belt *electrons* are distinguished into 'inner zone' and 'outer zone' populations. The $L = 2.8$ Earth radii line is being used to separate the inner and outer zone domains. So that only the outer zone populations of electrons affect the geostationary satellite. The energy ranges from a few hundreds of keV to a few MeV. Energetic Van Allen belt *protons* are usually contained within $L = 4$ Earth radii with maximum fluxes encountered at L of the order of 1.7 Earth radii. The energy ranges from 1 MeV to a few hundreds of MeV. Note that as a geostationary satellite orbit, at $L = 6.6$ Earth radii it remains outside the proton Van Allen belt, except during launch sequence (transfer orbit).

At the geostationary satellite orbit, the total flux of particles per cm^2 per year is given in Table 5.1.

The following formulae can be used to estimate the electron fluxes in the 'outer zone'.

$$N(> E) = 7.10^{14} \exp(-4E) \qquad \text{for active sun} \qquad (5)$$

$$N(> E) = 1.3 \times 10^{13} E^{-1.5} \qquad \text{for quiet sun} \qquad (6)$$

where $N(> E)$ represents the electron fluxes per cm^2 and per year with energy, expressed in MeV, greater than E.

Table 5.1   TOTAL FLUX OF PARTICLES PER CM^2 PER YEAR.

	SOLAR ACTIVITY	
	MAX	MIN
Trapped electrons ($E > 0.5$ MeV)	9.5×10^{13}	3.5×10^{13}
Trapped protons	Negligible effect	Negligible effect
High energy solar protons ($E > 40$ MeV)	6.5×10^9	1×10^7

Protons relevant to a geostationary satellite are mainly high energy solar protons generated by solar flares. Flux depends on the level of solar activity which conditions the occurrence of ordinary and anomalously large solar flares. One anomanously large solar flare is likely to occur during a period of seven years corresponding to the lifetime of present day satellites.

5.1.5 Terrestrial magnetic field

The Earth's magnetic field H is that of a magnetic dipole of moment $M_T = 7.9 \ 10^{15}$ Wb.m. This dipole forms an angle of $11.5°$ with the rotation axis of the Earth. Thus it creates an induction B which has two components:

(1) A radial component: $B_r = \dfrac{M_T \ 2 \cos \theta}{r^3}$ (weber/m^2)

(2) A normal component: $B_n = \dfrac{M_T \sin \theta}{r^3}$ (weber/m^2)

where r and θ are polar coordinates of the considered point in the dipole reference system.

For a geostationary satellite the normal component varies between $1.03 \ 10^{-7}$ and $1.05 \ 10^{-7}$ weber/m^2 and the radial component between $\pm 0.42 \cdot 10^{-7}$ weber/m^2. The component perpendicular to the orbital plane is practically constant and equals $1.03 \ 10^{-7}$ weber/m^2.

5.1.6 Meteorites

The Earth is surrounded by a cloud of meteorites (material fragments, rocks, stones . . .) the density of which decreases as the altitude increases and as the mass increases. Their velocity ranges from a few km/s to several tens of km/s. The following formulae give the flux N in particles of mass m or greater per square metre and per second.

$$\log_{10} N(\geqslant m) = -14.37 - 1.213 \log_{10} m \quad \text{in the range } 10^6 \text{g} < m < 1\text{g} \qquad (8)$$

$$\log_{10} N(\geqslant m) = -14.34 - 1.534 \log_{10} m - 0.063(\log_{10} m)^2$$
$$\text{in the range } 10^{-12} \text{ g} < m < 10^{-6}\text{g} \qquad (9)$$

5.2   EFFECTS OF THE SPACE ENVIRONMENT

Both mechanical and thermal effects will be discussed. Then the effects on satellite materials and components will be presented.

5.2.1 Mechanical effects (Legendre, 1980)

5.2.1.1 Gravitational torque

This torque is generated by the dependance on distance from the Earth's centre of the Earth's gravitational field and by inhomogeneity of the satellite structure. These effects may cause the satellite to rotate about its centre of mass unless the axis of smaller inertia of the satellite is aligned with the Earth local vertical. Assuming the z-axis to be of revolution for the satellite, the corresponding torque is given by:

$$T_g = 3\,\frac{\mu}{r^3}\,(I_z - I_x)\,\theta$$

where μ is the Earth gravitational constant, r is the distance from satellite to the centre of the Earth, I_z is the moment of inertia about the z-axis, I_x (smaller than I_z) is the moment of inertia about any axis perpendicular to the z-axis, and θ is the angular deviation, assumed to be small, between the z-axis and the perpendicular to the orbital plane.

This torque, which may be used to stabilize satellites placed in a low orbit, is rather inefficient for the stabilization of geostationary satellites which means that in the latter case it may easily be combated by making I_x and I_z not too different one from another. For instance, with $I_z = 180\ \text{m}^2\,\text{kg}$, and $I_x = 100\ \text{m}^2\text{kg}$, the maximum torque is $C_g = 2.2 \times 10^{-7}\ \text{N} \cdot \text{m}$, when θ is smaller than $10°$.

5.2.1.2 Influence of the asphericity of the terrestrial potential

The perturbing potential given in equation (3) creates a tangential acceleration Γ positively oriented according to increasing longitudes, so that:

$$\Gamma_T = -\frac{1}{r}\frac{dU_p}{dL} = \frac{\mu}{r^2}\left(\frac{R_e}{r}\right)^2 6\,J_{22}\sin 2(L - L_{22}) \qquad (10)$$

This acceleration generates a variation of the velocity V_s of the satellite in its orbit, which causes a drift in longitude

$$\Gamma_T = \frac{dV_s}{dt} = \frac{d}{dt}(r\omega) = \frac{dr}{dt}\omega + r\frac{d\omega}{dt} \qquad (11)$$

where ω represents the angular satellite velocity (rad/s) and can be calculated from:

$$\omega = \frac{2\pi}{T} = \sqrt{\frac{\mu}{a^3}} \qquad (12)$$

where a is the semi-major axis of the orbit. As the orbit is nearly circular one can consider that $r = a$ and $\omega = \mu^{1/2} r^{-3/2}$ resulting in:

$$\frac{d\omega}{\omega} = -\frac{3}{2}\left(\frac{dr}{r}\right)_{r=a} \tag{13}$$

From equations (11) and (13) it can be deduced:

$$\left(\frac{dr}{dt}\right)_{r=a} = -\frac{2}{\omega}\Gamma_T \tag{14}$$

so:

$$\frac{d^2L}{dt^2} = \frac{d\omega}{dt} = \frac{3}{a}\Gamma_T \tag{15}$$

and the longitudinal acceleration undergone by the satellite is expressed as follows:

$$\frac{d^2L}{dt^2} = k^2 \sin 2\,(L - L_{22}) \tag{16}$$

where $k^2 = (18\omega^2 R_e^2/a^2)\, J_{22} = 3.10^{-5}\ \text{rad}/(\text{day})^2 = 4.10^{-15}\text{rad}/\text{s}^2$
and $L_{22} = -15\,^\circ = 15\,^\circ\text{W}$

The above derivation is based on several approximations. The actual semi-major axis of the satellite orbit is $a = 42166.2$ km, which differs from the value of $a = 42164.2$ km as given by equation (12). This difference comes from the influence of the zonal harmonic J_2 of equation (1). The actual acceleration undergone by a geostationary satellite is shown on Figure 5.3. It can be seen that the acceleration depends on the location of the satellite. There are two stable equilibrium positions (105° longitude west and 76° longitude east) and two positions with unstable equilibrium (11° longitude west and 162° longitude east).

From (16) one sees that the motion of the satellite around a point of stable equilibrium is approximatively governed by the equation:

$$\frac{d^2\Lambda}{dt^2} = -k^2 \sin 2\Lambda \tag{17}$$

where Λ is the longitude of the satellite measured in relation to the nearest point of stable equilibrium ($\Lambda = L - L_{22} \models 90^\circ$).

A primary integral of (17) is:

$$\left(\frac{d\Lambda}{dt}\right)^2 - k^2 \cos 2\Lambda = \text{constant} \tag{18}$$

In Figure 5.4 the curves give the drift $d\Lambda/dt$ in terms of the longitude Λ about a point of stable equilibrium. The numbers in brackets give the period.

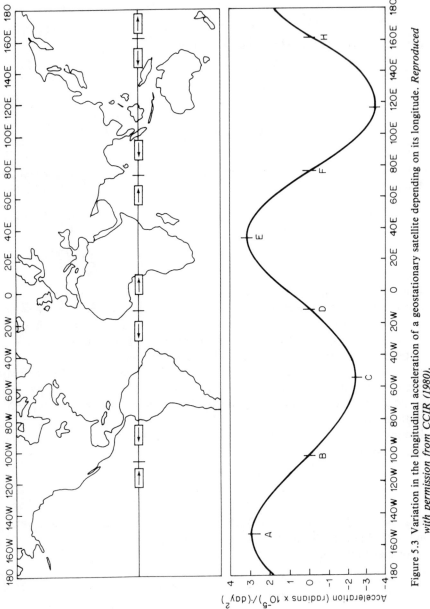

Figure 5.3 Variation in the longitudinal acceleration of a geostationary satellite depending on its longitude. *Reproduced with permission from CCIR (1980).*

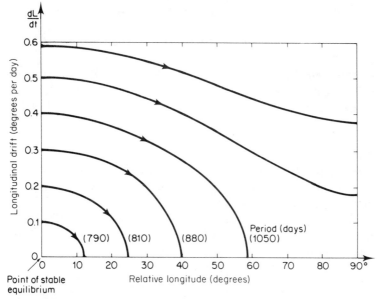

Figure 5.4 Longitudinal drift of a geostationary satellite.

5.2.1.3 Influence of the composite luni-solar attraction

This attraction mainly causes a satellite motion perpendicular to the equatorial plane. Hence the inclination of the orbital plane shifts at a rate of *about* $1°$ per year ($0.3°$ caused by the Sun and $0.7°$ caused by the Moon) towards the eclliptic plane. More precisely, this variation in inclination is between $0.75°$ and $0.95°$ per year according to the date (minimum in 1978 and maximum in 1987) as it is related to the 17 years periodicity of the motion of the ascending longitudinal mode of the Moon orbit. Moreover, the luni-solar attraction causes a small longitudinal drift of the satellite which is combated by a slight modification of the semi-major axis of the satellite orbit which now becomes $a = 42165.8$ km.

5.2.1.4 Influence of aerodynamic drag

Residual atmosphere at the geostationary satellite orbit has a negligible effect on the motion of geostationary satellites. However, its effect is of importance when satellites are on the transfer orbit prior to injection on the geostationary satellite orbit.

5.2.1.5 Influence of solar radiation pressure

An element of surface dS with normal n, making an angle θ with the direction of the Sun defined by a unit length vector $\mathbf{u}$ is acted on by a force:

$$\frac{dF}{dS} = -\frac{W}{c} [(1 + \rho) \cos^2 \theta \, \mathbf{n} + (1 - \rho) \cos \theta \, \mathbf{n} \times (\mathbf{u} \times \mathbf{n})] \qquad (19)$$

where c is the velocity of light, ρ the reflectance that is the ratio of the reflected flux to the incident flux, and W the solar radiation flux.

If element dS is totally reflecting ($\rho = 1$) the pressure is normal to the surface

$$\frac{dF}{dS} = -\frac{2W}{c} \cos^2 \theta \, \mathbf{n}$$

If element dS is totally absorbing ($\rho = 0$), the radiation pressure displays two components: one normal to the surface element $(dF/dS)_N = W/c \cos^2 \theta$ and one tangential to the surface element $(dF/dS)_T = -W/c \cos \theta \sin \theta$.

A satellite with an apparent surface area S_a oriented towards the Sun, with a reflectance $\rho = 0.5$ and with a mass m is subject to the following force normal to the surface:

$$F = \frac{1.5 \, W}{c} S_a$$

where $W/c = 4.51 \times 10^{-6} \, \text{N/m}^2$ at 1 IAU.

The acceleration due to the solar radiation pressure on the surface is therefore given by:

$$\gamma = \frac{F}{m} = 6.77 \times 10^{-6} \frac{S_a}{m} \, \text{m/s}^2$$

Solar arrays constitute practically the entire apparent surface area of satellites. Practically S_a/m is in the order of $10^{-2} \, \text{m}^2/\text{kg}$. For example, with INTELSAT V, $S_a = 18 \, \text{m}^2$ and $m = 1000 \, \text{kg}$, then $S_a/m = 1.8 \, 10^{-2} \, \text{m}^2/\text{kg}$. With such satellites the acceleration resulting from the solar radiation pressure is not large enought to bring a significant perturbation of the orbit compared with other causes of perturbation.

Satellites equipped with large solar arrays (for example, a surface of $100 \, \text{m}^2$ for a mass of 1000 kg), have a ratio S_a/m larger than or equal to 10^{-1}. For instance the direct broadcasting satellites TDF1, displays solar arrays of $70 \, \text{m}^2$ and a mass of 1100 kg. The main impact is on the eccentricity of the orbit which undergoes periodic disturbances (period of one year) with an amplitude in the order of 10^{-3}. In such circumstances specific station keeping operations are required to maintain the eccentricity within acceptable limits.

Moreover, the resultant of the forces applied on all the surface elements dS of a satellite generally does not contain the satellite centre of mass. Consequently there is a disturbing torque which acts on the satellite's attitude. As each elementary force is proportional to $W \cos \theta$, the torque applied to the satellite is proportional to $WS_a \cos \theta$, where S_a is the apparent surface area of the satellite as seen from the Sun. Hence the torque is dependent on the orientation of the satellite relative to the Sun. It should be recalled that due to the Sun declination the north–south axis of a geostationary satellite makes an angle between $67°$ and $113°$ with respect to the direction of the Sun. The torque tends to cause a drift of the north–south axis of the satellite.

5.2.1.6 Influence of the Earth's magnetic field

The Earth's magnetic terrestrial field creates a torque $\mathbf{C_M} = \mathbf{M} \times \mathbf{B}$ on a satellite having a magnetic moment M. Although the largest component of the field is perpendicular to the Equator, it produces the weakest effect if the satellite revolves around its north–south axis (spin stabilization) owing to the fact that the corresponding torque is in the equatorial plane and the resultant sum of the torque per turn is nil. If the satellite is 'body' stabilized it makes a complete turn per day around its north–south axis and the resultant of the torque is nil on a 24 h average.

The global magnetic moment of a satellite results from remanent moments, such as moments due to electric currents in the wires, induced moments, proportional to the terrestrial magnetic field. These moments can be reduced or compensated before the launch to achieve 10^{-4} N · m on the ground quite easily. As the Earth's magnetic field is inversely proportional to the cube of the distance to the centre of the Earth, the torque in geostationary orbit is divided by:

$$(42.166/6.380)^3 = 289.$$

Therefore the global magnetic moment of a geostationary satellite while orbiting is of the order of $3.5 \ 10^{-7}$ N · m

Actually, the launching conditions modify part of the adjustments made on the ground; it is advisable to establish a margin and to adopt a torque value of $C_M = 10^{-6}$ N · m

5.2.1.7 Influence of the meteorite impact

The momentum transmitted to a satellite by a meteorite collision is best evaluated statistically, namely by the probability of meteorites of given masses colliding into the satellite and by the momentum transfer resulting from the collision. Impacts between meteorites and satellites may be assumed to occur at random and the Poisson distribution may be applied. Thus the probability of n impacts in the mass range $M_2–M_1$ occurring over an area S in time t may be summarized as:

$$P(n) = \frac{(Sft)^n \exp(-Sft)}{n!} \qquad (20)$$

where f is the particle flux in the mass range $M_2 - M_1 = N(> m_1) - N(> m_2)$ with $N(> m)$ given by equations (8) and (9), S is the exposed area (m^2) and t is the time of exposure (s).

5.2.1.8 Self-generated torques

The displacement of the antennae, the solar arrays and the fuel cause torques which affect the main body of the satellite. Furthermore, the station keeping of the geostationary satellites periodically require corrective forces applied to the centre of mass of the satellite.

As propellent tanks are emptied during the mission, the centre of mass varies relative to the body of the satellite. Moreover, thrusters differ slightly one from another and may not be perfectly aligned. The corrective forces of station keeping are therefore not applied precisely to the centre of mass, and a station-keeping disturbing torque results during these corrections. For instance, assuming a thrust of 2 N and a displacement of the centre of mass of 5×10^{-3} m results in a perturbing torque C_p of 10^{-2} N $\cdot$ m.

5.2.1.9 Effect of radiated radio frequency power

Pressure caused by the radio frequency radiation of the antennae creates a force which cannot be neglected if the transmitted power is high.

For a very directive antenna radiating a power P_T, the produced force F is:

$$F = -\frac{dm}{dt} \cdot c = -\frac{P_T}{c} \tag{21}$$

where c is the light velocity.

For a satellite transmitting 1 kW within a beam of 1° (direct broadcasting satellite) the force is $F = 3 \times 10^{-6}$ N. If the torque arm length is 1 m, the torque is 3×10^{-6} N $\cdot$ m

This force is only of importance if the radiated power is large and is concentrated in a narrow beam. In that case the axis of the antenna must contain the satellite's centre of mass or the satellite must be designed with two antennas symmetrically disposed with respect to the centre of mass.

5.2.1.10 Conclusions

The geostationary satellite is subject to perturbations which cause it to drift away from its nominal position and which create attitude perturbing torques. The drift in latitude is the most significant. This drift results from the luni-solar attraction and is due to a variation in the inclination of the orbital plane of about 1° per year. The drift in longitude is basically due to the asymmetry of the terrestrial potential and depends on the nominal station of the satellite

Table 5.2 ATTITUDE PERTURBING TORQUES.

Origin of perturbation	Instantaneous torque (N · m)	Observations
Station keeping	10^{-2}	During corrections only
Radiation pressure	5×10^{-6}	Always present except at eclipses
Magnetic field	10^{-6}	Daily average small
Gravity gradient	10^{-7}	Always present

relative to the two points of equilibrium situated at $105°$ longitude west and $76°$ longitude east. An additional longitude drift is caused by the satellite orbit eccentricity variation due to the solar radiation pressure.

The perturbing torques affecting attitude are summarized in Table 5.2.

5.2.2 Thermal effects

The satellite is heated by solar radiation whilst the sides facing cold space are cooled. Heat transfers result only from conduction and radiation as the vacuum excludes any convection heat transfer. The temperature of a geostationary satellite is determined by the balance between the radiation it absorbs from the Sun and the heat it radiates.

5.2.2.1 Average temperature of the satellite

The average temperature is obtained from the following thermal balance

$$P_S + P_I = P_R + P_A \tag{22}$$

where P_S is the power absorbed from the sun ($P_S = \alpha W S_a$ is the result of solar radiation flux W on an apparent surface S_a with an absorptance α defined as the ratio of the absorbed flux to the incident flux), P_I is the power dissipated internally, P_R is the radiated thermal flux, and P_A is the accumulated power during transient heat transfers.

For a spherical perfect heat conductor with a radius r at the equilibrium temperature T, it follows that:

$P_S = \alpha W \pi r^2$
$P_I = 0$
$P_A = 0$
$P_R = \varepsilon \sigma T^4 \uparrow 4\pi r^2$ is the power σT^4 radiated (see section 5.1.4) by the total surface $4\pi r^2$ of emittance ε, which is the ratio of the radiant flux intensity from the considered body to that of a blackbody at the same temperature (σ is the Stefan-Boltzmann constant (5.67×10^{-8} W m^{-2} K^{-4}), and $W = 1353$ W/m^2 at 1 $\dot{I}AU$ distance).

Table 5.3 EQUILIBRIUM TEMPERATURE OF A SPHERICAL PERFECT HEAT CONDUCTOR
IN SPACE AT 1 IAU DISTANCE FROM THE SUN.

Shielding	Absorptance α	Emittance ε	$\dfrac{\alpha}{\varepsilon}$	Temperature (°C)
Cold: white colour	0.20	0.80	0.25	−75
Mean: black colour	0.97	0.90	1.08	+12.5
Hot: brilliant gold	0.25	0.045	5.5	+155

So the equilibrium temperature is:

$$T = \left(\frac{\alpha}{\varepsilon}\frac{W}{4\sigma}\right)^{1/4} \tag{23}$$

The equilibrium temperature of this spherical passive body ($P_I = 0$) only depends on the *optical* characteristics of the external surface, i.e. practically on its colour. For various shieldings, Table 5.3 indicates that the equilibrium temperature can vary between −75 to +155°.

The equipment of the satellite will operate satisfactorily in a narrower temperature range, for example from 0 to +45° C. The shieldings need therefore be selected and combined judiciously to satisfy these conditions. As the solar cells often comprise the largest part of the surface of the satellite, their optical characteristics are very important. Their absorptance lies between 0.7 and 0.8 (in unloaded conditions) and they have an emittance of between 0.80 and 0.85.

5.2.3 Effects on the materials

5.2.3.1 Vacuum effects

In vacuum, metals and semiconductors sublimate and outgas. The corresponding loss of mass depends on the temperature (1.000 Å/year at 110 °C, 10^{-3} cm/year at 170 °C and 10^{-1} cm/year at 240 °C for magnesium). As temperatures above 200° are easily avoided and on condition that the shieldings used are not too thin, these effects are not important. The possibility of condensation of gases on cold surfaces is more of a problem (e.g. short circuits on insulating surfaces); the use of materials which are too easily sublimable or become conductive, like zinc and cesium, should therefore be avoided. On the other hand, a big advantage of the vacuum is that it spares metals the problems associated with corrosion. Finally, polymers tend to decay into volatile products.

Some surfaces, when in contact, tend to diffuse into one another in a cold welding process and bind solidly. Bearings and rotary joints undergo an increased frictional effect. One can maintain these moving pieces in insulated pressurized enclosures or use lubricants with a low rate of evaporation or sublimation. Ceramic ball-bearings made from special ceramic metal compounds such as stellite or nickel-bonded titanium carbide may also be used.

5.2.3.2 Effects of Solar Radiation

The ultra-violet radiation whose spectrum stretches from 100 to 1000 Å, causes ionization in the materials. This results in:

(1) An increase in the conductivity of the insulators and the modifications of the absorptance and emittance of the shieldings.
(2) A decrease in the efficiency of the solar cells with time (30 per cent decrease for silicon cells at the end of seven years).

At wavelengths above 1000 Å, the solids can be ionized; the polymers are discoloured and their mechanical characteristics weakened. Above 3000 Å, the effects on metals and semiconductors are negligible.

5.2.3.3 Effects of meteorites

The probability of impact of a meteorite on the satellite has been already discussed in Section 5.2.1.7. The impact of meteorites creates an erosion of the materials of about 1 Å per year at geostationary satellites altitude (200 Å at low altitudes). For the heaviest meteorites these impacts can create a perforation of the sheet metals if these are too thin, which in some cases can be disastrous for the survival of the satellite. A protection can be obtained using bumpers which are several sheets of metal slightly separated. The first sheet fragments the meteorite and the following ones stops the debris.

5.2.3.4 Effects of cosmic particles

As a result of charged particles, metals and semiconductors undergo atomic excitations of their electrons. Plastics are even ionized. Mineral insulators undergo both effects. Solar flares in particular affect the minority carriers in the semiconductors, the optical transmission of the glass and some polymers. The main effect of high energy particles is to degrade the performance of the solar cells and modify surfaces designed for thermal control from their original finish. Galactic cosmic rays are usually neglected in solar array damage predictions.

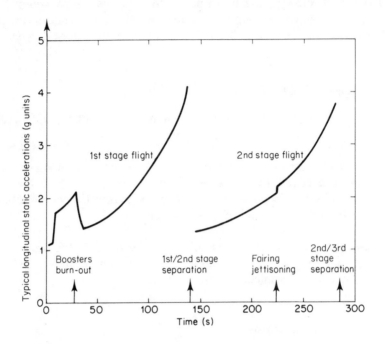

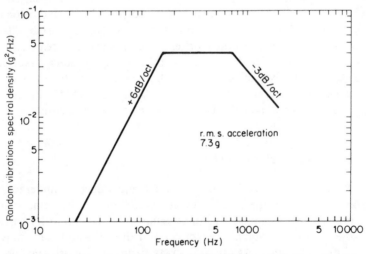

Figure 5.5 Environment during launch (ARIANE 3 launcher). *Reproduced by permission of the European Space Agency.*

5.3 ENVIRONMENT BEFORE THE INJECTION INTO GEOSTATIONARY ORBIT

The injection into geostationary orbit is preceded by two phases (See Chapter 7) in the course of which the environment deviates somewhat from what has been described for a geostationary satellite:

(1) The *launching phase*, which ends with the injection into the transfer orbit, and lasts several minutes,
(2) The *transfer orbit phase*, during which the satellite describes elliptical orbits (200–36 000 km). This phase lasts a few tens of hours.

5.3.1 Environment during launching

A fairing protects the satellite from aerodynamic overheating and pressure as it is propelled through the low atmosphere. Overheating of the fairing has little effect on the satellite. The greatest stresses are those shocks and vibrations communicated by the launcher during stage separation and powered flight (Figure 5.5)(European Space Agency, 1981)

5.3.2 Environment in the transfer orbit

During this phase, the satellite is generally stabilized by spinning and its configuration differs from operational configuration: full apogee motor, solar arrays and antennae folded.

The environment and the effects described in the preceding paragraphs are applicable with two differences:

(1) The thermal balance is affected by the Earth proximity, e.g. eclipses every few hours, earth radiation and albedo, and by the specific satellite configuration and attitude.
(2) At the perigee, the atmospheric drag is not negligible and can cause a lowering of the apogee.

REFERENCES

EUROPEAN SPACE AGENCY (1981) *Ariane user manual*, Paris.

KING, J. H. (1974) Solar proton influences for 1977–1983 space missions, *J. Spacecraft*, **11**(3), pp. 401–408.

LEGENDRE, P. (1980) Le maintien à poste des satellites geostationnaires. Evolution de l'orbite. *Le mouvement du véhicule spatial en orbite. Cours de technologie spatial.* CNES, pp. 583–607.

STASSINOPOULOS, E. G. (1980) The geostationary radiation environment, *J. Spacecraft*, **17**(2), pp. 145–152.

CHAPTER 6   Geostationary satellite construction: communication and bus sub-systems

The design of a geostationary satellite must take into account three main operational considerations:

(1) The primary telecommunications mission which requires that the RF signals are received and transmitted at a specified power level within the angles defined by the coverage zones on Earth.
(2) The space environment and its effects on materials, components and sub-systems.
(3) The environmental stresses during launch which vary according to the launcher used and the launch procedure.

The first point was covered in Chapters 2, 3 and 4, point (2) was examined in Chapter 5 and point (3) will be discussed in Chapter 7. This chapter is devoted to the various sub-systems which constitute a geostationary communications satellite.

6.1   SUB-SYSTEMS

A geostationary communication satellite can be divided into sub-systems each of which has a specific function. It is common practice to distinguish the *communication sub-systems* or payload (repeaters and antennas) that perform the primary mission, from the *bus or platform* which includes all other sub-systems (common sub-systems) and is designed to support the communication subsystems as illustrated in Figure 6.1. A list of the sub-systems is given in Table 6.1. The functions each carries out and their main characteristics are indicated therein.

The requirements common to all satellite projects that need emphasis are:

(1) Minimal mass.
(2) Minimal power consumption.
(3) High reliability.

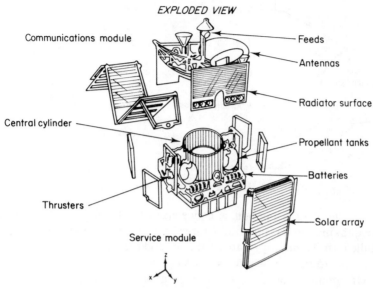

Figure 6.1   TELECOM 1 satellite (exploded view).

Table 6.1   SATELLITE SUB-SYSTEMS.

Sub-system	Function	Main characteristics
Attitude and orbit (AOCS)	Attitude stabilization Orbit determination	Accuracy
Propulsion	Provides velocity increments and torques	Specific impulse Mass of propellant
Telemetry, tracking and command (TTC)	Exchange of house-keeping data with control centre	Number of channels Security of communications
Thermal control	Temperature regulation	Heat dissipation capability
Structure	Supports equipments	Stiffness
Electric power supply	Provides electric energy at various voltage levels	Power Voltage regulation
Antennae	Receive and transmit RF signals	Coverage Gain
Repeaters	Amplify signals and change frequency	Noise figure Linearity Output RF power

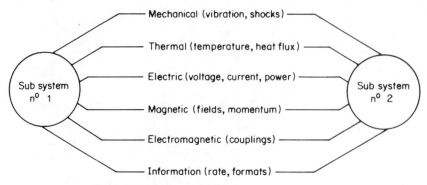

Figure 6.2 Various interfaces between sub-systems.

From the moment a mission is conceived and defined, each sub-system is designed taking into account these three criteria and the particular function to be fulfilled, the technology state-of-the-art and the characteristics of other sub-systems, as different sub-systems interact and interfere in many ways (Figure 6.2). Of importance are the Electromagnetic Interference (EMI) and Electromagnetic Compatibility (EMC) requirements.

6.2 ATTITUDE CONTROL (Wertz, 1978)

While correctly positioned at its nominal orbital location, a satellite may display motions about its centre of mass. These motions can be resolved into rotations about three reference axes: *yaw, roll* and *pitch axes*, as illustrated in Figure 6.3. The satellite attitude is determined by the angles that the satellite body axes make with these reference axes. Attitude control is vital if a satellite is to fulfil its function. The precision and reliability of this sub-system affects

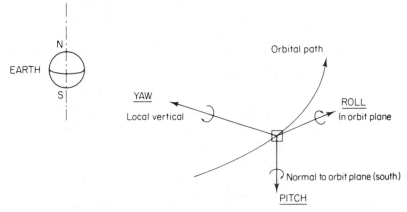

Figure 6.3 Reference (yaw, roll and pitch) axes about which a satellite spins.

the performance of most other sub-systems (e.g. spot beam antennae and solar arrays must be precisely oriented). As an example, satellite attitude must be controlled with an accuracy of about $0.1°$. This means that the axes of the satellite body must be aligned within a $0.1°$ tolerance with yaw, roll and pitch reference axes.

Attitude control consists of two separate functions:

(1) Rotating the part of the satellite aiming at the Earth around the pitch axis one revolution per day ($0.25°$ per minute).
(2) Stabilizing the satellite so as to combat the disturbing torques.

Occasionally, a distinction is made between *passive* and *active attitude control*. In the former case, the required attitude corresponds to a position of stable equilibrium of the satellite. In the latter, the satellite is unstable, or insufficiently stable, within the desired attitude configuration. Active attitude control is achieved by continually carrying out four operations as follows:

(1) Detect the satellite attitude.
(2) Compare with the reference axes.
(3) Determine the corrective torques.
(4) Correct the attitude by actuators mounted on the satellite.

The control loop is based on the use of error detection. If error detection and decision systems are on Earth, the control loop must use the telemetry, telecommand (TTC) sub-system for information exchange between the satellite and Earth. This is called open loop control. The load of the TTC sub-system can be reduced by confining the attitude control to the satellite (closed loop) (Lacombe and Havas, 1978). This solution is used, for instance, on INTELSAT V (Quaglione, 1980).

6.2.1 Sensors for detection of attitude (Duchon and Vermande, 1980)

Attitude detection requires precision which depends not only on the technique used, but also on sensor misalignment relative to the satellite body. The most commonly used sensors are *sun sensors* and *earth horizon sensors*, although stellar sensors and radiofrequency sensors can also be used.

6.2.1.2 Sun sensor

This is a photovoltaic device in which a current flows when it is illuminated by sunlight. These sensors can measure the angle between the direction of the Sun and an axis related to the satellite within a tolerance of about $0.005°$.

6.2.1.2 Horizon sensor

From space, the Earth and its atmosphere appears as a black spherical body at a temperature of 255 K when measured in the infra-red band at the carbon dioxide absorption wavelength of 14–16 microns. Its radiation contrasts strongly with the low temperature radiation (a few Kelvin) of the background outerspace. This allows the contour of the globe to be known, by using a heat sensitive device (bolometer, thermocouple or thermopile). The diffusion of reflected solar light can also be used (albedo of the Earth) and is detected by photoelectric cells or phototransistors. Precision of contour determination is around $0.05°$.

6.2.1.3 Stellar Sensor

It works like the sun sensor. This sensor is accurate (5×10^{-4} degrees), but is costly and heavy. There is a risk it can be saturated by the Sun, Earth, and interference from other bright sources.

6.2.1.4 Inertial sensors

These use gyroscopes for sensing satellite movement, or gyrometers to measure the angular velocity (Greensite, 1970; Schmidtbauer *et al.*, 1973).

6.2.1.5 Radiofrequency sensors

These sensors measure the characteristics of radio-frequency waves sent from beacons on the Earth to the satellite. Determination of the angle of the satellite antenna axis relative to the direction of radio-frequency waves can be achieved with high accuracy (about $0.05°$, depending on frequency and antenna diameter) while the rotation around the line of sight is difficult to evaluate. By measuring the rotation of a polarized signal coming from a single radio beacon, one can obtain a value of the attitude around the yaw axis. Unfortunately the orientation of the polarization is affected by Faraday rotation, and the precision is only about $0.5°$ (Tammes and Bleiweis, 1976).

6.2.1.6 Laser sensor

It has been suggested that a laser beam could be used to determine the orientation of the satellite. The anticipated precision is $0.006°$ around the roll axis and $0.6°$ around the yaw axis (Sepp, 1975; Aruga and Igarashi, 1977).

6.2.2 Attitude determination

This is achieved by defining the orientation of the satellite relative to the reference axes of Figure 6.3.

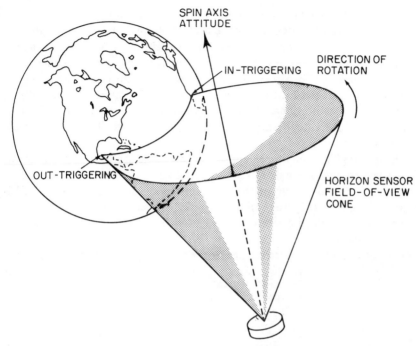

Figure 6.4 Earth   contour   detection. *Reproduced   by   permission   of D. Reidel Publishing Company from Wertz (1978).*

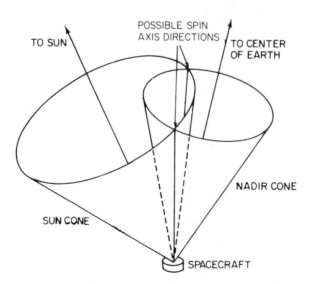

Figure 6.5   Determination of the two possible satellite spin axis directions. *Reproduced by permission of D. Reidel Publishing Company from Wertz (1978).*

For a spinning satellite (see Section 6.2.4.2), the aiming of a sensor at a specified radiating body defines a cone around the spin axis with a half angle at the apex equal to the angle between the boresight of the sensor and the spin axis of the satellite. The Earth and the Sun can be taken as examples of radiating bodies to be aimed at. As shown in Figure 6.4 sweeping the illuminated zone of the Earth produces a signal which allows the nadir angle to be known from the signal duration. The nadir angle is the angle between the spin axis of the satellite and the axis linking the satellite to the centre of the earth (nadir axis).

A sun sensor gives a measurement of the half angle at the apex of the second cone associated with the axis linking the satellite to the Sun. The spin axis of the satellite is found at one of the intersections of the two cones defined by the two aiming lines (Figure 6.5). The choice between one of the two intersections necessitates a third orientation measurement or a prior knowledge of the satellite orientation.

For a body stabilized satellite (see Section 6.2.4.4), which does not benefit

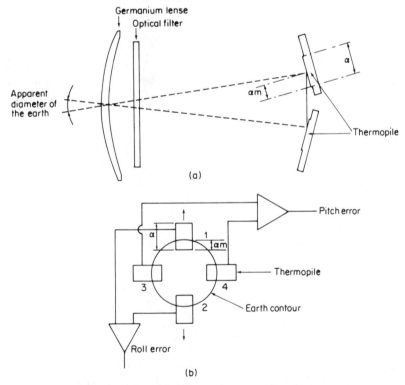

Figure 6.6   Determination of the Earth centre direction: (a) earth image generation; (b) wiring of thermopiles. *Reproduced by permission of Centre National d'Etudes Spatiales from Bourcier et al. (1980).*

from the satellite's rotation, one solution is to combine an aim at the Sun with an aim at the Earth, the latter being carried out either by a mechanically sweeping sensor, or by a horizon sensor using thermopiles (Figure 6.6) which delivers a signal related to the angle between the direction of the centre of the Earth and the direction of the sensor's optical axis (Bourcier *et al.*, 1980). New concepts are based on the use of gyroscopic devices associated with a sun sensor (Hsing *et al.*, 1978).

6.2.3 Actuators for attitude control

Once the attitude of the satellite is known, attitude corrections can be initiated. The required correcting torques are of the order of 10^{-4} to 10^{-1} N · m and are generated by controlling actuators.

The torque generating devices, or *actuators*, consist of:

6.2.3.1 Momentum devices

Momentum devices are *reaction wheels* and *control moment gyroscopes* (CMG).

Varying the speed of rotation ω of a wheel with inertia momentum I alters the kinetic momentum $H = I\omega$ of the wheel and induces a correcting torque T given by:

$$T = \frac{dH}{dt} = I\frac{d\omega}{dt}$$

The control moment gyroscopes (CMG) have a gimballed constant speed wheel, which allows the axis of the kinetic momentum of the wheel to be changed. As the kinetic momentum axis moves, a correcting torque T equal to the derivative with time of the kinetic momentum vector is generated. Due to low reliability, CMG are seldom used on geostationary satellites for active torque generation.

The above devices are well suited to controlling the attitude when the satellite undergoes alternate periodic (quasi-zero mean) disturbing torques (e.g. the effect of the oblateness of Earth).

Random disturbing torques with non-zero mean values (e.g. the effect of solar wind), or large amplitude disturbing torques can introduce a variation of the kinetic momentum which exceeds the limits fixed for the rotation speed of the wheel or the control moment gyroscope orientation. Saturation occurs and therefore an unloading torque is required using another actuator (thruster, for example).

6.2.3.2 Thrusters

Thrusters produce forces by expelling material through nozzles. If the lever arm is 1 m long, the thrust should be of about 10^{-4} to 10^{-1} N to achieve the required correcting torque value. The thrust F is related to the expelled material mass per time unit dm/dt, and depends on the specific impulse I_{sp} of the expelled material (see Section 6.4.1.2), as follows:

$$F = gI_{sp} \frac{dm}{dt}$$

where g is the gravitational acceleration at the Earth's surface. For reasons of weight and simplicity, these thrusters are usually part of the group of thrusters used for station keeping control (see Section 6.3).

6.2.3.3 Magnetic coils

Magnetic coils create a magnetic moment M when driven by an electrical current I. The interaction of the coil's magnetic momentum with the Earth's magnetic field B generates the desired torque, expressed as: $\mathbf{T} = \mathbf{M} \times \mathbf{B}$. These devices are electrical power consuming and only generate torques of low magnitude.

6.2.3.4 Solar sails

Solar sails use the effect of solar radiation pressure to create a torque (Section 5.2.1.5).

6.2.4 Attitude control techniques

6.2.4.1 Stabilization by gravitational gradient

This example illustrates passive attitude control largely used on low orbiting satellites when they do not require precise pointing. As discussed in Section 5.2.1.1., theory (Lazennec, 1966) shows that Earth's gravitational torques lead the satellite to align its axis of lowest inertia with the local vertical. But this technique is difficult to use on geostationary satellites for two reasons:

(1) The Earth's gravitational field varies as the square of the inverse of the distance to the Earth's centre. At an altitude of 36 000 km the gravitational torque is not strong enough unless the satellite is in two parts whose centres of mass, aligned on the local vertical, are far enough apart.
(2) Station keeping requires corrections which greatly disturb this type of stabilization.

Moreover as the movement around the position of equilibrium is oscillatory and undamped this technique by itself does not allow precise pointing of the antennae.

In order that the technique be efficient, large inertial momentums must be provided, using deployable arms (about 100 m long) equipped with masses at the ends. Further, oscillatory movement must be damped using energy dissipating devices, such as magnetic coils coupled with the Earth's magnetic field. This technique has been used for the Dodge experiment in the US, and led to attitude control accuracy of $\pm 3°$ in roll and pitch and $\pm 7°$ in yaw (Mobley, 1968). Another example is the Radio Astronomy Explorer 2 (RAE2) satellite.

6.2.4.2 Spin stabilization

Spin stabilization is a means of maintaining orientation of a space vehicle because a rotating body offers an inherent *gyroscopic stiffness* to torques tending to disturb the orientation of the rotational axis. Where there is no disturbing torque, the kinetic momentum maintains a fixed direction indefinitely, relative to an absolute reference axis system. For a geostationary satellite the axis to be maintained in a fixed direction is the axis parallel to the Earth's rotational axis, and therefore the satellite spin axis is North/South oriented.

Spin stabilization is achieved by rotation of the geostationary satellite body between 30 and 120 r.p.m., creating an inertial stiffness which maintains the satellite spin axis perpendicular to the equatorial plane. Hence this simple technique benefits from the properties of a gyroscope, but has the inconvenience of either accepting a toroidal radiation pattern antenna and therefore low gain, or imposing the use of a de-spun antenna or communication payload

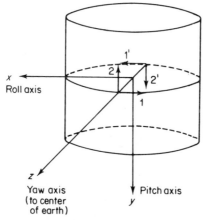

Figure 6.7 Actuators implementation on a spinning satellite: (1) spin speed control; (2) attitude control about the roll and yaw axes.

which requires a specific technology (see Section 6.9). In addition, this stabilization technique is most often used during the transfer orbital phase of the satellite (Chapter 7).

As the moments of inertia of the satellite body about the spin axis may not be very large relative to those about any orthogonal axis, oscillations of the spin axis are possible. These oscillations must be damped by internal dissipation of kinetic energy (nutation damper).

The disturbing torques discussed in Chapter 5 have two effects:

(1) They reduce the actual speed of rotation about the axis.
(2) They 'depoint' the spin axis from the North/South direction.

Therefore, the rotation speed must be maintained by periodic use of thrusters (No. 1 in Figure 6.7), and depointing must be corrected.

The interest of spin stabilization is better understood if one compares the action of a disturbing torque T_d on a satellite without kinetic momentum and on a spin-stabilized satellite. Figure 6.8 shows the action of the disturbing

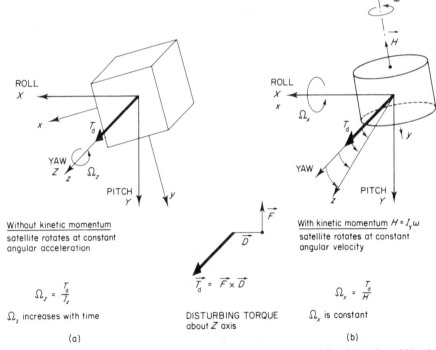

Figure 6.8   Action of a disturbing torque on a satellite (a) without, and (b) with on-board kinetic momentum.

torque T_d applied about the z-axis: the satellite without kinetic momentum rotates about the z-axis at constant angular acceleration $d\Omega/dt$, given by:

$$d\Omega/dt = T_d/I_z$$

Considering the same disturbing torque T_d, the spin-stabilized satellite with kinetic momentum H tends to rotate as a result of the gyroscopic effect about the x-axis at constant angular velocity Ω, given by:

$$\Omega = T_d/H$$

Correction is achieved by applying a torque which neutralizes the drift. This is generally applied periodically, when depointing reaches the upper bound using a thruster such as 2 in Figure 6.7. It works by impulses synchronized to the satellite's spin. As examples of spin stabilized satellites one may cite the INTELSAT I, II and III satellites (Figure 1.5).

6.2.4.3 Dual spin stabilization

To allow for spot beam coverage, large satellite antennae are required and spin stabilization is poorly adapted. It is preferable to mount the antenna on a part of the satellite body which spins at an angular velocity of 1 revolution per day ($0.25°$/min) about the pitch axis (the axis perpendicular to the orbital plane). This implies that the platform be despun relative to the main satellite body which rotates so as to create as above the gyroscopic stiffness. INTELSAT IV and VI (Figure 6.9) are examples of this technique.

6.2.4.4 Body-fixed satellite stabilization

This stabilization technique, commonly referred to as 3-axis, stabilization refers to a situation where the body of the satellite is maintained at a fixed attitude relative to the earth. It should be noted that as long as attitude control is performed all satellites are *three-axis stabilized*, although the misleading common usage restricts the use of this term to body-fixed satellites. With body-fixed satellites the implementation of large antennae and solar arrays is not so much of a problem since the body of the satellite does not rotate. However, the gyroscopic stiffness inherent to body-spun satellites, commonly called spinners, is lost. This gyroscopic stiffness can be restored using one or several *momentum wheels* on-board the satellite which bring a kinetic momentum about the pitch axis.

6.2.4.4.1 Body-fixed satellite with single momentum wheel

The satellite is equipped with a single momentum wheel whose axis at nominal attitude is the pitch-axis. The attitude control subsystem operates as follows:

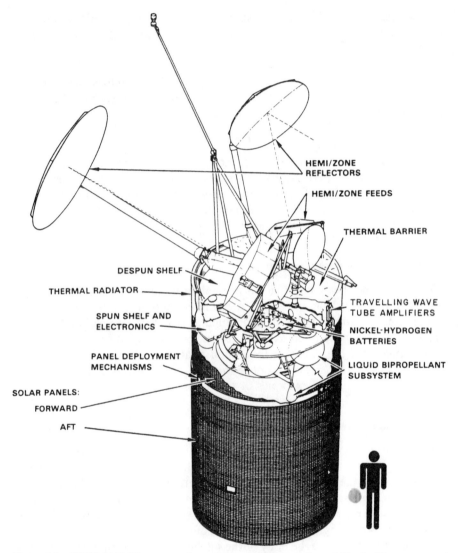

HEMI/ZONE
REFLECTORS

HEMI/ZONE FEEDS

THERMAL BARRIER

DESPUN SHELF

THERMAL RADIATOR

TRAVELLING WAVE
TUBE AMPLIFIERS

SPUN SHELF AND
ELECTRONICS

NICKEL-HYDROGEN
BATTERIES

PANEL DEPLOYMENT
MECHANISMS

LIQUID BIPROPELLANT
SUBSYSTEM

SOLAR PANELS:

FORWARD

AFT

Figure 6.9   INTELSAT VI spacecraft configuration (dual spin stabilized) (Thompson and Johnston, 1983).

Pitch and roll angles are detected by IR sensors aiming at the Earth. The gyroscopic stiffness of the momentum wheel limits the motions about the yaw and roll axes. During orbital motion, yaw and roll axes interchange every six hours, which allows control of the yaw from the measurement of the roll. During phases with large disturbing torques (e.g. during station-keeping manoeuvres) roll and yaw axes are separately controlled as variations about the yaw axis are measured by an integrating gyrometer.

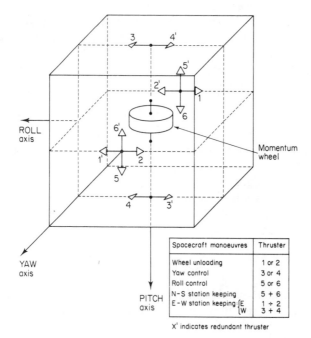

Figure 6.10   Attitude control of a bias momentum type body-fixed satellite.

Controlling torques about the pitch axis are generated by varying the wheel rotating speed. Controlling torques about the yaw and roll axes are generated by thrusters (Figure 6.10).

Pointing accuracy obtained with a stabilization system of this type is of the order of $0.05°$ about the roll axis, $0.02°$ about the pitch axis, and $0.3°$ about the yaw axis.

Attitude control about the pitch axis. Assume a disturbing torque T_d applied about the pitch axis. Let I_S and I_W be the inertial momentums of the satellite and the momentum wheel, respectively, ω the angular velocity of the wheel, and ϕ the satellite angle about the pitch axis. The kinetic moment theorem results in:

$$I_S \ddot{\phi} + I_W \dot{\omega} = T_d$$

As the satellite must revolve at a constant angular velocity of one revolution per day around the pitch axis, $\dot{\Phi} = 0.25°/\text{min}$ and $\ddot{\phi} = 0$.

$$\text{Hence: } \dot{\omega} = \frac{T_d}{I_W}$$

If T_d is constant, the momentum wheel must be accelerated by its motor. When maximum or minimum speed ω_M is reached, it must be off-loaded; the speed must be adjusted to the nominal speed by applying an opposite *unloading torque* T_u. This is done with two thrusters (one for each direction) oriented perpendicularly to the satellite axis (1 or 2 in Figure 6.10). If at the instant $t = 0$ the wheel rotates at average velocity ω_0, it reaches velocity ω_M, when constrained by torque T_d, at a time t_1, such that:

$$\frac{\omega_M - \omega_0}{t_1} = \frac{T_d}{I_W}, \quad t_1 = \frac{\Delta H}{T_d} \tag{1}$$

where ΔH is the variation of the kinetic momentum H of the wheel between its average velocity and its extreme velocity ω_M.

The unloading torque T_u should be equal to the torque generated by the variation in the rotation speed of the wheel during unloading, so as not to disturb the satellite attitude. When this condition has been met, the unloading time t_u is such that $T_u \cdot t_u = T_d \cdot t_1$. It follows, therefore:

$$T_u = t_1 \cdot \frac{T_d}{T_u} \tag{2}$$

Attitude control about the roll and yaw axes. Only one of these axes has to be considered here as the same considerations apply to the other. The gyroscopic stabilization about these axes is produced by the momentum wheel. As the satellite's kinetic momentum is negligible compared with that of the wheel, a disturbing torque T_d applied about the yaw axis generates a movement about the roll axis at constant angular velocity. If $H = I_W\omega$ is the wheel's kinetic momentum (ω being at its lowest value corresponding to the worst case) the drift is $\Omega = T_d/H$. For maximum depointing ε, the time between two corrections will be:

$$t_2 = \frac{\varepsilon}{\Omega} \tag{3}$$

The correcting torque opposing the disturbing one is applied by one of the two thrusters (one for each direction): 3 or 4 in Figure 6.10 for correction about the yaw axis, and 5 or 6 for correction about the roll axis. Each thruster acts for a duration t_c such that $T_t \times t_c = T_d \times t_2$, where T_t is the torque generated by the thruster.

$$t_c = t_2 \cdot \frac{T_d}{T_t} \tag{4}$$

Required propellant mass. Let F be the thrust, I_{sp} the specific impulse, and t the total operating time, the propellant mass is given by (see Section 6.4.1.2):

$$m = \frac{F \cdot t}{g \cdot I_{sp}}$$

Within a one-year time interval, the number of thruster operations is

$365 \times 24 \times 3600/t_1$ for unloading, $365 \times 24 \times 3600/t_2$ for corrections about yaw axis, and the same value for correcting about the roll axis.

The *total* operating time t is then:

$$t = 365 \times 24 \times 3600 \times \left(\frac{t_u}{t_1} + 2\,\frac{t_c}{t_2}\right)$$

The *annual* propellent mass is given by:

$$m = \frac{F}{g \cdot I_{sp}} \left(\frac{t_u}{t_1} + 2\,\frac{t_c}{t_2}\right) \times 365 \times 24 \times 3600 \tag{5}$$

Example. Given a satellite configured as in Figure 6.10 (i.e. with peripheral thrusters). Its characteristics are as follows:

Thrusters:

Thrust: $F = 1\text{N}$

Specific impulse of propellent (nitrogen): $I_{sp} = 70$ s

Thruster lever arm: $D = 0.75$ m

Inertia wheel:

Nominal speed: $\dfrac{\omega}{2\pi} = 125$ turns per second (7500 r.p.m.)

Nominal kinetic momentum: $H = 50\ \text{N} \cdot \text{m} \cdot \text{s}$

Kinetic momentum variations: $\Delta H \pm 5\ \text{N} \cdot \text{m} \cdot \text{s}$

Disturbing torques (about each axis): $T_d = 5 \times 10^{-6}\ \text{N} \cdot \text{m}$.
Stabilization must be ensured within $0.1°$ about any axis.
Pitch control. The time t_1 between two unloadings is:

$$t_1 = \Delta H/T_d = 1 \times 10^6\,\text{s} \ (11.6\text{ days}).$$

The unloading torque generated by a thruster is $T_u = F \cdot D = 0.75\ \text{N} \cdot \text{m}$. The unloading duration t_u is:

$$t_u = t_1\,\frac{T_d}{T_u} = 6.7\text{ s}$$

Yaw (or roll) control. The drift about the yaw axis $\Omega = T_d/H = 1 \times 10^{-7}$ rad/s. The time t_2 between two corrections is $t_2 = \varepsilon/\Omega = 1.7 \times 10^4$ s (4.7 h).
The correcting torque generated by a thruster is $T_t = F \cdot D = 0.75\ \text{N} \cdot \text{m}$.
The operating time t_c of a thruster is $t_c = t_2 \cdot T_d/T_t = 0.12$ s.
Propellent mass (7 years). For seven years' use, the propellent mass is:

$$m = \frac{7F}{gT_{sp}} \left(\frac{t_u}{t_1} + 2\,\frac{t_c}{t_2}\right) \times 365 \times 24 \times 3600 = 6.7\text{ kg}$$

One should add to this value a margin to take into account the thruster dispersal and the undrawn propellant mass from tanks and pipes at depletion. If, instead of nitrogen, hydrazine is used, which has a specific impulse four times greater, the mass would only be 1.7 kg.

6.2.4.4.2   Body-fixed satellite with several momentum wheels

Using a single momentum wheel the direction of the kinetic momentum wheel is constrained to be that of the wheel axis. Should there be a misalignment between the wheel axis and the pitch axis then one must compensate the misalignment using thrusters. Propellent consumption is also required if one wishes to offset the satellite body axis relative to the pitch axis so as to adjust the antenna pointing if necessary. To avoid propellent consumption it is possible to generate a variable direction kinetic momentum from the combined action of two or three momentum wheels. These wheels have their axes inclined relative to the pitch axis, and the direction of the resulting kinetic momentum is determined by adjusting the rotation speed of each wheel. Moreover, this technique offers a possible redundancy by implementing a spare wheel which can replace any wheel in case of failure.

6.2.4.4.3   Body-fixed satellite without on-board kinetic momentum

The previous techniques do not differ fundamentally from a spin stabilized satellite, in that the rotating sections (momentum wheels, or satellite body) have a moment of inertia high enough for the gyroscopic effect to limit the number of corrections necessitated by disturbing torques acting on the yaw and roll axes. With a body-fixed satellite, instead of using an on-board kinetic momentum (bias momentum type), satellite stabilization can be achieved using three reaction wheels along the three principal axes which cancel the effects of these torques, by exchange of kinetic momentum between the vehicle body and each of the wheels. As most of the disturbing torques are zero mean value, the rotation speed of the wheels oscillates about a zero value. The gyroscopic stiffness created by the wheels is low and because of this, the satellite must be permanently attitude controlled (zero momentum type).

The discussion concerning the satellite attitude control about the pitch axis can be applied to all three axes, but the complete equation for the system is much more complex because of the terms of gyroscopic coupling due to the interaction of movement between the satellite and the wheels. An on-board computer controls the rotation speed of the wheels from information on the satellite attitude. One example of implementation of this technique is the Japanese BSE satellite (broadcasting satellite for experimental purpose) (Figure 6.11). Figure 6.12 shows the variations of kinetic momentum of the three wheels (Shimoseko and Matsumoto, 1980). It can be observed that the wheel momentum, i.e. the wheel rotation speed, is oscillatory about zero

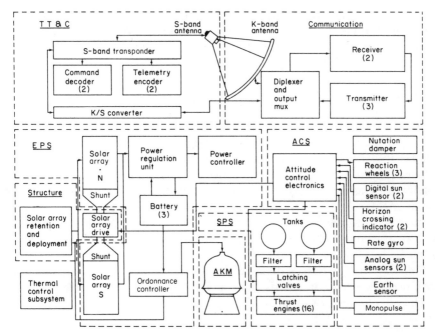

Figure 6.11   Organization of a zero momentum type body-fixed satellite (BSE satellite).
Reproduced with permission from Ichikawa (1978). Copyright © 1978 IEEE.

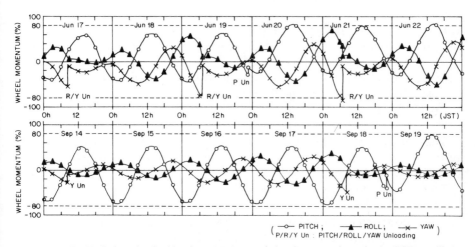

Figure 6.12   Variations in the kinetic momentum of the three reaction wheels (BSE satellite).
From Shimoseko and Matsumoto (1980). *Copyright American Institute of Aeronautics and Astronautics and reprinted by permission of the AIAA.*

value. As not all the disturbing torques are zero mean value, the wheel momentum must increase to compensate for such torques and the rotation speed of the wheel reaches occasionally its maximum permissible value. Unloading must then take place.

6.3   STATION KEEPING

A geostationary satellite should remain stationary relative to the Earth, and occupy a specific position on the geostationary satellite orbit. It is, however, subject to drifts in longitude and latitude owing to various disturbances (see Chapter 5). As it is impossible in practice to maintain perfect stationarity, the satellite is constrained to remain within a *'window'* whose limits are defined by an angular shift as seen from the centre of the Earth around the required nominal position. For example, a window with sides of $0.1°$, corresponds to a drift of about 75 km on the sphere containing the geostationary satellite orbit.

Station-keeping consists of maintaining the satellite within this 'window' with periodic orbit corrections using as less propellent as possible.

The dimensions of this window are specified according to the mission and are subject to the following:

(1) The smaller these dimensions are, the simpler the tracking required by the earth station antennae.
(2) When earth station antennae have large beamwidth, or when the station is on-board a vehicle (aeroplane, boat, truck) which, as it moves, necessitates a tracking system, 'windows' of larger dimensions can be accepted.
(3) Geostationary satellites with multiple spot beams antennae pointed at precise points on the Earth, require precise positioning in order to avoid a displacement of the coverage area.
(4) Adopting a strict tolerance on satellite position maintenance allows better use of the geostationary satellite orbit and of the radiofrequency spectrum. (CCIR, Rep. 453–3, 1982).

The revised WARC 79 Radio Regulations require a station keeping accuracy of $\pm 0.1°$ in longitude for the fixed-satellite and broadcasting-satellite services. (CCIR, Rec. 484-2, 1982)

Factors influencing orbital parameters are (see Chapter 5):

(1) The attraction of the Moon and the Sun. This results in a variation of the inclination of the orbital plane in the order of $1°/year$;
(2) Asphericity of the Earth's gravitation field. This results mainly in a longitudinal drift depending on the satellite station.

(3) The pressure of solar radiation, which increases the eccentricity of the orbit, and causes an apparent oscillatory longitudinal drift about the nominal satellite station.

Station-keeping is based on a strategy, which takes into account constraints such as propellent consumption (Legendre, 1980). The strategy can be summarized as follows:

(1) Determine the direction and speed of drift;
(2) Predict by extrapolation which day the satellite will escape the window;
(3) Determine precisely the true orbit by a new series of measurements, a few days before that date;
(4) Calculate the date, the amplitude and duration of the velocity increments required to modify the orbital parameters (optimized for the chosen strategy and according to the principal disturbing effects);
(5) Monitor the effects of the correction.

Generally, corrections are carried out by *velocity increments* applied perpendicular to the orbital plane at orbital nodes using thrusters 5 and 6 of Figure 6.10 simultaneously (*north–south station keeping*) and tangential to the orbit (*east–west station keeping*) using either the pair of thrusters 1 and 2 (correction acting eastward on the satellite) or 3 and 4 (correction acting westward on the satellite).

6.3.1 Determination of position (Saint-Etienne, 1973; Cariou, 1980)

For a geostationary satellite, position determination depends on the measurement of angle and range.

6.3.1.1 Angle measurement

Angle measurement can be achieved by varying the ground antenna pointing and searching for a maximum reception gain or using monopulse techniques (see Chapter 8). Depending on the characteristics of the antenna, the precision of this measurement is between $0.05°$ and $1°$.

Interferometry can also be used. Here, two stations A and B separated by a distance L, receive a signal from the satellite, say the telemetry carrier at frequency f (Figure 6.13).

The difference in distance ΔR between the satellite and each station is proportional to the difference in propagation time $\Delta t = \Delta R / c$ and is measured by a phase shift $\Delta \phi = 2\pi f \Delta t$ between the received signals

$$\Delta\phi = \frac{2\pi L \cos E}{\lambda} \tag{6}$$

where E is the antennae elevation angle.

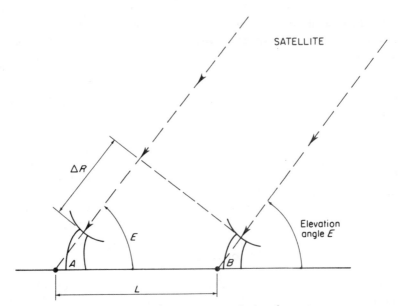

Figure 6.13　Angle measurement by interferometry.

From this formula, the value of the elevation angle is obtained. The satellite is located on a cone of axis AB with a half-angle at the apex E. The combination of two pairs of stations allows localization of the satellite on the intersection of the two related cones. The measurement precision is around $0.01°$. Although it is useful during launch and orbital station acquisition, it is not precise enough for determination of the final orbit, which is performed from range measurements.

6.3.1.2　Range measurement

The distance R between a transmitter and a receiver is obtained by a measurement of the phase shift $\Delta\phi$ between the transmitted and received signals:

$$\Delta\phi = 2\pi f \, \frac{D}{c}$$

In practice, the phase shift is measured between a sinusoidal signal modulating the telecommand carrier, and this same signal after retransmission by the satellite in the form of a modulation of the telemetry carrier (see Section 6.5).

6.3.2　North–south station keeping

North–south station keeping consists of changing the orbital plane inclination. An increment of velocity ΔV must be applied perpendicular to the orbital

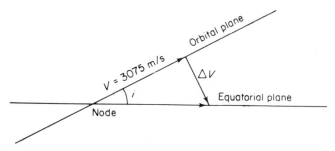

Figure 6.14   Inclination correction.

plane at the nodes (Figure 6.14). Each correction has an amplitude of $\Delta V = iV$ where V is the velocity (3074 m/s) of the geostationary satellite, and i the inclination to be corrected.

As the luni–solar attraction reflects in a change of inclination, an annual velocity increment must be applied equal to

$$\Delta V = 3075 \frac{\pi}{180} \left(\frac{di}{dt}\right) m/s$$

where di/dt is the mean annual inclination variation (in degrees). Table 6.2 gives precise values of the required velocity increment considering exact values of orbit plane inclination variation depending on the year, as it relates to the 17 years periodicity of the moon orbit characteristics. 50 m/s is an order of magnitude to be kept in mind.

Table 6.2   REQUIRED VELOCITY INCREMENT
TO COMPENSATE FOR ORBIT INCLINATION.

Year	Velocity increment (m/s)
1980	41.24
1981	42.58
1982	44.19
1983	45.99
1984	47.70
1985	49.16
1986	50.20
1987	50.72
1988	50.62
1989	49.96
1990	48.77
1991	47.21
1992	45.48
1993	43.71
1994	42.15

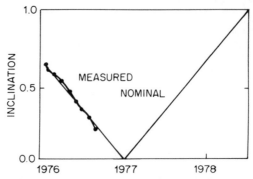

Figure 6.15   Variation of the orbital inclination of the
CTS satellite with time.

If the 'window' is large enough, corrections during the satellite's life may not be necessary. Indeed, where the maximum tolerable inclination is i degrees during the satellite's life, it is acceptable to position the satellite at the start of its mission on a slightly inclined (e.g. by i degrees) geosynchronous orbit, with a correctly placed ascending node. The inclination therefore decreases, then becomes zero, and then increases to reach i degrees after $2i/(di/dt)$ years.

Figure 6.15 shows the inclination of the Canadian CTS satellite launched in 1976. It was placed in orbit with an inclination of $i_0 = 0.7°$. So at the end of its expected life of two years, the inclination becomes $0.8°$.

6.3.3   East–west station keeping

Movement in the equatorial plane due to the asphericity of the Earth's gravity obeys equation (18) of Chapter 5, which follows:

$$\left(\frac{d\Lambda}{dt}\right) - k^2 \cos 2\Lambda = \text{constant} \tag{7}$$

where Λ is the satellite longitude measured relative to the closest stable point of equilibrium.

The satellite must be kept at the nominal longitude Λ_N, measured relative to the closest stable position of equilibrium, allowing a maximum deviation $\varepsilon/2$ on both sides of Λ_N.

If Δ is the longitude measured from $\Lambda_N (\Delta = \Lambda - \Lambda_N)$, it follows:

$$\cos 2\Lambda = \cos 2(\Lambda_N + \Delta) = \cos 2\Lambda_N - 2 \Delta \sin 2\Lambda_N$$

and Equation (7) above is written:

$$\left(\frac{d\Delta}{dt}\right)^2 + 2k^2 \Delta \sin 2\Lambda_N = \text{constant} \tag{8}$$

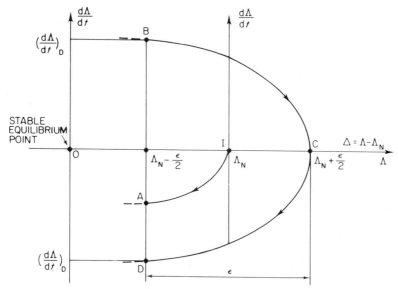

Figure 6.16   Longitude station keeping.

The curve representing the drift $d\Delta/dt$ in terms of Δ is a parabola defined by the nominal longitude Λ_N and by the initial conditions. If the satellite starts at point I (nominal position) with zero drift (Figure 6.16), it will drift towards the stable point of equilibrium according to the parabola determined by equation (8) which has its apex at I.

6.3.3.1   Strategy 'of the inclined plane'

When the longitude reaches the value $\Lambda_N - \varepsilon/2$ (point A) a velocity increment is applied to the satellite, so that the parabola which has its apex at C $(\Lambda = \Lambda_N + \varepsilon/2)$, the second limit of the 'window', is described. From C, the satellite moves back towards the stable point of equilibrium. At point D $(\Lambda = \Lambda_N - \varepsilon/2$, a suitable velocity increment δv_L is applied and the representative point in Figure 6.16 passes from D to B. Then, the cycle begins again.

East–west station keeping, therefore, imposes that the satellite representative point follows curve BCD of Figure 6.16. This is known as the '*inclined plane strategy*'.

6.3.3.1.1   Required velocity increment

In equation (8) the constant is determined considering point C $(\Delta = \varepsilon/2, d\Delta/dt = 0)$. It follows therefore:

$$\text{Constant} = 2k^2 \frac{\varepsilon}{2} \sin 2\Lambda_N$$

Equation (8) is written:

$$\left(\frac{d\Delta}{dt}\right)^2 = -2k^2\left(\Delta - \frac{2}{2}\right)\sin 2\Lambda_N \tag{9}$$

At point $D(\Delta = -\varepsilon/2)$

$$\left(\frac{d\Delta}{dt}\right)^2_D = 2k^2\varepsilon \sin 2\Lambda_N \tag{10}$$

Whence

$$\left(\frac{d\Delta}{dt}\right)_D = -k\sqrt{(2\varepsilon \sin 2\Lambda_N)} \tag{11}$$

To pass from point D to point B, a variation of the drift $d\Delta/dt$ equal to $-2(d\Delta/dt)_D$ must be taken into consideration. A variation of drift $d\Delta/dt$ is equivalent to a velocity increment $\delta v_L = \Gamma_T \, dt$, where Γ_T is given by equation (15) in Chapter 5.

$$\delta v_L = \Gamma_T \, dt = \frac{a}{3}\left(\frac{d^2L}{dt} \, dt\right) = \frac{a}{3}\left(\frac{dL}{dt}\right) = \frac{a}{3}\left(\frac{d\Delta}{dt}\right)$$

that is, for $d\Delta/dt = -2(d\Delta/dt)_D$

$$\delta v_L = \frac{2a}{3} k\sqrt{(2\varepsilon \sin 2\Lambda_N)} \tag{12}$$

$$\delta v_L = 2.5\sqrt{(\varepsilon \sin 2\Lambda_N)} \text{ m/s} \ (\varepsilon \text{ in radians}) \tag{13}$$

This velocity increment is always applied in the same direction, as the satellite always drifts in the same direction.

6.3.3.1.2 Correction periodicity

The periodicity at which velocity increments must be applied is equal to the time t required to run through the parabola curve. This time is calculated by integrating formula (9) from $\varepsilon/2$ to $-\varepsilon/2$:

$$t = \frac{2\sqrt{2}}{k}\sqrt{\frac{\varepsilon}{\sin 2\Lambda_N}} = 516\sqrt{\frac{\varepsilon}{\sin 2\Lambda_N}} \text{ days}$$

where ε is in radians.

The velocity increment and duration of the cycle depend on the satellite position relative to the stable equilibrium point and the longitudinal dimension of the 'window'.

6.3.3.1.3 Annual velocity increment

The annual velocity increment to apply is:

$$\delta V_L = \delta v_L \cdot \frac{365}{T} \qquad \delta V_L = 1.77 \sin 2\Lambda_N \text{ m/s} \tag{15}$$

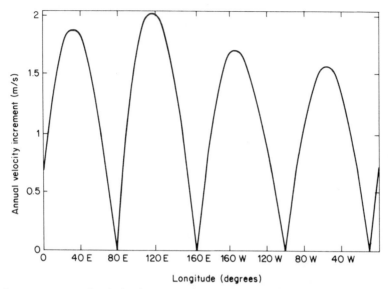

Figure 6.17   Annual velocity increment for east–west station keeping versus station longitude.

It depends on the longitude of the point at which the satellite is maintained, and not on the dimension of the window. At most it is 1.8 m/s, (or say 2 m/s).

The above calculations are based on the formulas given in Chapter 5.2.1.2 which were derived using simplifying assumptions. When considering the actual acceleration undergone by the satellite, as given in Figure 5.3, the actual required annual velocity increment is that displayed in Figure 6.17.

6.3.3.2   'Centre tracking' strategy

Formula (15) and Figure 6.17 show that the annual velocity increment depends on the longitude of the satellite's nominal position in orbit. Around a point of equilibrium, this is very weak. In practice, it is not the drifts due to asphericity of the Earth's potential which must be compensated, but the east–west component of velocity increments induced by corrections perpendicular to the equatorial plane ('north–south station keeping').

The strategy here is to place the satellite at the *centre of the window* before carrying out the north–south correction, so as to give the maximum margin over longitude. Then a longitudinal correction is applied before the satellite reaches the limit of the window that nudges it back to the centre of the window. A second such correction then neutralizes the drift caused by the first.

6.3.3.3   Examples of the above strategies (Figure 6.18)

Figure 6.18(a) shows variations of the satellite longitude according to the

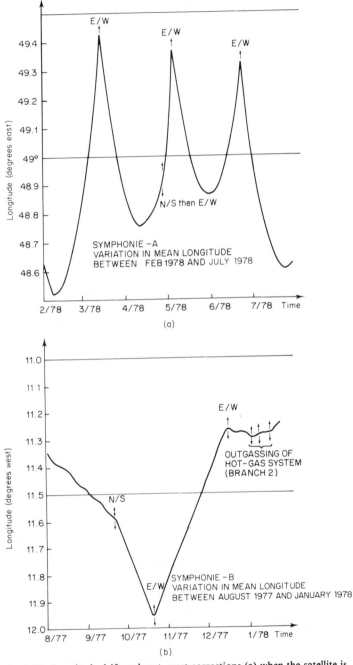

Figure 6.18 Longitude drift and east–west corrections (a) when the satellite is located far away from the equilibrium point, and (b) when the satellite is located near the equilibrium point. *Reproduced by permission of Centre National d'Etudes Spatiales from Legendre and Deixelberger (1980).*

inclined plane strategy. The satellite's nominal station is at a longitude of 49°E. At this longitude the natural drift of the satellite is eastward and westward corrections must be applied to the satellite as it is about to reach the eastern boundary of the window.

Figure 6.18(b) illustrates the centre tracking strategy. Now the satellite's nominal station is at a longitude of 11.5°W, that is very near to an equilibrium point on the geostationary orbit. The effect of a north–south correction on the longitudinal drift can be observed, and a subsequent east–west correction is applied as the satellite reaches the window limit. The satellite then moves back towards the centre of the window and an additional east–west correction is performed to compensate for the drift caused by the previous correction.

6.3.3.4 Correction of eccentricity

The previous strategies are used when eccentricity variations in the orbit are small enough not to require specific corrections. However, for large satellites, sensitive to solar radiation pressure (e.g. direct broadcasting satellites), the eccentricity variations can become excessive. Then *specific corrections* are required to maintain the *orbit eccentricity* within specified limits. The applied corrections may serve both to correct eccentricity and to compensate for longitudinal drift.

6.3.4 Station keeping velocity increment budget

This budget is established considering:

(1) *Inclination corrections* (about 50 m/s annually).
(2) *Drift corrections* in longitude (a few m/s annually).

The actual budget depends on four factors:

(1) The date when station keeping is initiated
(2) The longitude of orbital positioning;
(3) The width of the window;
(4) The satellite's S/m ratio.

The budget is in the order of 50 to 60 m/s for one year. The greater part is due to inclination corrections.

6.4 PROPULSION SUB-SYSTEM

The propulsion sub-system aims at generating forces acting on the body of the satellite. It comprises:

(1) Low thrust (10^{-3} N to 20 N) actuators (Reaction Control System: RCS) devoted to attitude and orbit corrections which provide an annual velocity increment of the order of 50 m/s.
(2) A high thrust (400 N to 50 000 N) motor (Apogee Kick Motor: AKM or Apogee Boost Motor: ABM) which provides the velocity increment required for the geostationary satellite orbit injection at the transfer orbit apogee (about 1500 m/s, see Section 7.1.3.)
(3) With Space Shuttle launched satellites, eventually a Perigee Kick Motor (PKM) which provides the velocity increment required to inject the satellite into the transfer orbit (about 2430 m/s, see Section 7.1.6).

6.4.1   Characteristics of thrusters

There are two families of thrusters:

(1) Chemical, which have a thrust level of between 0.5 N and a few 10 000 N;
(2) Electrical, which with current technology can produce between 2 and 10 mN.

Specific features of RCS thrusters are:

(1) Low thrust levels (a few mN to a few N);
(2) Cumulative operating times from several hundreds to several thousands of hours made up of many short operating cycles (a few milliseconds to a few hours);
(3) A long life, about seven to ten years for telecommunication satellites.

The thrusters are characterized by their *thrust* and the *specific impulse* of the propellent they use. The mass of propellent necessary to achieve a given velocity increment is dependent on the specific impulse of the propellent: the higher the specific impulse, the smaller the mass. High values of specific impulses are therefore of interest as weight on-board satellites is limited.

6.4.1.1.   Expression of the velocity increment

The required mass of propellent is calculated by the fundamental equation:

$$M \, dV = w \, dM \qquad (16)$$

this states that between the time t and the $t + dt$, the satellite with the initial mass M and velocity V has lost a mass $- dM$ and has increased its velocity by dV. The velocity of the expelled mass dM with respect to the satellite is w. On integrating between the instant t_0 (mass of satellite $M + m$) and the instant t_1

(mass of satellite M), it follows that:

$$\Delta V = w \, \mathrm{Log}_e \frac{(M+m)}{M} \qquad (17)$$

where m is the mass of the propellent consumed and M is the mass of the satellite after burn-out of the propellent.

6.4.1.2 Specific impulse

The specific impulse is the impulse which is communicated to the satellite during a time dt (for instance, $F \, dt$ if F is the thrust) by unit weight $g \cdot dm$ of propellent during this time:

$$I_{sp} = \frac{F \, dt}{g \cdot dm} = \frac{F}{g \cdot \frac{dm}{dt}} \qquad (18)$$

The specific impulse is therefore also the thrust by weight unit of propellent consumed per second. As dm/dt is the rate ρ at which the propellent is consumed

$$I_{sp} = \frac{F}{\rho g} \qquad (19)$$

The equation (16) can also be written $M \, dV/dt = w(dM/dt)$, i.e. $F = w\rho$ which is

$$I_{sp} = w/g \qquad (20)$$

So the specific impulse has the dimensions of time and is expressed in seconds. The magnitude of the specific impulses of various propellents are given in Table 6.3

Table 6.3 SPECIFIC IMPULSES.

Type of Propellent	$I_{sp}(s)$
Cold gas (nitrogen)	70
Hydrazine	220
Heated hydrazine	300
Bipropellent	290[a]
	310[b]
Electric ions	1000 to 10 000
Solid	290

[a] Blowdown operation mode (i.e. pressurant gas is stored in the same tank as propellent and pressure decreases as propellent is consumed).
[b] Regulated pressure operation mode (i.e. a regulator maintains a constant gas pressure).

6.4.1.3 Mass of propellent

Equation (17) can be written as:

$$m = M\left(\exp\frac{\Delta V}{w} - 1\right) \tag{21}$$

Using (20):

$$m = M\left(\exp\frac{\Delta V}{gI_{sp}} - 1\right) \tag{22}$$

It may be of special interest to relate the required mass of propellent to the total initial mass $M_s = M + m$ of the satellite (this concerns orbit injection):

$$m = M_s\left[1 - \exp\left(\frac{-\Delta V}{gI_{sp}}\right)\right] \tag{23}$$

6.4.2 Chemical thrusters

The principle of chemical thrusters is to generate gases at high temperature by chemical reaction of propellents. The gases are then accelerated by a nozzle. Propellents can be either *solid* or *liquid*.

Solid propellent motors are used to provide the required velocity increment for orbital injection. A solid mixture of oxidant and fuel is contained within a case ended by a nozzle. Firing of the propellent grain is performed using an igniter. Specific impulse values are in the order of 290 s. (see Chapter 7.1.4)

The most popular *liquid propellent* is hydrazine. Hydrazine vaporizes and decomposes when in contact with a catalyst, producing a hot gas which is a mixture of nitrogen, hydrogen and ammonia at temperatures of 900 °C. Figure 6.19 displays the cross-section of a catalytic thruster. Electro-mechanical valves control the propellent flow in the decomposition chamber. Heaters prevent hydrazine from freezing in the valve and elevate the catalyst bed temperature to minimize thruster degradation.

To reduce the part of the satellite's mass taken up by the propellent there is a tendency to increase the specific impulse using electrothermal hydrazine thrusters or bipropellent systems (Owens, 1976; Sackeim *et al.*, 1980; Fritz *et al.*, 1983).

Electrothermal hydrazine thrusters (EHT) (Free *et al.*, 1978; Valentini *et al.*, 1979) use electrical energy to increase the temperature of the gas after catalystic decomposition to about 1900 °C before exiting through the nozzle. This allows specific impulse values to reach around 300 s.

Bipropellent systems are based on spontaneous combustion resulting from

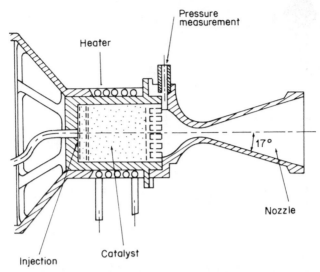

Figure 6.19 Configuration of a hydrazine thruster.

the contact of two propellents, for instance monomethyl hydrazine (combust-
ible) and nitrogen tetroxide (oxidant) (Pfeffer and Viellard, 1970).

There is a gradual movement towards the use of *bipropellent spacecraft propulsion systems* for both geostationary satellite orbit injection and on-orbit
control as an *unified propulsion system* offers significant mass savings over the
conventional combination of a solid apogee kick motor and a hydrazine
reaction control system. Indeed, storable bipropellent spacecraft unified
propulusion systems take advantage of the inherently higher specific impulse
offered by bipropellents and of the facility of storing in common tanks
propellent required for both orbit injection and on-orbit control. Moreover,
any propellent mass unused for orbit injection is still available for on-orbit
control, and results in an increased satellite lifetime.

Figure 6.20 shows an example of a storable bipropellent spacecraft unified
propulsion system (Moseley, 1984). This integrated bipropellent system has
twelve 22-Newton bipropellent thrusters and one 490 N apogee thruster. A
separate helium pressurant sphere (isolated from the propellent system by a
pyrotechnic valve during launch) provides helium through a single pressure
regulator during apogee manoeuvres. The pressurization sub-system is then
isolated for the remainder of the mission, resulting in a blowdown mode of
operation of the propulsion system.

Latching isolation valves in the propellent lines can isolate either redundant
set of attitude and orbit control thrusters. Positive, redundant propellent isolation is provided by pyrotechnic valves for ground handling and launch safety.
Filters and pressure transducers are included to ensure propellent cleanliness
and to monitor sub-system pressures.

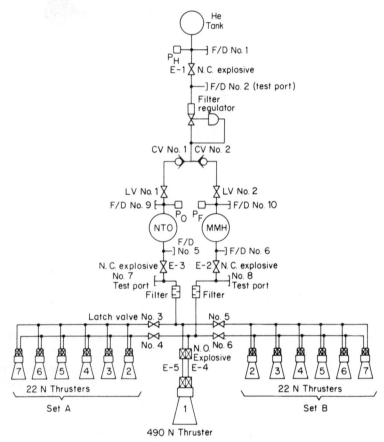

Figure 6.20   INSAT-1 bipropellent unified propulsion system (Moseley, 1984).

6.4.3.   Electric thrusters

In an electric thruster, ionised mass is accelerated in an electromagnetic or electrostatic field and serves to generate the thrust.

Electric propulsion is an advanced technology (Loeb *et al.*, 1979; Wayne and Finke, 1980; CCIR, Rep 843, 1982), compared to chemical propulsion. It gives *low thrust* (less than 0.1 N) with a *high specific impulse* (1000 to 10 000 s). As electric propulsion needs a high voltage power supply, not only should the specific impulse of the propellent be considered, but also the specific power P_s, equal to the ratio of electric power to thrust. This is in the order of 25 to 50 W/mN, depending on the type of thruster.

Electric thrusters are mainly: (Pfeffer *et al.*, 1976; Fearn, 1977; CCIR, Rep. 843, 1982):

Table for figure 6.20.

Item	Qty per S/S	Nominal mass (kg) INSAT-1	Supplier
1. Propellant tanks	–	30.672	Lockheed
A. Propellant (MMH & NTO) tank	2	(29.443)	
B. Auxiliary fuel tank	–	(1.229)	
2. Pressurant (helium) tank	1	11.620	Brunswick
3. Apogee (490N) thruster	1	3.647	Marquardt
4. AOCS (22N) thruster	12	7.838	Marquardt
5. Pressurant (helium) module	1	2.840	Marquardt
A. Pyro valve (NC: normally closed)	1	(0.132)	Pyronetics
B. Regulator	1	(0.762)	Consolidated
C. Check valve (CV)	2	(0.308)	HTL
D. Latch valve (LV)	2	(0.572)	Marquardt
E. Pressure transducer LP	2	(0.379)	Paine
F. Mounting brackets/manifolds	A/R	(0.460)	FACC
G. Pressure transducer HP	–	(0.227)	Teledyne
6. Propellant (MMH & NTO) modules	2	2.832	Marquardt
A. Pyro valve (NC: normally closed)	2	(0.263)	Pyronetics
B. Filter	2	(0.282)	Wintec
C. Latch valve (LV)	4	(1.143)	Marquardt
D. Pyro valve (NO: normally open)	2	(0.254)	Pyronetics
E. Mounting brackets/manifolds	A/R	(0.890)	FACC
7. Fill/drain valves (F/D)	8	0.768	Pyronetics
Total subsystem mass		60.217	

(1) *Plasma thrusters* (I_{sp} = 1000 to 5000 s) which employ a solid fuel rod (Teflon) that is spring fed to the thruster. A capacitor periodically releases its stored energy across the face of the rod. A few surface layers of the rod become ionized and accelerated by the contained self-generated field produced by the electric current (Free *et al.*, 1978). This gives the advantage of mechanical simplicity and avoids the need for charge neutralization as the plasma created is electrically neutral. The drawback is the potential RF interference problem.

(2) *Ion thrusters* (I_{sp} = 2 000 to 10 000 s) where charged particles (mercury or cesium ions) are accelerated to high velocities by means of a high voltage grid system. Downstream of the accelerator grid, a neutralizer provides a source of electrons for *ion beam neutralization*. This function is necessary

to maintain the space vehicle at a constant potential relative to the local space plasma. Two techniques have been developed:

(a) *mercury ionization* by electronic bombardment (Fearn, 1977; 1978) or by radio-frequency excitation (Bassner, 1979) prior to acceleration of mercury ions by the electric field. Specific impulse values are in the range 2 000 to 3 000 s.

(b) *field emission* (Bugeat and Valentian, 1982) where the atoms of a liquid metal (cesium) are ionized and then accelerated under the action of a high strength electric field achieved by peak effect at the extremity of a sharp electrode. This process does not require an external power source for ionization. Such thrusters have been developed by the SEP under contract of the ESA and display values of specific impulses as high as 8000 to 10 000 s.

Taking into account the high potential specific impulse, the on-board propellent mass of electric thrusters should be low compared with chemical thrusters. Nevertheless, electric thrusters demand additional electrical energy, obtained by an increase in the size of the solar arrays. They are only really useful for *extended satellite lifetime* (more than seven years). Finally, the low levels of thrust of electric thrusters are of some concern for the correction strategies (Free, 1972).

6.4.4   Example of calculation of the mass of the propulsion subsystem

The mass m of propellent (specific impulse I_{sp}) required to provide a velocity increment ΔV to a satellite of mass M is given by formula (22). For a 500 kg (dry mass) seven-year lifetime satellite equipped with a hydrazine propulsion sub-system ($I_{sp} = 220$ s) the north–south correction requires $\Delta V_1 = 50$ m/s $\times$ 7 years $= 350$ m/s and therefore a propellent mass m_1 of 88 kg. The longitudinal correction, for which $\Delta V_2 = 1$ m/s $\times$ 7 years $= 7$ m/s gives a mass $m_2 = 1.6$ kg. The station acquisition and attitude control would require about 10 kg more. In addition to the total mass of hydrazine, there are tanks, pipes, filters and thrusters, giving an additional mass of about 30 kg. The total mass is then 130 kg.

6.5   TELEMETRY, TRACKING AND COMMAND (TTC)

This sub-system is used to:

(1) *Transmit house-keeping information* and status of the satellite to the ground control station.
(2) *Provide angular* and *range measurements* to permit *localization* of the satellite.

Table 6.4 TELECOMMAND LINK CHARACTERISTICS. *From ESA (1980), reproduced by permission of the European Space Agency.*

VHF System	
Frequency band: Up-link	148 to 149.9 MHz
Down-link	136 to 138 MHz
Number of sinusoidal tones	6
Range instrumental accuracy (includes ground and on-board hardware,	At threshold S/N: 141 m
excludes ionospheric propagation error)	At infinite S/N: 21 m
Ionospheric propagation error	Varies from 0.1 to more than 2 km
Maximum range ambiguity resolution	1.5×10^4 km
S band system	
Frequency band: Up-link	2025 to 2120 MHz
Down-link	2200 to 2300 MHz
Number of sinusoidal tones	7
Range instrumental accuracy (includes ground and on-board hardware,	At threshold S/N: 174 m
excludes tropospheric propagation error)	At infinite S/N: 3.75 m
Tropospheric propagation error	Varies from 0 to 300 m
Maximum range ambiguity resolution	1.8×10^4 km

(3) *Receive command signals* from the ground control station to initiate attitude and station keeping manoeuvres and operations of the on-board equipments.

The frequencies generally used are either at VHF or in S band (see Table 6.4), or in the band allocated for the main satellite telecommunications mission (e.g. 6/4 GHz) depending on whether the satellite is in the launch phase or on station.

6.5.1 Command

The number of remote controlled operations is generally small except during critical phases (orbital positioning, orbital correction), but the number of different commands to be transmitted, and decoded on-board without error is high (a few hundreds). Besides this, there may be interference between the remote control systems of several satellites. Finally, reception of noisy signals leads to some probability of error. Hence *protection* against errors and jamming is required. This is achieved by sending the received signal back to the ground control stations (acknowledgement) with address and data coding and repetition of orders. Protection against interference and jamming is obtained using either narrow band reception, or wide band reception combined with spectrum spreading of signals and code division multiple access (see Chapter 3, Section 3.4.6).

6.5.2   Telemetry

Data supplied by the various sensors on-board the satellite is generally transmitted in digital form in a time frame. The bit rate is low, around 150 to 1000 bits/s. To adapt telemetry to the different phases of the satellite mission, the multiplex format or the bit rate can be modified by remote control.

6.5.3   Tracking

Tracking is obtained from telemetry signals and ranging measurements. The principle of tracking was described in Section 6.3.1 above.
  Ranging measurements are performed as follows:

A series of sinusoidal signals called *tones* successively phase modulate the up-link telecommand carrier which nominal frequency is f_U. These signals are demodulated on-board the satellite, and phase remodulate the down-link telemetry carrier of frequency f_D (f_U and f_D having a fixed ratio, that is $f_U/f_D = 221/240$; NASA-ESA standards) (*ESA*, 1980). The phase of the received tones at the ground control station are compared to that of the transmitted tones in order to determine the satellite range. The precision of measurement is set by the highest frequency ('major tone', 100 kHz in S band). The series of other sinusoidal tones ('minor tones') resolves the ambiguity resulting from the fact that the station-to-satellite distance is greater than the wavelength of the major tone. The ambiguity can only be resolved if the sinusoidal tone frequency is less than 8 Hz. The minor tones transmitted successively are obtained by division of the major tone (20 kHz, 4 kHz, 800 Hz, 160 Hz, 32 Hz, 8 Hz) (Carton, 1980). Thus, the relative dephasing of the sinusoidal tone of 8 Hz frequency can be determined with the temporal precision of the major tone. The precision of the measurement is within a few metres (see Table 6.4). It depends on the signal/noise ratio, and on dephasing due to tropospheric propagation. For a major tone of frequency f_S, the r.m.s. value of the distance error is:

$$\sigma_d = \frac{c}{2} \frac{\sigma_\phi}{2\pi f_s} \qquad (24)$$

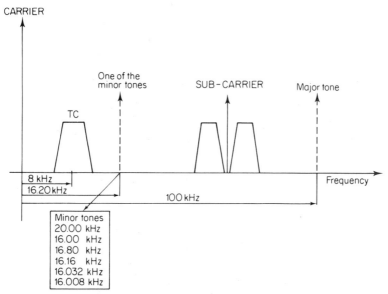

Figure 6.21   Spectrum of TT and C signals. *Reproduced by permission of the European Space Agency.*

where σ_ϕ is the r.m.s. value of the phase error, c the light velocity

$$\sigma_\phi = \frac{k}{\sqrt{(S/N)}} \qquad (25)$$

The value of the constant k depends on the type of receiver used.

Figure 6.21 shows an example of the spectrum of the telemetry and command carriers where minor tones are up frequency shifted to reduce the signal bandwidth. The modulation generally used is of the PCM/PSK/PM type.

6.6   THERMAL CONTROL

Thermal control is required to maintain equipments and structure of the satellite within *specified temperature ranges*. This must be ensured whether the equipments are *operating or not*. Structural deformations must be minimized to ensure correct performance of the attitude control and antenna sub-systems. The thermal control sub-system is to be designed considering the different requirements of the *operational and transfer orbit phases*, which involve different attitudes and orbits and different status of the apogee motor (Rolfo, 1981). Also the acceptable temperature range varies widely from one satellite equipment to another, e.g.:

Batteries: $0\,^\circ\text{C}$ to $+20\,^\circ\text{C}$.

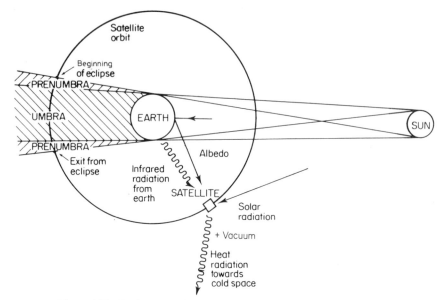

Figure 6.22 Environment factors of importance to the thermal control.

Solar cells: $-100\,^{\circ}$C to $+50\,^{\circ}$C.
Electronic equipment: $-10\,^{\circ}$C to $60\,^{\circ}$C.
Propellent tanks: $+10\,^{\circ}$C to $+50\,^{\circ}$C.
Infra-red sensors: $-20\,^{\circ}$C to $+45\,^{\circ}$C.

The mean temperature of a satellite is the result of the *thermal balance* between the energy received by radiation from external sources plus the energy dissipated inside the satellite mainly due to electrical losses and the energy radiated from the satellite to outer space. There is no convection heat exchange due to vacuum.

Figure 6.22 describes the environmental factors of importance to the thermal control subsystem and underlines the three sources of radiation a satellite is subject to (Sun, Earth, albedo of the Earth) each with differing characteristics which are absorbed in different ways by the satellite coating.

Thermal control can be *active* or *passive*, the latter being preferred for reasons of simplicity, lower cost and reliability.

6.6.1 Passive thermal control

This is based on the absorptance α and emittance ε of the surface finish. The ε and α parameters take values lying in the range zero to unity, and the ratio α/ε is of prime importance in determining the surface mean temperature when exposed to the Sun. Main finishes are:

(1) *White paint* which absorbs infra-red radiation (earth flux) and reflects solar flux. White paint is *cold* in the Sun ($-$ 150 °C to $-$ 50 °C), as the ratio α/ε is small (ε as high as 0.9, α in the order of 0.17).

(2) *Aluminium paint* which has a low ε (around 0.25) and a low α (around 0.25). The temperature of equilibrium when exposed to the Sun is around 0 °C. On the other hand, as its emittance is lower than that of the black paint, the aluminium paint is warmer in the dark than the black paint.

(3) *Black paint* which has a high ε (around 0.89) and very high α (around 0.97). When exposed to Sun, the temperature is above 0 °C.

(4) *Polished metal surface*, which absorbs the visible part of the solar spectrum, but reflects infra-red radiation. These coatings are *hot* in the Sun (50 °C to 150 °C) as the ratio α/ε is high (for instance, with gold $\varepsilon = 0.04$ and $\alpha = 0.25$).

Since communication satellite payloads need to dissipate *heat, radiators* with a very low α/ε finishes (α about 0.08, ε about 0.75) are employed and consist of fused silica mirrors with silver backing, called *optical solar reflectors* (OSR), or *second surface mirrors* (SSM) made from plastic (Teflon, Kapton or Mylar) sheets with silver or aluminium backing. They are located on the north and south panels so they may efficiently radiate their heat towards outer space since the north and south panels are least affected by daily variations in solar incidence.

A rapid estimation of the required radiator surface area S can be performed using the following equation, which expresses that the power P (heat) to be evacuated plus the power absorbed by the radiator from the Sun is equal to the power radiated by the radiator at the temperature of equilibrium T:

$$P + \alpha\phi S = \varepsilon S\sigma T^4$$

where: α and ε are the absorptance and the emittance of the radiator, ϕ is the power flux (W/m^2) received from the Sun (depending on distance and orientation of the surface relative to the direction of the Sun).

Assuming a maximum temperature for the radiator of a few tens of degrees (say 40 °C), S is calculated in the worst case (end of life when α is degraded, and at the time of year when the power received form the Sun is maximum for a given orientation of the surface, see section 5.1.4.1).

For a sun oriented surface lying in a plane perpendicular to the orbital plane:

$$\phi = \cos \delta \geq d^{-2} \times 1353 \text{ W/m}^2$$

where δ is the sun declination and d the distance to the sun, expressed in IAU. Worst conditions occur just before the spring equinox and from figure 5.1, $\delta = 4.3°$ and $d^{-2} = 1.011$, hence $\phi = 1.008 \times 1353$ W/m^2.

For a surface parallel to the orbital plane:

$$\phi = \cos(90° - \delta) \times d^{-2} \times 1353 \ \text{W/m}^2$$

Worst conditions occur:
—just before winter solstice for a surface on the south side of the satellite; then from figure 5.1, $\delta = 23.5°$ and $d^{-2} = 1.033$, hence $\phi = 0.947 \times 1353 \ \text{W/m}^2$,
—at summer solstice for a surface on the north side of the satellite; then from figure 5.1, $\delta = 23.5°$ and $d^{-2} = 0.965$, hence $\phi = 0.885 \times 1353 \ \text{W/m}^2$.

One sees that values of ϕ are smaller for surfaces lying parallel to the orbital plane. This explains why, with a body-fixed satellite, radiators are mounted on the north and south sides of the satellite.

When the radiator is not illuminated by the Sun (e.g. at summer solstice for the radiator mounted on the south side of the satellite), its equilibrium temperature is to be calculated with the above equation, taking $\phi = 0$. This temperature must not be too low to avoid thermal stresses.

Example of calculation for a body-fixed satellite:
—Heat to be radiated = 800 W (equally divided between north and south sides)
—α of the radiator after apogee motor firing and 7 years = 0.17 (value before launch = 0.06)
—ε of the radiator = 0.75
—Equilibrium temeprature of the radiator = 32°C

Deriving from the above equation:

$$S = P/(\varepsilon\sigma T^4 - \alpha\phi)$$

The surface area of the radiator on the south side of the satellite should be:

$$S = 400/(0.75 \times 5.67 \ 10^{-8} \times 305^4 - 0.17 \times 0.412 \times 1353) = 1.46 \ \text{m}^2$$

For the north side radiator, a similar calculation delivers 1.43 m².

The equilibrium temperature of the radiator on the south side at summer solstice is calculated using:

$$T = (P/S\varepsilon\sigma)^{¼} = (400/1.46 \times 0.75 \times 5.67 \ 10^{-8})^{¼} = +10°C$$

All the above assumes that the heat to be dissipated is uniformly distributed over the surface of the radiator, and hence that the temperature of the radiator is the same at all points. On the other hand, heat is generated mainly by TWTs through their small mounting surface. Then a power dispatcher is mandatory to distribute the heat over the whole surface of the rear side of the radiator.

Most passive thermal control systems also rely on *multilayer superinsulation blankets* composed of alternate layers of thin Mylar and Kapton films coated with vacuum deposited aluminium which ensure high isolation between the satellite inside parts and space.

6.6.2 Active thermal control

This is used as a supplement to passive means and comprises:

(1) *Heat pipes* (Figure 6.23) ensuring adiabatic treansfer of heat from heat sources to radiators by successive vaporization and condensation of a fluid at the two extremities of a pipe. These devices ensure a high capacity for heat transfer with small temperature differences because of the high latent heat value of the fluids used.
(2) *Hinged flaps* and *multiple blade louvres* mechanisms arranged to expose or to cover radiator areas.
(3) *Electrical heaters* activated either by thermostats or telecommand.

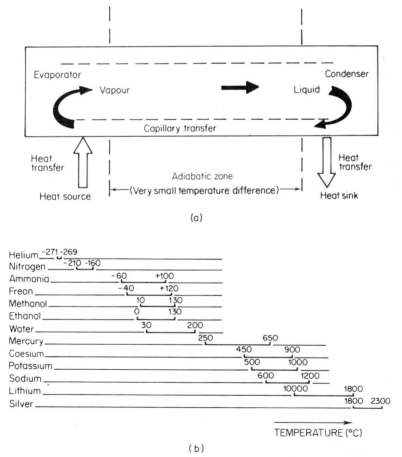

Figure 6.23   Heat pipe: (a) principle of operation; (b) temperature operating range.

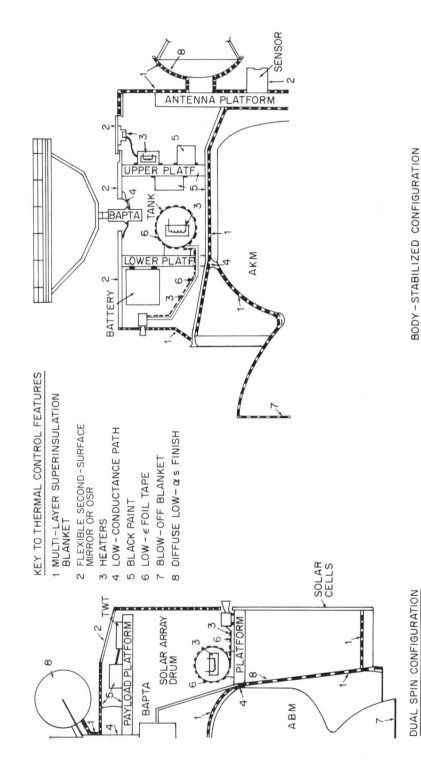

KEY TO THERMAL CONTROL FEATURES

1 MULTI-LAYER SUPERINSULATION BLANKET
2 FLEXIBLE SECOND-SURFACE MIRROR OR OSR
3 HEATERS
4 LOW-CONDUCTANCE PATH
5 BLACK PAINT
6 LOW-ε FOIL TAPE
7 BLOW-OFF BLANKET
8 DIFFUSE LOW-α s FINISH

BODY-STABILIZED CONFIGURATION

DUAL SPIN CONFIGURATION

Figure 6.24   Thermal control for dual-spin and body-stabilized configurations.

6.6.3 Conclusion

The above shows the objectives and means of thermal control. It should be noted that the stabilization sub-system concept greatly influences the design of the thermal control sub-system. Obviously, the problems are very different for a spin stabilized satellite than with a body-fixed satellite (Figure 6.24): a spin stabilized satellite has similar exposure conditions over the lateral faces of the satellite body.

6.7 STRUCTURE (Figure 6.25)

The functions of this subsystem are to:

(1) *Support the on-board equipment*, especially during the launch phases where the mechanical stresses are highest.
(2) *Ensure correct positioning of this equipment*: alignment of sensors, thrust axes of the thrusters, axes of the antennaw, relative to the axes of the satellite body, despite mechanical and thermal constraints.
(3) *Allow the various separations and deployments*, the satellite's configuration is subject to from launch to its operational station.
(4) *Avoid electrical charge accumulation*.

Structural materials must have resistance to deformation and be lightweight. These are conflicting requirements. Current techniques allow about 6 per cent of the satellite's total mass only to be used for its structure (Colette and Herdan, 1977). This has been achieved by use of aluminium and magnesium alloys, honeycomb panels, bonded assemblies (Chanteranne, 1979), and carbon fibre reinforced plastic materials (solar panels, antenna towers) (Giraubit, 1979; Giraubit *et al.*, 1979). The use of beryllium is restricted owing to its prohibitive cost.

6.8 ELECTRIC POWER SUPPLY: GENERATION, STORAGE AND CONDITIONING

The electric power supply of a satellite poses severe problems mainly because of mass and volume limitations. Direct broadcast satellites can require a few kilowatts of electric power (power transmitted in the order of a kilowatt, mass of 1000 kg). Electric power is directly related to the transmitted RF power, through the RF/DC efficiency of the high power amplifiers. The power sub-system consists of:

(1)   A *primary energy source* which converts into available energy some energy available other way (currently sun energy transformed into electrical energy by solar cells).

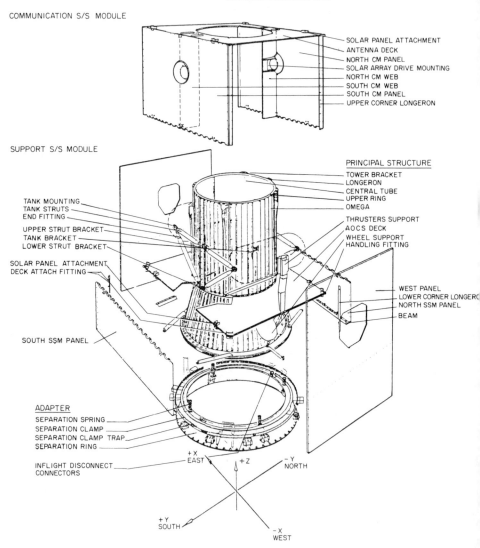

Figure 6.25   Example of a satellite structure (INTELSAT V)

(2) *A secondary energy source* which can be substituted when needed, e.g. during an eclipse period (battery of electrochemical accumulators).

(3) *Power conditioning* (regulation, distribution) and *protection* circuits.

6.8.1 Primary energy source

The only external energy source is *solar radiation*. On-board energy source (nuclear generators), although envisaged (Kaplan *et al.*, 1978), are not yet satisfactory for geostationary telecommunication satellites. Nevertheless, during the first few hours following the injection into transfer orbit and prior to the deployment of the solar panels, the electrochemical accumulators are the primary energy source.

6.8.1.1 Solar cells

Solar cells work on the principle of the photovoltaic effect (current appears at an NP junction when subjected to a flow of photons).

(*a*) *Solar radiation.* Above the atmosphere the average Sun power flux is 1353 W/m^2 (see Chapter 5.1.4.1). When the normal to the surface makes an angle θ with respect to the direction of the Sun, the received power, and therefore the output, are multiplied by cos θ. The surface of cells to be taken into account in the design is therefore the *projection on a plane* normal to the solar rays.

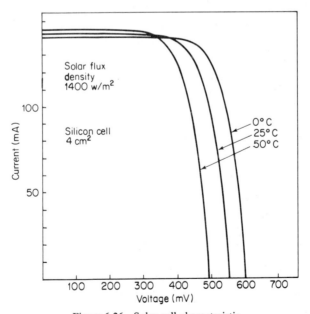

Figure 6.26   Solar cell characteristic.

(*b*) *Characteristics of a solar cell. Silicon* is the most widely used material in solar cells. The current to voltage characteristic (Figure 6.26) depends on the *temperature* (voltage falls by 50 per cent if the temperature varies from 27 °C to 150 °C).

The average output *efficiency* of a cell subjected to solar radiation in vacuum at a temperature of 27 °C is 10–14 per cent. This efficiency *decreases* under the effect of the *radiations* and drops by about 30 per cent in seven years. The major contributor in geostationary orbit is from protons during solar flares. In order to reduce the damage, solar cells are covered with shields from fused silica or cerium doped microsheets.

The surface area of a cell is between 4 and 8 cm^2. Currently silicon cells use a relatively thick monocrystalline chip (200 to 250 μm). Work is in progress to produce ultra thin (50 μm) low mass cells not very strongly affected by radiation. Other research aims to replace silicon with more efficient materials (18 per cent efficiency), e.g. aluminium and *gallium arsenide* (Sparks, 1980; Wheelon 1983).

6.8.1.2 Solar panels

Cells are connected in series and parallel to deliver a few tens of volts (e.g. 42 V, INTELSAT V (Brodersen and Rizos, 1978). They are bonded on panels ensuring the required rigidity and thermal regulation.

On spin stabilized satellites, the solar panels form the exterior envelope of the satellite body (INTELSAT I, II, III, IV). INTELSAT IVA and VI are equipped with a dual telescopic solar panel, the outer panel (skirt) of which is deployed after the satellite is on orbit so the area of solar panels is increased and full power is obtained.

Body-fixed stabilized satellites require panels that are *folded* during the launch phase and then deployed. If these are fixed relative to the satellite, which turns round once daily relative to the Sun, several panels must be employed so that sufficient surface area is illuminated, irrespective of the Sun's direction (the SYMPHONIE and AMSAT-OSCAR 10 satellites had three panels parallel to the satellite axis, with a 120° angle between them). More evolved systems permanently orient panels so that they intercept the maximum solar flux. For this, two movements must be initiated:

(1) A daily rotation at orbital speed (one turn per day).
(2) An annual movement of ± 23° to follow the apparent movement of the Sun on the Equator.

It is simpler and sufficiently effective to orient the solar panels according to the *daily movement*. Therefore the panels are in a plane containing the pitch axis and following the daily rotation, but only perpendicular to the solar rays at the equinoxes. The solar energy delivered at the solstices will in this instance be $100 \ (1 - \cos 23°) = 8$ per cent lower.

This requires:

(1) *Sun sensors.*
(2) *Electronics* for measurement and control.
(3) A *servo motor.*
(4) A *bearing assembly and power transfer assembly* (BAPTA). This can be avoided by using instead a flexible link and revolving the panels rapidly to their start positions every 24 h.

Body-fixed stabilized satellites are more *efficient* in their use of solar cells, compared with spinning satellites. At *end of life* body-mounted solar cells on spinning satellites provide 6 to 10 W/kg, whereas deployable panels supply 15 to 25 W/kg. The solar array mass in each case is in the order of 3 to 5 kg/m^2 (INTELSAT VI: 59 m^2, 250 kg, 2 kW; Telecom 1: 16 m^2, 47 kg, 1 kW). Figure 6.27 shows the advantage of deployable and oriented panels relative to those mounted on the body in terms of economy of mass.

Current research is aimed at increasing deployable solar panel dimensions

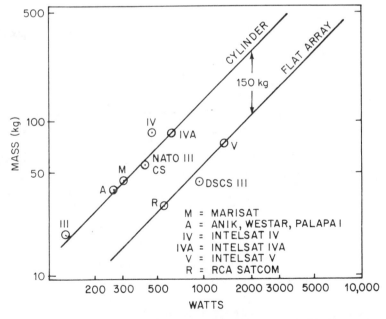

Figure 6.27   Mass of solar array (including substrate) versus electrical power (at beginning of life). *Reproduced by permission of Centre National d'Etudes Spatiales from Templeton and Cuccia (1979).*

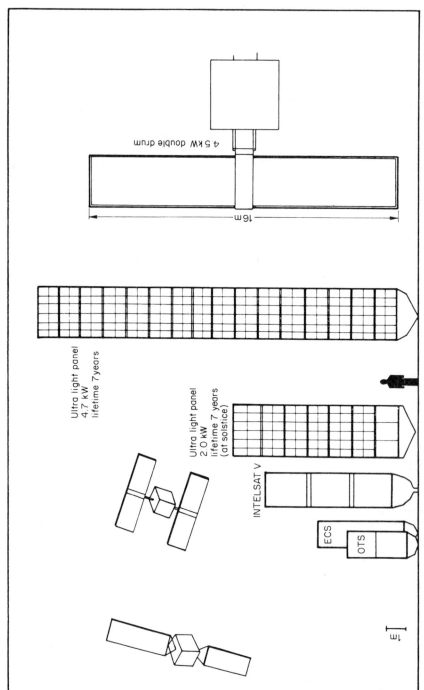

Figure 6.28 Evolution in size of flat solar arrays. *Reproduced by permission of Association Aeronautique et Astronautique de France from Lacombe and Havas (1978).*

and reducing weight. Values are reaching 50 to 60 W/kg, and power levels of a few tens of kilowatts (flexible deployable solar panels for direct broadcast satellites). Figure 6.28 shows the evolution in size of flat solar arrays for different body-stabilized satellites (Lacombe and Havas, 1978).

6.8.2 Secondary energy sources

The secondary energy source *stores* energy from the primary source and provides this stored energy when the primary source stops functioning. *Electrochemical accumulators* are the most appropriate for this purpose especially in the case of telecommunications satellites, when the mission must be performed during an *eclipse*.

These eclipses take place 84 days per year and can reach 70 minutes per day (see Figure 4.24).

Characteristics which are important for a secondary energy source are (Sparks, 1980):

(1) A sufficient lifetime, which depends on the depth of discharge and the temperature.
(2) A high specific energy in Wh/kg (Watt-hours per kg).

Table 6.5 shows the characteristics of different types of electrochemical accumulators. The values shown are for a depth of discharge of 100 per cent. The *depth of discharge* (DOD) represents the percentage of the maximum stored energy which is actually used. Practically, to guarantee a sufficient

Table 6.5 CHARACTERISTICS OF ELECTROCHEMICAL ACCUMULATORS.

Type	Energy per mass unit (Wh/kg)	Observations
Nickel—Cadmium (Ni—Cd)	20/40	Most used as secondary source. Large number of charging cycles with depth of discharge of 50 to 60 per cent.
Silver—Cadmium (Ag—Cd)	50/60	Not magnetic.
Nickel hydrogen (Ni—H$_2$)	60/80	Low energy density (Wh/cm^3 low) Long lifetime even with depth of 70 to 80 per cent.
Silver hydrogen (Ag—H$_2$)	100	Under development

lifetime, the depth of discharge *should* not exceed 60 to 80 per cent depending on type during the longest eclipse.

Until recently, nickel-cadmium accumulators have been mostly used (McKinney and Briggs, 1978). Nickel-hydrogen accumulators were used for the first time on the NTS 2 satellite (Dunlop and Stockel, 1980), and are replacing Ni-Cd on satellites because of their higher specific energy and longer life (Susplugas, 1979). Research is underway on silver-hydrogen accumulators.

6.8.3   Power conditioning and protection circuits

The power delivered by the primary energy source depends on the selected operating voltage and the extent of irradiation degradation of the solar cells. The cell voltage is also strongly dependent on *temperature* as during an eclipse the solar cell temperature can drop to $-180\ ^{\circ}$C. So just after an eclipse, during the illuminated phase when the temperature is in order of a few tens of degrees, the voltage delivered by the cells is about 2.5 times the nominal operating value. Therefore, it is necessary to provide *voltage regulation* or

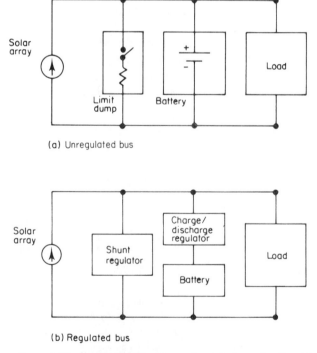

(a) Unregulated bus

(b) Regulated bus

Figure 6.29   Power conditioning and protection circuits: (a) unregulated bus; (b) regulated bus.

damping by a parallel resistor. In addition, one needs control circuits for charging the accumulators and for supplying energy to the on-board equipment.

Two main principles are used:

(1) *Unregulated bus* (Figure 6.29(a)) where the battery is placed in parallel with the solar generator, and defines the operating voltage of the solar array, altogether with the user load characteristic. Then some kind of damping facility dissipates any surplus energy.
(2) *Regulated bus* (Figure 6.29(b)) in which a shunt regulator defines the solar generator's voltage. The accumulators are connected to the power supply through a charge/discharge regulator.

The unregulated supply has the advantages of simplicity and a lower mass. However, the equipments are subject to variations in voltage, depending on the state of charge of the battery (e.g. INTELSAT V (Rintala *et al.*, 1979)). The regulator avoids this problem but at a cost to the mass budget and output power and less reliability (e.g. Symphonie (Debay *et al.*, 1979), OTS).

Apart from the circuits above, this sub-system has *protection circuits* (e.g. limitation of battery discharge to avoid inversion of polarization of the elements), and *regulation* and *conversion circuits* to supply the different on-board equipment with the continuous regulated voltages and the high voltages for the TWT.

6.8.4 Spinned satellite and body-fixed satellite power supply design

For this example of calculation, the required power at the end of the seven years' life is $P = 1200$ W. This requirement must be fulfilled in the worst case, that is at summer solstice when the solar radiation flux is $\phi = 0.89 \times 1353$ W/m^2 (see Figure 5.1).

6.8.4.1 Spin stabilized satellite example

The solar array is shaped as a cylinder, hence the true illuminated surface is $\pi/2$ times the equivalent surface normal to the radiation. Taking into account the dark side of the satellite, the total surface is twice larger. The geometric *shape factor* is then $F = \pi$. As the satellite rotates, solar cells successively face the Sun and the cold space. Hence the average temperature is relatively low, say 25 °C. At 25 °C the solar cell efficiency is $e = 14$ per cent. The expected degradation at end of life reduces the efficiency by 22 per cent. Some losses, such as those due to cover glasses, panel wiring, etc., are included in a factor $f = 0.9$. The solar cell area is $A = 2$ cm $\times$ 2 cm $= 4$ cm^2 and the filling factor is

0.85. The end of life power expresses as:

$$P = (1 - 0.22)\phi efnA/F \tag{26}$$

where n is the number of solar cells. Hence the required number of cells is:

$$n = 1200 \ \pi/(1 - 0.22) \times 1353 \times 0.89 \times 0.14 \times 0.9 \times 4.10^{-4} = 79\,750$$

Taking into account the filling factor, the solar array surface is $79\,750 \times 4.10^{-4}/0.85 = 37.5$ m^2 which represents a 2.16 m diameter cylinder, 5.5 m high.

Assuming an average solar cell mass (including wiring, bonding and covers) of 0.8 g, and a substrate mass of 1.6 kg/m^2, the mass of the solar panel is $79\,750 \times 0.8 \ 10^{-3} + 37.5 \times 1.6 = 124$ kg. The power per mass unit of the array is then $1200/124 = 9.7$ W/kg.

6.8.4.2 Body-fixed satellite example

The satellite has two rectangular Sun oriented solar arrays with axes along the pitch axis. As the solar arrays always face the Sun the degradation is increased, and the average solar cell temperature is higher than with the spin stabilized satellite case, say about 60 °C. Solar cell efficiency reduces to $e = 11$ per cent and the degradation is now 28 per cent. Loss factor is slightly less due to shadowing of the solar array by the antennae, say $f = 0.88$. The solar cell area is $A = 2$ cm $\times$ 4 cm $= 8$ cm^2 and the filling factor is 0.75. The end of life power expresses as:

$$P = (1 - 0.28)\phi efnA \tag{27}$$

and the number of solar cells is:

$$n = 1200/(1 - 0.28) \times 1353 \times 0.89 \times 0.11 \times 0.88 \times 8.10^{-4} = 17\,870$$

The solar array surface is $17\,870 \times 8.10^{-4}/0.75 = 19$ m^2.

The average solar cell mass is 1.6 g. The substrate mass is 1.8 kg/m^2 as the solar array needs its own structure. The mass of the two solar panels is $17\,870 \times 1.6 \ 10^{-3} + 19 \times 1.8 = 63$ kg. The power per mass unit of the arrays is then $1200/63 = 19$ W/kg.

The above design is similar to that of the INTELSAT V satellite power subsystem (McKinney and Briggs, 1978).

6.8.4.3 Battery

An element of the battery is characterized by its capacity C(**A.h**) and by the mean voltage during discharge V_d. Given the depth of discharge (DOD) α and the discharge efficiency η_d, the energy available from this element is:

$$E = CV_d\alpha\eta_d \ (\text{W.h})$$

Let P be the power to be supplied to the satellite during eclipse. Considering that the maximum eclipse duration is 70 mn or 1.2 h, the energy to be supplied is $1.2P$ (Wh). The required number of elements is then:

$$n = 1.2P/CV_d\alpha\eta_d$$

Example of calculation: The power to be supplied on a geostationary satellite during eclipse is $P = 800$ W. NiCd elements with capacity $C = 30$ Ah are used, and DOD is $\alpha = 55\%$ to guarantee a 7 years lifetime. The voltage during discharge is $V_d = 1.2$ volt per element and the discharge efficiency is $\eta_d = 0.9$. The number of elements is:

$$n = 1.2 \times 800/30 \times 1.2 \times 0.55 \times 0.9 = 54 \text{ elements}$$

To improve the reliability, the battery is often divided in two parts and one more element is added to each part to allow for nomial voltage operation even though any one of the elements undergoes a short-circuit failure. Diodes are also placed in parallel with each element to by-pass them in case of open-circuit failure.

A rapid estimation of the mass of the battery can be performed from the energy per mass unit E_m. The stored energy must be $1.2P/\alpha\eta_d$. The mass of the battery is then:

$$M(\text{kg}) = 1.2P/\alpha\eta_d E_m$$

Assuming energy per unit mass $E_m = 30$ Wh/kg, the mass of the battery for the above example can be estimated as:

$$M = 1.2 \times 800/0.55 \times 0.9 = 65 \text{ kg}$$

The charge of the battery is generally performed at constant current, the value of the intensity expressed in amperes being equal to $C/10$, where C is the capacity in Ah. Protection against overcharge is achieved by monitoring the battery voltage and ending charge when this voltage oversteps a given value. Once the battery is charged, an ideal charge current is maintained with an intensity equal to $C/50$ in order to withstand the battery self-discharge.

6.9   ANTENNAE

The design of a satellite antenna is conditioned by the required *coverage*. Chapter 4.7 has indicated methods to determine the value of the angle from the satellite subtending the serviced zone on the Earth. This determines the *beamwidth* of the antenna which is usually taken as the 3dB beamwidth (*half-power beam*), although other values of the gain fall-out can be adopted.

Considering the *3dB beamwidth*, the *boresight antenna* gain is equal to (Chapter 2.1.1.3):

$$G = \eta(\pi D/\lambda)^2$$

where the satellite antenna diameter D is related to the 3 dB beamwidth by the following:

$$D = 70 \, \lambda/\theta_{3 \, \text{dB}}$$

Should the *service area* be the total surface of the Earth as seen from one geostationary satellite, the required beamwidth is $17.5°$. Taking the 3 dB beamwidth equal to this angle, then the maximum antenna gain at the sub-satellite point on the equator (boresight of the antenna) is 20 dB. Note that even though the satellite *antenna gain* varies at most 3 dB within the coverage, the *received power* is 4.3 dB less at a station located on the boundary of the zone compared with a station located at the subsatellite point, due to *range variation* (see Section 4.4).

Section 4.7.2 has indicated that the satellite antenna coverage area were often restricted to *specific zones* on the Earth. Hence the antenna beamwidth is smaller. Such *narrow beams* display either a simple circular cross-section, and are then called *spot beams*, or an elliptic or more complicated cross-section (in order to fit at best the zone to be serviced), and are then called *shaped beams* (CCIR, Rep. 676, 1982).

The smaller the beamwidth the higher the gain, so with spot or shaped beams one benefits from a higher satellite antenna gain. For instance a $1°$ antenna beamwidth offers a 45 dB gain, that is a 25 dB gain improvement over a global coverage antenna. The resulting increase in the satellite EIRP, with same satellite transmitter power, allows for the implementation of smaller and cheaper earth stations. If a wider region must be covered, for instance a few degrees in size, the advantages of spot beams, with beamwidth of the order of $1°$ or even less, can be retained by dividing the coverage into several spot beams. This concept is that of a *multibeam antenna* (Foldes and Dienemann, 1980).

Finally, as discussed in Section 3.6, *frequency reuse* with spot beams and multibeam satellite systems allow for an increase of the system capacity. Frequency reuse is achieved either by *spatial separation* of the beams, that is beams at same frequency covering different parts of the Earth, or by *polarization discrimination,* that is two beams at the same frequency but with orthogonal polarizations covering the same part of the Earth.

The efficient use of the *radiofrequency spectrum* and of the geostationary satellite orbit also implies appropriate *radiation patterns* with highly *attenuated sidelobes,* thus reducing beam spillover (CCIR, Rep. 558–2, 1982), and *precise pointing* of the beam which depends on the satellite attitude and station keeping control performance (CCIR, Rep. 453–3, 1982).

6.9.1   Spot beams

A spot beam is a *circular cross-section* beam produced by a narrow beamwidth

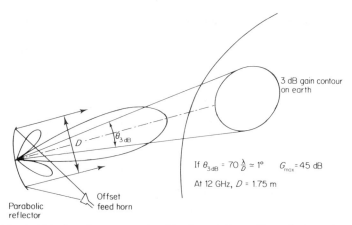

Figure 6.30   Spot beam: example of implementation with a reflector antenna.

(about 1°) satellite antenna. The coverage zone on the Earth has the shape of a circle or of an ellipse (see Figure 6.30).

6.9.1.1.   3 dB coverage

The antenna gain on the boresight is expressed as:

$$G = 48\ 360\ \eta/\theta_{3\ \mathrm{dB}}^2$$

where η is the antenna efficiency and $\theta_{3\ \mathrm{dB}}$ is the antenna 3 dB antenna beamwidth expressed in degrees.

6.9.1.2   5 dB coverage

For an antenna of diameter D_0, with a 3 dB beamwidth $\theta_{3\ \mathrm{dB}} = \theta$, the curves in Figure 6.31 shows how the gain in the axis and the gain in the direction $\theta/2$ varies with the diameter (Hatch, 1969). It appears that with D between D_0 and 1.3 D_0, the gain in the direction $\theta/2$ remains *higher* than its initial value and the gain in the axis increases by 2 dB.

So by replacing the antenna of diameter D_0 by one of 1.3 D_0, stations previously at the periphery of the 3 dB coverage zone are now at the periphery of the 5 dB coverage zone. As a consequence of the increase in antenna gain these stations suffer no disadvantage while stations at the centre of the zone benefit from 2 dB more power. The inconvenience is that the satellite has a more bulky antenna.

6.9.2   Elliptically shaped beams

An elliptically shaped beam is a narrow beam with an *elliptic cross-section*.

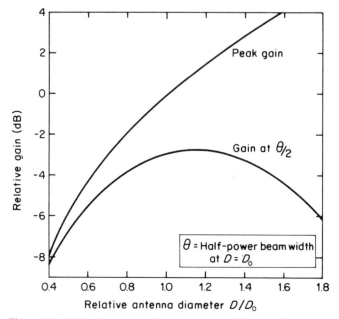

Figure 6.31   Off-axis antenna gain versus antenna diameter. *Reproduced with permission from Hatch (1969). Copyright © 1969
IEEE.*

Such a beam is characterized by two beamwidth angles in two perpendicular
planes. The coverage zone on the Earth has the shape of a circle or of an ellipse
and is designed to fit at best the contour of the zone to be serviced (see Figure
6.32).

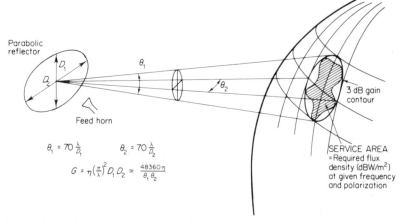

Figure 6.32   Elliptically shaped beam: implementation with an elliptically shaped
reflector antenna.

6.9.2.1 3 dB coverage

The antenna gain on the boresight expresses as: $G = 48\,360\,\eta/\theta_1\theta_2$ where η is the antenna efficiency and θ_1, θ_2 are the antenna half-power beamwidths in two orthogonal planes (expressed in degrees).

6.9.2.2 Influence of antenna beam depointing

All the above assumes perfect pointing of the beam. However, depointing reduces signal power at the boundary of coverage. To compensate for this, the beam must be widened such that at maximum depointing ε, the required minimum power is received at the zone boundary. The antenna gain in the axis for a 3 dB coverage is therefore:

$$G = \frac{48\,360\,\eta}{(\theta_1 + 2\varepsilon)(\theta_2 + 2\varepsilon)} \tag{29}$$

instead of

$$G' = \frac{48\,360\,\eta}{\theta_1 \cdot \theta_2}$$

The gain loss A is:

$$A = \frac{G}{G'} = \frac{\theta_1 \cdot \theta_2}{(\theta_1 + 2\varepsilon)(\theta_2 + 2\varepsilon)} = \frac{1}{\left(1 + \frac{2\varepsilon}{\theta_1}\right)\left(1 + \frac{2\varepsilon}{\theta_2}\right)} \tag{30}$$

and the loss is expressed in decibels:

$$A_{\mathrm{dB}} = 10 \log\left(1 + \frac{2\varepsilon}{\theta_1}\right) + 10 \log\left(1 + \frac{2\varepsilon}{\theta_2}\right) \tag{31}$$

The above loss in gain results from the 2ε widening of the 3 dB antenna coverage. To maintain nominal power at the zone boundary, transmitter power must be increased by A_{dB}. As ε is small compared to θ_1 and θ_2, it follows therefore:

$$A_{\mathrm{dB}} = \frac{10}{2.3}\left(\frac{2\varepsilon}{\theta_1} + \frac{2\varepsilon}{\theta_2}\right) \qquad A_{\mathrm{dB}} = 8.7 \frac{\theta_1 + \theta_2}{\theta_1\theta_2}\varepsilon \tag{32}$$

To put this into perspective, if $\theta_1 = \theta_2 = \theta_{3\,\mathrm{dB}}$

$$A_{\mathrm{dB}} = 17.4\left(\frac{\varepsilon}{\theta_{3\,\mathrm{dB}}}\right) \tag{33}$$

therefore if $\varepsilon < \theta_{3\,\mathrm{dB}}/10$ the loss is less than 2 dB.

The preceding case, which is for coverage of a zone requiring minimum power at all points within the zone, must be distinguished from the case where the service must be ensured with minimum required power at a unique earth station nominally located on the antenna axis and where satellite antenna depointing causes attenuation in the received signal according to the direction θ relative to the antenna axis.

$$G_{dB} = G_0 - 12\left(\frac{\theta}{\theta_{3\,dB}}\right)^2 \tag{34}$$

whence loss in gain for a depointing ε:

$$A_{dB} = 12\left(\frac{\varepsilon}{\theta_{3\,dB}}\right)^2 \tag{35}$$

The above loss in gain results from variations in the satellite antenna direction caused by attitude variations of the satellite. To maintain the signal power at the earth station, nominally situated in the antenna axis, transmitter power must be increased by A_{dB} as given by (35).

6.9.2.3 Example

With an antenna of $1°$ beamwidth and attitude control of $0.3°$, the loss in the gain as given by formula (33) is 5.2 dB and, 1.1 dB as given by formula (35).

This shows that the margin of power required to ensure a 3 dB zone coverage is larger than that required to ensure the minimum required power at a unique earth station for the same performance satellite attitude control. If this margin appears too high, taking into account the best achievable performance of satellite attitude control ($0.05°$), a *control of the satellite antenna pointing* must be considered: for instance the TDF satellite is equipped with a gimballed antenna reflector relative to the satellite platform (Arnaud *et al.*, 1980).

6.9.3 Shaped beams (any contour)

The elliptically shaped beam described above is obtained by shaping the antenna reflector and achieves simple circular or elliptical contours on the Earth. Moreover, these contours can not be modified during the lifetime of the satellite to cope, for instance, with possible modifications of the traffic pattern or a change of the satellite orbital position. More *sophisticated* contours can be obtained using *multiple feed antennae* where the individual radiation patterns of each feed are combined to generate the desired beam shape. In a given direction the electric field results from the addition of the fields of the individual feeds depending on their *relative power and phase*. The radio-frequency power at the transmitter output is split and phased among the various feeds by means of sets of power dividers and phase shifters which constitute a so-called *beam forming network* (see Figure 6.33). By changing

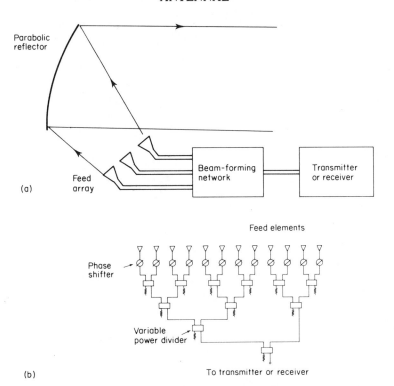

Figure 6.33 Shaped beam antenna: (a) implementation with a reflector antenna and a multiple feed array; (b) beam forming network.

through the TTC link the relative power and phase between feeds the beam shape can be modified during the satellite lifetime. Examples of shaped beams are found on the INTELSAT V satellite (Figure 4.13). The ultimate concept is that of changing the amplitude–phase distribution of the feeders in a time-sequenced manner so as to generate a *scanning shaped beam* which is able to hop from one part of the service area to another.

6.9.4 Multiple beams

The coverage of the service area is achieved by several spot beams.

6.9.4.1 Separated beams

The spot beams can be widely separated as, for instance, where the service area consists of the set of the main cities of a country between which heavy route communications requirements exist. Figure 6.34 illustrates the coverage of the heavy traffic zones of Europe with six spot beams using the same frequency

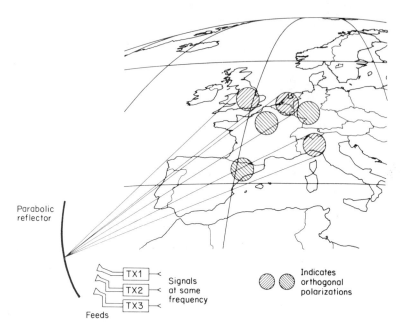

Figure 6.34   Multibeam antenna for the coverage of the European main traffic zones with six spot beams operating at the same frequency on both polarizations (only the set of feeds and transmitters for one polarization is shown).

band (De Montlivault and Hagenbucher, 1982). Note that where the distance between two beams is not large enough, additional isolation is obtained using orthogonal polarization.

6.9.4.2   Contiguous beams with a small number of beams

The spot beams can be *contiguous* where the service area can be approximated by a cluster of a few circles. In this case the contiguous beams operate at *different frequencies* so as to avoid *interference* (see Figure 6.35) and hence the total available bandwidth is divided into as many sub-bands as beams.

6.9.4.3   Beam lattice

With the previous scheme as the number of beams increases, the frequency band devoted to each spot beam becomes smaller, hence reducing the capacity of the link. One can combine the previous concept with that of *frequency reuse* to achieve a *continuous beam lattice* by repeating the beam pattern of Figure 6.35 with reuse of the same frequency sub-bands from one beam pattern to

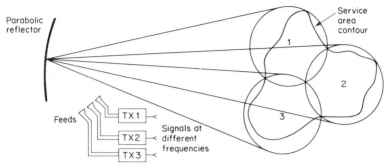

Figure 6.35   Multibeam antenna with three contiguous beams operating at different frequencies.

another. Figure 6.36 illustrates an example of coverage of Europe with a beam lattice using three different frequency sub-bands (Lopriore *et al.*, 1982).

Due to frequency reuse, *co-channel interference* is generated within each beam of the pattern by the surrounding beam patterns. Figure 6.37 shows that the value of the interference within a beam depends on the *antenna radiation pattern* associated with any spot beam and on the *distance* between centre of beams reusing the same frequency. With a triangular lattice such as the one displayed in Figure 6.37 the frequency reuse scheme induces six main co-channel interferences in each beam from the vertices of a hexagon. Augmenting the number of beams (i.e. of frequencies) within the beam pattern increases the distance from one beam to another at the same frequency and hence reduces the interference, but this also reduces the useable bandwidth within each beam.

6.9.5   Different types of satellite antennae

6.9.5.1   Rotating antenna platform (spin stabilized satellites)

6.9.5.1.1   *Toroidal radiation pattern antenna* (Banget *et al.*, 1963)

The simplest satellite antenna that can be used on a *spin stabilized satellite* is a fixed antenna revolving with the satellite around the spin axis with a *toroidal radiation pattern* of approximately 18° beamwidth (Figure 6.38(a)). Unfortunately the gain of such an antenna does not exceed a *few decibels*. Such a technique was used on the first operational satellites such as INTELSAT I ('Early Bird') and II which had antenna gains of 4 to 5 dB at reception and about 9 dB at transmission.

6.9.5.1.2   *De-spun antenna* (Donnelly *et al.*, 1969; Matthews *et al.*, 1978)

In order to concentrate the transmitted power on the area to be covered, one

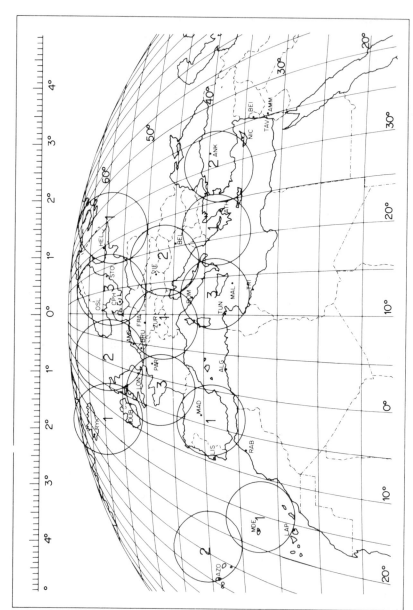

Figure 6.36 Coverage of Europe using a beam lattice with a three-frequency (1, 2, 3) beam pattern analogue to that of Figure 6.35 (Lopriore *et al.*, 1982).

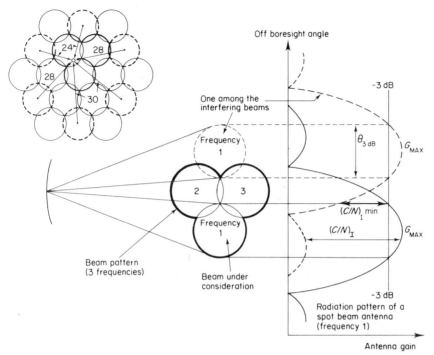

Figure 6.37 Construction of interference within a beam lattice with three frequencies. In the upper part, values indicate the six main components of the carrier-to-interference power ratio for beam at frequency 1 (worst case).

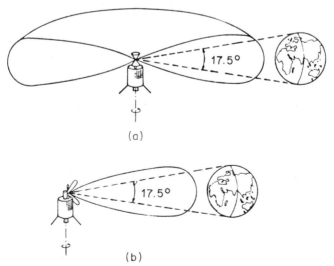

Figure 6.38 Rotating antennae: (a) toroidal radiation pattern antenna; (b) de-spun antenna.

can mount a *de-spun antenna* on the satellite (Figure 6.38(b)), that is a reflector antenna which spins around the satellite's rotation axis, in the opposite direction, thus maintaining the antenna axis pointed towards the Earth.

6.9.5.1.3   Electronically de-spun antenna (Rankin et al., Rosenthal et al., 1969).

Electronically de-spun antennas offer an alternate solution which avoids a mechanical bearing assembly. The antenna consists of an *array* of several antennas which are sequentially *switched* to de-spin the radiation pattern (Drabowitch and Morlon, 1967). Unfortunately, as electronically de-spun antennas scan the radiated beam in a stepwise fashion, a significant 1–3 dB variation in the antenna's directive gain is observed. Moreover, some loss is to be expected in the feed system.

6.9.5.2   Stabilized antenna platform (dual spin or body-stabilized satellites)

The antennae are mounted on a *stabilized platform* which is a despun platform in the case of a *dual-spin satellite* [e.g. COMSTAR (Abutaleb et al., 1977), Intelsat IV (Jilg, 1972) and Intelsat VI (Thompson and Johnston, 1983), (see Figure 6.9) or the body of the satellite itself (INTELSAT V, TELECOM 1). The despun platform implies a bearing assembly and a rotary joint or a power transfer assembly between the two sections that are in relative rotation. This presents difficulties with lubrication over the lifetime of the satellite (evaporation of lubricant), and mechanical friction disturbing the gyroscopic effect.

The antennae are rigidly fixed, and their pointing accuracy is the same as the attitude control (up to 0.05°). Higher accuracy requires an antenna tracking system. (TDF satellite, (Arnaud, 1980)).

6.9.5.2.1   Horn antenna

The horn antenna is the simplest directive antenna and is widely used for global coverage of the Earth which implies a 3 dB beamwidth equal to 17.5°. Such a beamwidth value is achieved at 4 GHz using a 0.3 m horn aperture diameter. Smaller beamwidth would imply larger aperture and related length of the horn, which may be difficult to install on a satellite. Moreover, horns suffer from poor sidelobes characteristics, which, however, can be improved *using corrugated horns* entailling annular grooves on the inner wall.

Although not used for spot beam generation, horns are frequently employed as a primary feed of reflector antennae.

6.9.5.2.2   Reflector antenna (Clarricoats and Poulton, 1977)

Most commonly used on communications satellites, the reflector antenna consists of a *parabolic reflector* with a *feed* located at the focal point. The antenna can also be arranged in a dual-reflector system where the main reflector is illuminated by a combination of a primary feed and a subreflector (Cassegrainian or Gregorian systems). Axisymmetric configurations suffer from *aperture blocking* effect of the radiated field by the feed or the subreflector, which results in a loss in the antenna gain and building of side-lobes in the radiation pattern.

Aperture blocking can be avoided using an *offset reflector* configuration, where an offset portion of the parabolic reflecting surface is used instead of an on-axis portion. Moreover, an offset configuration can more easily be accommodated in satellite design especially when it is necessary to deploy the reflector after launching.

Unfortunately, as a result of the *asymmetry* of the offset reflector configuration a cross-polarized component in the antenna radiation field is generated when the reflector is illuminated by a linearly polarized field or, when circular polarization is employed, the antenna beam is squinted from the electrical boresight. This is of some concern with systems employing frequency reuse by orthogonal polarization.

Reflector antennae may provide *shaped beams* or *multiple beams* when the primary feed is composed of an array or a set of feeds located in the vicinity of the focal point (see Figure 6.39) (Rudge, 1975) (Matthews *et al.*, 1976).

If the primary feed consists of an array of feeds driven at same frequency by a beam forming network, the combination of the electric fields radiated by the individual feeds results in a *shaped beam*. For instance, an array of 88 feeds is used on the INTELSAT V satellite to produce the hemispheric and zone coverage of Figure 4.13.

Separate offset feeds generate distinct beams at a specific frequency and in a given direction, resulting in a multi-spot beam coverage, as illustrated in

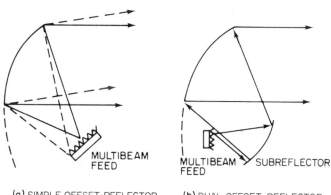

(a) SIMPLE OFFSET-REFLECTOR SYSTEM

(b) DUAL OFFSET-REFLECTOR GREGORIAN SYSTEM

Figure 6.39   Offset reflector antennae. *Reproduced with permission from Matthews* et al. (1976). Copyright © 1976 IEEE.

Figure 6.34. This technique can also provide a beam lattice coverage when the number of beams (and hence of feeds) is not too large. When the number of feeds increases, feeds tend to be located far away from the focal point which results in higher radiation pattern sidelobes. With a beam lattice comprising a large number of beams the number of feeds can be lowered by sharing feeds between beams: each beam at a specific frequency and in a given direction is then generated by proper split and phasing of the radiofrequency power between common feeds (Scott *et al.*, 1982).

Large communications satellites will tend to operate at several frequency bands and with different coverage zones. To avoid the installation of several antennae, each one with its own reflector, several primary feed arrays can share one reflector. However, this poses mechanical problems as many feeds are candidate to occupy the focal point region. Some flexibility is achieved by the use of *polarization* and/or *frequency sensitive surfaces* which allow orthogonally polarized waves, or waves at different frequencies, to be focused at distinct locations. Figure 6.40 shows an example of implementation where a frequency sensitive surface is used as the subreflector of a dual reflector

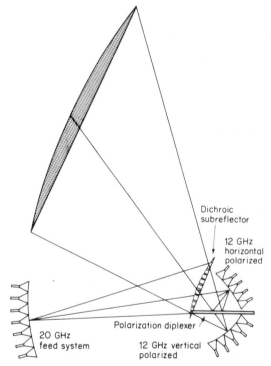

Figure 6.40   Dual frequency and polarization multibeam reflector antenna (Lopriore *et al.*, 1982).

system at 30/20 GHz and is transparent to waves at 14/12 GHz. The polarization sensitive surface separates the orthogonal polarized waves at 14/12 GHz (Lopriore *et al.*, 1982).

6.9.5.2.3 Lens antennae (Matthews *et al.*, 1976; Scott *et al.*, 1976)

The primary source consists of one or several radiating feed elements which are associated with a *lens* which focuses the radiated electromagnetic energy. These antennae have the advantage over reflector antennae of having the feed behind the aperture, which avoids blocking. This feature is of special interest for *multiple beam* antennae which require a bulky feed array and beam forming network. Lens antennae have the advantage over offset reflector antennae to be axisymmetric, hence offering large cross-polarization isolation.

The lens is composed of either:

(1) An *assembly of conducting wave guides* (zone lens) whose length is designed so as to produce the necessary phase advance to transform the incident spherical wave front into a plane wave front (Figure 6.41). These lenses are light but suffer from relatively small frequency bandwidths (around 5 per cent).
(2) *A homogeneous dielectric medium.* This gives a large frequency bandwidth but is heavier.
(3) An *assembly of transmission lines* ('TEM' or 'bootlace' lens). A ray-path through the lens is constrained to follow an RF transmission line interconnecting small pick-up and reradiating elements (Figure 6.42). Bandwidth is fairly large, and weight is between the previous two lenses.

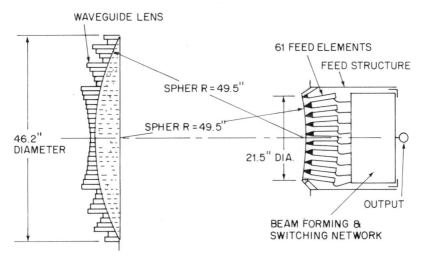

Figure 6.41 Waveguide lens multiple beam antenna (Scott *et al.*, 1976).

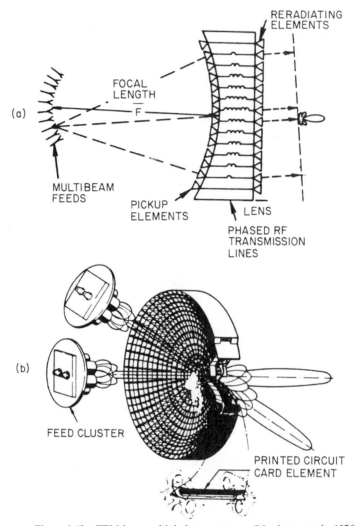

Figure 6.42   TEM lens multiple beam antenna (Matthews *et al.*, 1976).

6.9.5.2.4   *Phased array antennae* (Stark, 1974; Coirault, 1980)

A phased array antenna is a group of radiators which are spread out and fed from a single source and are therefore coherently excited. The phasing of the separate radiators is such that their contributions are additive on a plane wavefront at some chosen angle, which can be varied by changing the phasing (Figure 6.43). Re-use of frequency by polarization is also possible.

6.9.6   Conclusion

Ideally, the radiation pattern of a satellite antenna mainlobe should be such that energy is concentrated towards the earth station hence, maximizing the

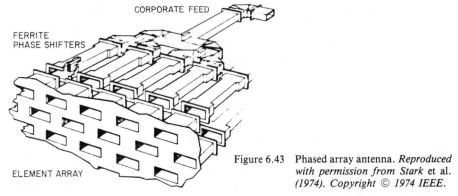

Figure 6.43   Phased array antenna. *Reproduced with permission from Stark et al. (1974). Copyright © 1974 IEEE.*

EIRP. But main lobe gain considerations lead towards the design of satellite antennae which may result in inefficient orbit utilization. Orbit utilization and interference reduction are conditioned by the *sidelobe radiation patterns*. Typical functions for the far side envelope can be found in the CCIR (Figure 6.44) (CCIR, Rep. 558–2, 1982).

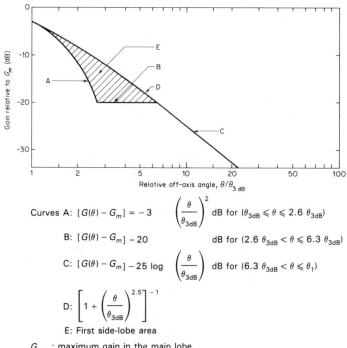

Curves A: $[G(\theta) - G_m] = -3 \left(\dfrac{\theta}{\theta_{3dB}}\right)^2$ dB for $(\theta_{3dB} \leqslant \theta \leqslant 2.6\ \theta_{3dB})$

B: $[G(\theta) - G_m] - 20$ dB for $(2.6\ \theta_{3dB} < \theta \leqslant 6.3\ \theta_{3dB})$

C: $[G(\theta) - G_m] - 25 \log \left(\dfrac{\theta}{\theta_{3dB}}\right)$ dB for $(6.3\ \theta_{3dB} < \theta \leqslant \theta_1)$

D: $\left[1 + \left(\dfrac{\theta}{\theta_{3dB}}\right)^{2.5}\right]^{-1}$

E: First side-lobe area

G_m : maximum gain in the main lobe

θ_{3dB} : 3 dB half beam width in the considered plane (3 dB below G_m)

θ_1 : value of θ when $G(\theta)$ in (C) is equal to -10 dB

Figure 6.44   Satellite radiation pattern envelope function. *Reproduced with permission from CCIR Report 558–2 (1980).*

For the foreseeable future, *multiple beam antenna* technology is likely to find increasing use as it allows for *frequency reuse* either by *spatial separation* and/or by orthogonal *polarization*. Multiple beam antennae provide the desired coverage either by *separate spot beams* or a *lattice of beams*, or *scanning spot beams* (Acampora *et al.*, 1979; Durrani and Keblawi, 1980; Reudnik and Yeh, 1980). Multibeam antenna technology requires the combination of many techniques such as reflectors with feed array (INTELSAT V (Nygren, 1980)) or lenses. High directivity implies large reflectors which must be *deployable*. These large reflectors for satellites (Powell, 1979; Foldes and Dienemann, 1980) can be made of fine mesh (mesh reflectors) so as not to offer too much resistance to solar radiation pressure (Figure 6.45).

Improved directivity and *pointing accuracy* go hand in hand. So in addition to the limitations associated with attitude control imprecision already mentioned, errors due to the unknown angular deviation between the electromagnetic and mechanical axes of the antennae, misalignment of the mechanical axes relative to the reference axes of the satellite, and thermal

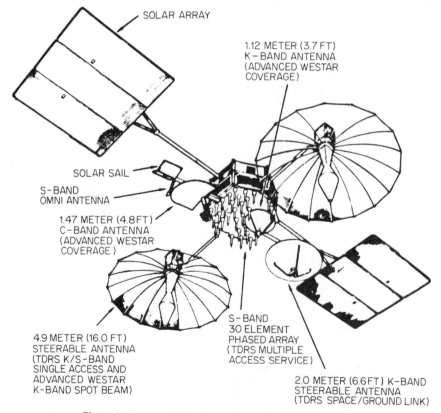

SOLAR ARRAY

1.12 METER (3.7 FT)
K-BAND ANTENNA
(ADVANCED WESTAR
COVERAGE)

SOLAR SAIL
S-BAND
OMNI ANTENNA

1.47 METER (4.8 FT)
C-BAND ANTENNA
(ADVANCED WESTAR
COVERAGE)

4.9 METER (16.0 FT)
STEERABLE ANTENNA
(TDRS K/S-BAND
SINGLE ACCESS AND
ADVANCED WESTAR
K-BAND SPOT BEAM)

S-BAND
30 ELEMENT
PHASED ARRAY
(TDRS MULTIPLE
ACCESS SERVICE)

2.0 METER (6.6 FT) K-BAND
STEERABLE ANTENNA
(TDRS SPACE/GROUND LINK)

Figure 6.45   Deployable antennae (Advanced Westar satellite).

deformation of the vehicle structure must be taken into account. Finally, the complexity of the satellite antennae causes many problems related to the mission analysis, interference reduction and losses, and in the development and manufacture of a communications satellite (English, 1980; Chang and Taormina, 1980).

6.10   REPEATER

A satellite repeater is the electronic assembly which ensures the following functions:

(1) *Amplification of signals* from an input power in the order of -100 dBW, to an output power of about 10 dBW (approximately 110 dB gain).
(2) *Frequency down conversion* which avoids interference between the powerful transmitted signal and the weak incoming signal.

These two basic functions refer to a *conventional satellite repeater*. A *regenerative repeater* also provides *on-board detection* of the received signal prior to baseband processing and remodulation for down-link transmission.

6.10.1   Repeater architecture

The design of a repeater is conditioned by the following considerations. The receiver allows for amplification and down conversion. Post-conversion frequency in the receiver is either the down-link frequency or an intermediate frequency (IF) depending on whether the repeater is of the *single* or *dual*

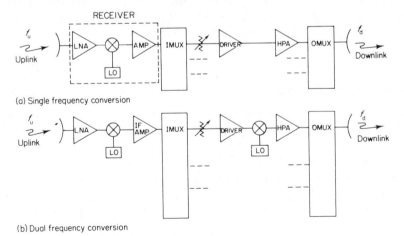

Figure 6.46   Repeater architecture: (a) single frequency conversion; (b) dual frequency conversion.

frequency conversion type (Figure 6.46). The down converter's high noise figure may make it necessary to place a low noise amplifier (LNA) before the down converter. Down-link signals are amplified by a high power amplifier (HPA) which provides the required output transmitted power. Efficient use of the HPA requires that it is driven in a non-linear part of the input/output characteristic. If the transponder is to amplify several carriers simultaneously, undesirable *intermodulation products* are generated. Moreover, off the shelf amplifying devices are power limited. So dividing the available bandwidth into several channels called *channelization*, is adopted. With multibeam satellites another problem to be considered is the beam interconnectivity.

Finally, some *redundancy* must be incorporated to achieve the required reliability. This is obtained by using redundant elements and switches (see Chapter 9).

6.10.1.1 Single and dual frequency conversion

The single frequency conversion repeater requires a local oscillator (LO) whose frequency is the difference between the received and transmitted frequencies (Figure 6.46a). For a 6/4 GHz commercial link, the local oscillator frequency is in the order of 2.2 GHz and at 14/12 GHz it is 1.5 GHz. Military satellites at 8/7 GHz have local oscillators at about 0.7 GHz. Most repeaters (for instance, INTELSAT IV, IVA and Telecom I) use single conversion.

When down-link frequency is high, the available technology may not provide amplifiers with sufficiently large bandwidth and required power gain. This constraint makes single frequency conversion impossible to implement. In that case *dual frequency conversion* is used (Figure 6.46(b)); the down and up conversions define an *intermediate frequency* (IF) which can be chosen independently of the up- and down-links. The intermediate frequency is from around a hundred MHz to a few GHz (4 GHz for INTELSAT V with a 14/12 GHz link).

6.10.1.2 Channelization

The *channelization* is ensured by a *set of filters* called *input multiplexer* (IMUX), and *output multiplexers* (OMUX). It offers two main advantages:

(1) The *number of carriers* per channel can be *reduced.* As non-linearity generates more intermodulation products if many carriers are amplified simultaneously, intermodulation noise can be reduced by assigning fewer carriers to each channel, possibly even only one carrier per channel.
(2) It enables the *required down-link power* level to be obtained on *each channel,* as each channel benefits from the maximum power available from

the present technology (Valentini *et al.*, 1979). Prior to transmission all channels are recombined together by the output multiplexer (OMUX), and it may be necessary to implement a variable attenuator on each channel, switchable by remote control, in order to equalize gains of the different channels.

Note that the term *transponder* is often associated with *one channel* of the *channelized satellite repeater*.

Channelization became a reality as weight, volume and power loss of the channel elements, e.g. filters, driver, attenuator, TWT, etc. were reduced. So, whereas INTELSAT II had only one 125 MHz bandwidth transponder and INTELSAT III had 2 (225 MHz bandwidth), INTELSAT IV has 12 (each of 36 MHz bandwidth), INTELSAT IVA has 20 (36 MHz bandwidth) INTELSAT V has 27 (bandwidths are 36–41–72–77–241 MHz), INTELSAT VI has 48 (bandwidths are 36–41–72–77–150 MHz).

Parallel channels create a problem in that a small amount of energy can interfere with channels other than the one being used. This causes a distortion of the output signal ('ACI' *adjacent channel interference*) caused by different paths having different group delays. The greater the spacing between channels, and as the filter response gets closer to the ideal bandpass filter shape, the lower this distortion becomes. The solution lies in optimizing between technological possibilities for filters (Kudsia *et al.*, 1980), and acceptable quality loss because of distortion in the link budget. Figure 6.46a shows an example of channelization where single frequency conversion is used. If *dual frequency conversion* is adopted the *up conversion* can be implemented at two different locations:

(1) In the *receiver*, that is simultaneously for all the channels.
(2) In the *transmitter* with a mixer for each channel (Figure 6.46b). The IMUX filters work at intermediate frequency which permits use of lighter and smaller coaxial TEM 1 resonators rather than waveguide resonators needed at microwave frequencies. The TWT driver stage also operates at intermediate frequency and can employ a bipolar transistor amplifier. But the mixer works at high level. In any event, the linearity and dynamics of the mixer and driver stages are not critical as their effect is only applicable for a reduced number of carriers. While more complex because of numerous up conversions, this implementation reduces intermodulation noise generation.

More flexibility in the design of the output multiplexer is obtained by separating the channels into two groups: the *odd* and *even channels* according to frequency allocation, which are recombined on two separate OMUX. Hence the guardband between adjacent channels on OMUX is larger.

6.10.1.3   Beam interconnectivity

Multiple beam satellites have channels that are allocated to the different pairs of beams for up-links and down-links. Hence, the communications payload has *several repeaters* that ensure the links between the specific zones covered by the antennae. The *connection* between *up-links* and *down-links* of the dedicated zones may be established one of three ways (in a fixed, semi-fixed or dynamic interconnection).

(1) *Fixed interconnection* is achieved by mere cable routing between repeaters and beams.
(2) *Semi-fixed interconnection* implies use of switches commanded from ground through the TTC sub-system, which modifies the receiving to transmitting beam interconnection scheme according to major traffic pattern changes during the satellite mission.

   Both fixed and semi-fixed schemes require one repeater for each connection. So as the number of beams increases, the number of repeaters required for complete interconnectivity of the network, which is equal to the square of the number of beams, becomes prohibitively large. The solution then is to implement a dynamic interconnection.
(3) The *dynamic interconnection* sequentially routes the traffic between one zone and another in real time by means of a switch matrix (Figure 3.33). This is called a *satellite switched (SS) multiple access* scheme. It is most often combined with time division multiple access to give a SS/TDMA system (see Section 3.6.3). One example of such a system is INTELSAT VI.

6.10.2   Main components

The overall performance of the communciation payload sub-system is closely related to the characteristics of the repeater components. For instance, the receiver figure of merit G/T of the satellite depends on the receiver input amplifier noise figure. The down-link frequency stability depends on that of the local oscillator. Out-of-band signals are generated by the repeater non-linearities and are sensitive to filter characteristics. Finally the EIRP of the satellite depends on the saturated power of the HPA and the loss of the output devices (filters, feeders, etc.).

6.10.2.1   Low noise amplifier (LNA)

Basically the low noise amplifier (preamplifier) conditions the *repeater noise figure*. High gain (5 to 20 dB) and low noise figure for the first stage and second stage are necessary to preserve the total noise figure.

Advances in technology have made tunnel diode amplifiers obsolete and field effect transistors are more and more replacing parametric amplifiers at increasingly higher frequencies. Noise figures of 3 dB at 6 GHz, 4 dB at 14 GHz, 8 dB at 30 GHz, are typical values obtained with field effect transistor amplifiers, and 3 dB at 14 GHz, 4 dB at 30 GHz, with parametric amplifiers.

6.10.2.2 Down converter

The down converter includes a mixer, filters and a local oscillator. Typically mixer conversion loss is 5 to 7 dB. *Long term frequency stability* of the local oscillator is 10^{-5}, and *short term frequency shift* is of the order of a few Hz r.m.s. within a 5 to 100 Hz bandwidth.

Mixers typically use Schottky diodes and local oscillators commonly have quartz crystals combined with multipliers.

6.10.2.3 Post conversion amplifiers

After the frequency conversion *high gain amplification* is necessary to bring the signals to a high enough level for the input of the transmitter power stages. Depending on the frequency, this amplification is by bipolar transistors (4 GHz), FETs (4 GH and 12 GHz), or TWTs. As the level of the signal increases, non-linear effects tend to become important.

6.10.2.4 Input and output multiplexers

These devices determine the input and the output of the *channelized part* of the repeater. They make use of high Q's bandpass filters which impose the *transponder bandwidth*.

The *input multiplexer* (IMUX) splits into separate channels the total bandwidth. Typical implementation associate *circulators and a set of bandpass filters*. Figure 6.47 shows an example where channels are split into two sets of odd and even channels. The IMUX is then divided into two parts, each one sharing the receiver output power through an hybrid. The second input port of the hybrid is often used, as indicated in Figure 6.47, to provide an alternative path to the IMUX for the wideband signal when a redundant receiver is installed, avoiding the need for a switch at the output of the receivers.

The *output multiplexer* (OMUX) combines the channels after power amplification. Severe requirements are imposed concerning the insertion loss of the OMUX as any power loss reduces the satellite EIRP and generates heat. As circulators are bulky and suffer from losses, they are replaced in typical implementations by a short-circuited *waveguide manifold* which connects the bandpass filters, as shown in Figure 6.48. However, the interaction between

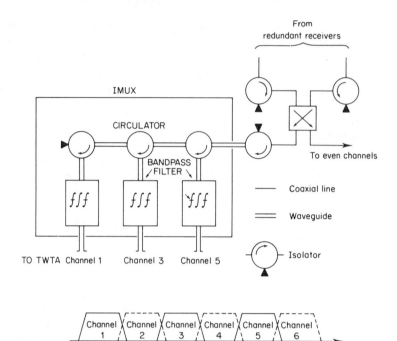

Figure 6.47   Implementation of an input multiplexer (IMUX) using circulators
to interconnect the filters. Isolators limit the standing wave ratio
(SWR) on the lines which arise from load mismatch at
terminations.

filters and manifold leads to a more complicated design and optimization than
with circulators.

The filters implemented in the IMUX and the OMUX must fulfil severe
requirements on *amplitude and group delay variations versus* frequency (*filter
mask*) such as: small *in-band amplitude* peak-to-peak and *slope variations* to
avoid signal distortion, specially at the IMUX as it feeds a non-linear power
amplifier; high *out-of-band attenuation* to avoid channel overlapping; constant *group delay* within the bandwidth to avoid signal distortion. These
requirements imply multipole Chebyshev or elliptic filters (Figure 6.49)
possibly associated with group delay equalizers. Figure 6.50 shows a typical
realization of filters based on waveguide cavities coupled by irises. An important feature is the *mechanical stability* to limit relative frequency response drifts
caused by ageing and thermal deformation, to typically less than $2.5 \ 10^{-4}$ over
the seven to ten years of satellite lifetime. Candidate materials are: *aluminium*
with a thermal expansion factor of $22.4 \ 10^{-6}/°C$, *Invar* (36% steel, 64%
nickel) with a thermal expansion factor of $1.6 \ 10^{-6}/°C$, *graphite fibre reinforced plastic* with a thermal expansion factor of $-1 \ 10^{-6}/°C$. Although

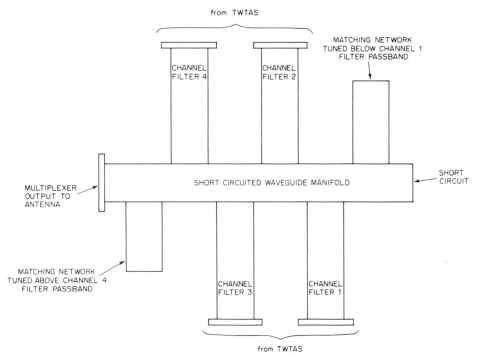

Figure 6.48   Implementation of an output multiplexer (OMUX) using a waveguide manifold to interconnect the filters.

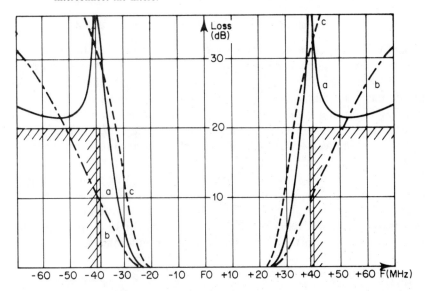

Figure 6.49   Comparison of insertion loss versus frequency with four-pole elliptic filters (a) and four-pole (b) or six-pole (c) Chebyshev filters.

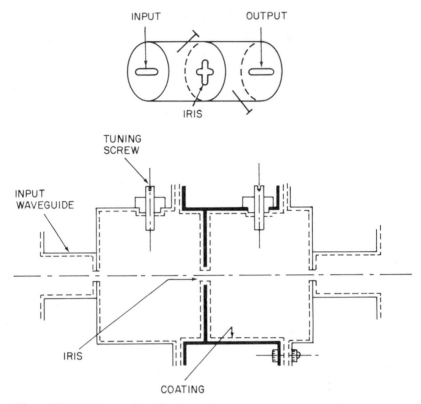

Figure 6.50   Implementation of a four-pole dual mode elliptic filter with two cavities.

aluminium shows a large value of the thermal expansion factor its specific gravity is low: 2.7. Invar has a specific gravity of 8.05, but its stiffness allows for thin-wall manufacturing to reduce weight. The graphite fibre reinforced plastic has a specific gravity as low as 1.6 and a stiffness twice that of steel, but manufacturing is delicate. Silver or gold coating provides high electrical and finish of the surface inside the cavities (Kudsia *et al.*, 1980).

As an example, INTELSAT VI has 10-section coaxial resonator input multiplexer filters with cross coupling to achieve group delay equalization. Each filter is about $7 \times 10 \times 3$ cm. The 50 bandpass, when integrated, amount to a total mass of 28.3 kg. Contiguous output multiplexers are used at both 4 and 11 GHz with triple cavity dual mode cylindrical resonators mounted on the output waveguide manifold. The total mass of the 10 output multiplexers, with a total of 48 ports, is 26.1 kg (Thompson and Johnston, 1984).

6.10.2.5   Switch matrix

With multibeam satellites *switches* are required to modify the beam interconnectivity (see Chapter 3.6). Switches operated through ground commands allow for a semi-fixed interconnection scheme (changed every few months or so). When SS/TDMA is involved the high rate of switch reconfiguration (several times within a millisecond) implies fast acting switch matrix. The switch matrix sequentially interconnects receivers associated with the up-link beams to the down-link beam transmitters. Rapid switching implies: (1) *solid state switches* (2) *on-board control* of the switch state sequence by a distribution control unit (DCU).

Former switching matrices used PIN diodes as switching devices. The trend is now to replace PIN diodes by dual gate field effect transistors which offer comparable isolation but a lower switching time (0.1 to 1 ns instead of 10 to 100 ns).

Of all the architectures which are candidates for a $N \times N$ switch matrix, where N is the number of beams, only two are of interest if the information on one up-link beam is to be distributed to several down-link beams (*broadcast mode*). These are: the *power divider-combiner architecture*, and the *coupler cross-bar architecture* (Figure 6.51). The power divider-combiner switch matrix entails N input N-way dividers and N output N-way combiners. Any output of a divider is connected to one output of a combiner via a switch. The coupler cross-bar switch matrix has N input lines and N output lines with cross-point consisting of directional couplers and one switching element placed between the two couplers. In order to equally distribute the power among the N cross-point on the same input row, couplers must have coupling coefficients of $1/(N+1)$, $1/N$, . . ., $1/2$, in the 1st, 2nd, . . ., Nth cross-points. Table 6.6 compares the performance of several switch matrix implementations.

6.10.2.6 Channel power amplifier

The amplifier in the channelized part of a satellite repeater typically comprises one or several stages of *power amplification (driver)*, and the *output stage*.

For a satellite repeater the EIRP requirement constrains the output stage power. Power amplifiers are non-linear and, as discussed in Section 3.4.3.3, non-linear operation generates undesirable signals called *intermodulation products*. Figure 6.52 shows an example of intermodulation products generated when a non-linear amplifier is driven by two equal power carriers at frequencies f_1 and f_2. Only *third order* products at frequencies $2f_2 - f_1$ and $2f_1 - f_2$, and *fifth order* products, i.e. all combinations such as $3f_2 - 2f_1$, are significant.

Wideband operation with a large number of carriers implies a satellite repeater as *linear* as possible, or else intermodulation noise is generated and this reduces the overall carrier-to-noise power ratio at the earth station receiver front end. The required overall carrier-to-noise power ratio imposes a *minimum value of the carrier-to-intermodulation noise power ratio* $(C/N)_{IM}$.

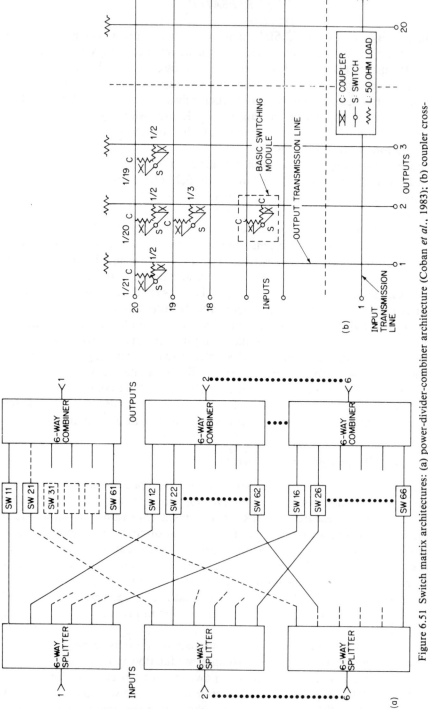

Figure 6.51 Switch matrix architectures: (a) power-divider-combiner architecture (Coban et al., 1983); (b) coupler cross-bar architecture (Ho et al., 1982).

Table 6.6 PERFORMANCE OF SEVERAL SWITCH MATRIX IMPLEMENTATIONS.

	Comsat[a]	Thomson[b]	General Electric[c]	Ford[c]	NTT[d]
Array size	8 × 8	8 × 8	20 × 20	20 × 20	4 × 4
Type of architecture	Divider combiner	Divider combiner	Coupler crossbar	Coupler crossbar	Divider combiner
Switching device	Pin diode	Pin diode	Dual gate FET	Dual gate FET	Dual gate FET
Bandwidth	3.5–6.5 GHz	3.7–4.2 GHz	6–7 GHz	3.5–6 GHz	1.8 GHz ± 140 MHz
Switching time	$\leqq 60$ ns	$\leqq 50$ ns	25 ns	15 ns	$\leqq 100$ ns
Insertion loss	$\leqq 23$ dB	$\leqq 28$ dB	18 dB	20.7 dB	$\leqq 17$ dB
Ins. loss variation (any 500 MHz)	$\leqq 1$ dB	$\leqq 1$ dB	1 dB	1 dB	—
Path to path insert. loss scatter	$\leqq 1.7$ dB	$\leqq 1.5$ dB	—	—	$\leqq 3$ dB
Path isolation	$\geqq 50$ dB	$\geqq 50$ dB	$\gg 50$ dB	$\geqq 45$ dB	$\geqq 53$ dB
Intermodulation $(C/N)_{IM}$	$\geqq 45$ dB	$\geqq 45$ dB	—	—	—
Group delay variation	$\leqq 0.5$ ns	$\leqq 1$ ns	—	$\leqq 0.5$ ns	—
Switch matrix mass	2.95 kg	2.3 kg	11 kg	8.7 kg	—
Switch matrix size (cm)	15 × 16 × 11	10.5 × 12 × 12	47 × 48 × 17	47 × 48 × 8.5	27 × 20 × 15
Power consumption	<7.5 W	8.5 W	33 W	5.7 W	1.7 W

[a] Assal *et al.* 1982. [b] Rozec and Assal, 1976. [c] Spisz, 1983. [d] Kato *et al.*, 1983.

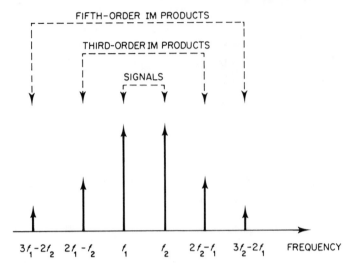

Figure 6.52   Intermodulation products generated in a non-linear device driven by two equal power carriers at frequencies f_1 and f_2.

Radiofrequency to direct-current (RF/DC) power *efficiency* of the output stage must be high to reduce thermal dissipation and electric consumption, so the output stage is operated near saturation where efficiency is maximum, but in a highly non-linear region. This leads to some value of $(C/N)_{IM}$ assuming that negligible intermodulation noise is generated within the previous amplifier stages, and specifically in the non-channelized wideband parts of the repeater.

A low value of the intermodulation noise power is obtained when the carrier *power level* at the output of the satellite post-converstion *wideband amplifier* is limited to a *maximum value*, depending on technology, so that it operates in a sufficiently linear region. The *remaining power amplification* gain from the level of power at the output of the post-conversion wideband amplifier to the power level required to drive the output stage must then take place in the *channelized part* (the output stage has typically a gain of about 55 dB with technology in common use and hence the required output power determines the input drive power). Indeed, operation at a higher power level is possible in the channelized part while generating little intermodulation noise power due to the fact that the number of carriers per channel is small.

6.10.2.6.1   Driver

The driver is a *narrowband amplifier* which amplifies signals at the IMUX output to the power level required to drive the output stage. The above considerations serve as guidelines for the design of the driver (power gain and number of stages) within the channelized part. The driver may comprise a step-by-step variable attenuator, switchable by remote control in order to equalize gains from one channel to another. Drivers are typically implemented using solid state components (Field effect or bipolar transistors).

6.10.2.6.2 Output stage

Figure 6.53 shows typical *output power, gain* and *phase shift* curves in terms of *input power* when the amplifier is driven by a *single carrier*. The variations in phase shift of the output carrier when the input power varies turns into *phase modulation* when the input carrier is amplitude modulated. This is quantified by the AM/PM *conversion coefficient* K_p, expressed as:

$$K_p = \Delta\phi/\Delta P_i (^\circ/\mathrm{dB})$$

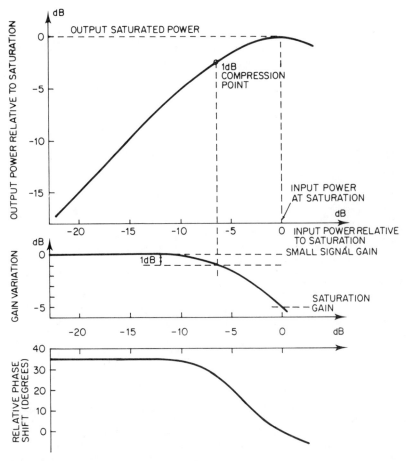

Figure 6.53 Travelling wave tube output power, gain and phase shift versus input power.

where $\Delta\phi$ is the phase shift variation expressed in degrees, and ΔP_i is the relative variation of the input power, expressed in dB.

6.10.2.6.3 Multi-carrier operation

Figure 6.54 shows typical relative output power curves in terms of relative input power when the amplifier is driven by two equal carriers. Curves represent the *output power* of *one out of the two carriers* and of *one out of the two third order intermodulation products* versus the power of *one of the two input carriers*. The *third order interception point* is the intersection of the linear parts of the carrier output power curve with that of the intermodulation products curve. The higher the interception point, the more linear the device is.

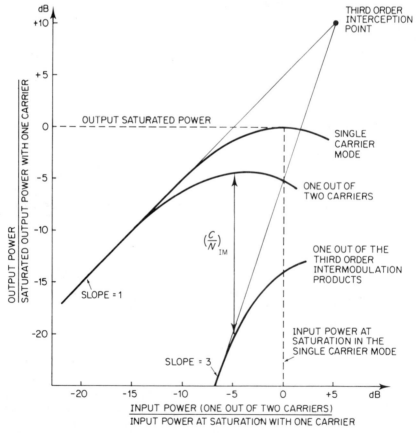

Figure 6.54 TWT output power of one out of two carriers and of one third-order intermodulation products versus input power.

With *multi-carrier operation*, phase variations of one of the output carriers at constant input level appears when the input power level of the other carrier varies, causing *phase modulation* on this carrier as a result of amplitude modulation of the other carrier. This is quantitied by the AM/PM *transfer coefficient* K_T, expressed in $°/dB$.

It should be noted that predistortion techniques are available to linearize power amplifiers (Bremenson and Jaubert, 1974; Czech, 1984; Gibson *et al.*, 1984).

6.10.2.6.4   Output stage technology

Nowadays output stages are implemented mostly with travelling-wave tubes (TWT) but solid state power amplifier (SSPA) technology is emerging. (Freeling and Weinrich, 1984; Gibson *et al.*, 1984).

A TWT works by the *interaction* between a *beam of electrons* and a *radio wave*. Figure 6.55 illustrates the construction of a TWT. The electron beam is generated by an heated cathode, focused by the shadow grid, accelerated by an anode and kept focused within the helix by permanent magnets. The RF wave travels along the helix so that its axial velocity is smaller than that of the electron beam, allowing for energy transfer from the kinetic energy of the electrons to the RF wave. The multistage voltage depressed collector increases the TWT efficiency by recovering residual energy from the electron beam.

Typical values for satellite TWTs are:

Output saturated power: from 8.5 to 20 W (150 to 250 W for direct broadcasting applications)
Saturated gain: about 55 dB
$(C/N)_{IM}$ at saturation: 10 to 12 dB
AM/PM conversion coefficient K_p: about $4.5°/dB$

An electronic power conditioner (EPC) generates the high voltage (up to 4000 V) required for operation of the TWT. The overall efficiency is about 32 per cent (TWT:40 per cent, EPC = 80 per cent) and the total mass is about 2.2 kg (TWT:0.7 kg, EPC:1.5 kg).

SSPAs make use of GaAs FET devices. Typical values at the moment are:

Output power: nearly 10 W at C-band
$(C/N)_{IM}$ at saturation: 14 to 18 dB
AM/PM transfer coefficient K_p: about $2°/dB$

The EPC associated to the SSPA converts the variable spacecraft bus voltage to a regulated set of bias voltages (about 10 V) and comprises a temperature compensation network. The overall efficiency is about 28 per cent (SSPA:33

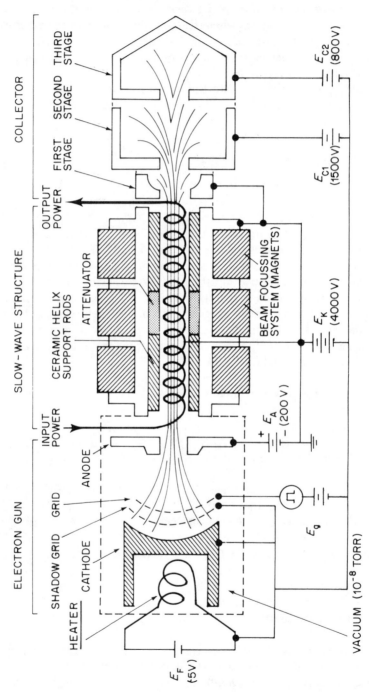

Figure 6.55 Construction of a travelling wave tube (TWT).

per cent, EPC:87 per cent) and the total mass is about 900 g (SSPA:400 g, EPC:500 g).

6.10.2.6.5 *TWTA versus SSPA*

Table 6.7 establishes a comparison of a TWTA and a SSPA with 8.5 W saturated output power operating in the 4 GHz frequency range (La Prade, 1983). At higher values of frequency only TWTs are presently used for satellite communications.

SSPA is *lighter in weight*, *more linear* and offers a significant *improvement in reliability*. Improved linearity translates into *increased transponder capacity* (Freeling and Weinrich, 1984).

6.10.3 Regenerative repeater

On-board regeneration in a digital satellite communications system is an attractive and rapidly developing option for future SS/TDMA systems. An SS/TDMA system with on-board regeneration performs the required *switching at baseband* instead of microwave frequencies. A simplified block diagram of a regenerative repeater is shown in Figure 6.56. Compared with a conventional one a regenerative repeater has in addition a demodulator, remodulator, and a baseband switch matrix replaces the microwave switch matrix.

As coherent demodulators have complex implementations, and as spacecraft hardware must be reliable and lightweight with minimum consumption, a *differentially coherent demodulator* appears to be desirable, even though it does not perform as well. The penality is about 2.5 dB. The differentially coherent demodulator is based on correlating the present RF waveform with the previous one delayed by one symbol time. This approach eliminates the need for carrier recovery, but necessitates a temperature-stable

Table 6.7 PERFORMANCE COMPARISON OF THE SSPA VERSUS THE TWTA.

	TWTA	SSPA
Operating frequency range	3.7–4.2 GHz	3.7–4.2 GHz
Saturated power output	8.5 W	8.5 W
Gain at saturation	58 dB	58 dB
Third order intermodulation product relative level $(C/N)_{\text{IM}}$[a]	11 dB	15 dB
AM/PM conversion coefficient K_{p}	$4.5°/\text{dB}$	$2°/\text{dB}$
DC to RF efficiency including EPC[b]	32%	28%
Mass including EPC	2.2 kg	0.9 kg
Failure in 10^9 hours	>2000	<500

[a] Close to saturation.
[b] EPC: Electric power conditioning.

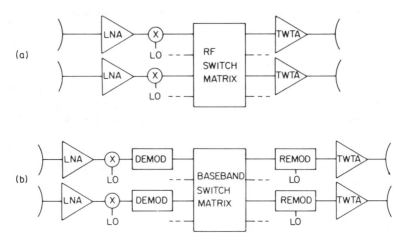

Figure 6.56   Architecture of a conventional multibeam satellite (a) compared with that of a regenerative one (b).

RF time delay. Several delay line technologies have been developed—MIC filters on fused silica (Lee, 1978; Childs et al., 1981), transmission lines on high dielectric substrates (Lee and Childs, 1979) or in waveguide techniques (Ohm, 1981). However, coherent demodulator implementations have also been reported (Stilwell, 1982; Reisenfeld, 1982; Izumisawa et al., 1984).

The *modulator* should be as simple as possible. Modulation can be performed directly at the transmit frequency (Koga et al., 1977; Ohm and Alberty, 1982; Ananasso et al., 1984) or at a lower frequency which then implies a subsequent up-converter (Stilwell, 1982).

Baseband switching can be accomplished by storing the incoming data from each up-link channel into an input data memory and routing the data to an output memory associated with the corresponding down-link channel by means of a non-blocking switch. An alternative scheme is to switch the incoming data into a single memory whose output is connected to each down-link channel by means of another switch. A baseband switch works with specified baseband frames. Each frame time data is written into one data memory and at the same time the previous frame of data is read out of another data memory. An example of implementation is given by Moat et al. (1982). Some *retiming circuits* may be necessary. Simple retiming circuits employ elastic buffers (Inukai et al., 1983). Baseband switching can be combined with *rate conversion* (Moat et al., 1982; Inukai et al., 1983).

Finally, to combat rain attenuation, *coding* may be employed. An LSI convolutional on-board decoder has been developed and implemented for applications at 30/20 GHz (Clark and McCallister, 1982).

REFERENCES

ABUTALEB, G. *et al.* (1977) The Comstar satellite system, *Comsat Tech. Rev.*, **7**(1), 35–83.

ACAMPORA, A., C. DRAGONE and D. O. REUDNIK (1979) A satellite system with limited scan spot beams, *IEEE Trans. on Comm.*, **27**(10), pp. 1406–1415.

ANANASSO, F., L. LO PRESTI, M. PENT and E. SAGGESE (1984) Simulation-aided design of the ITALSAT satellite regenerative transmission channel, *AIAA, 10th CSSC*, Orlando, Paper 84-0709.

ARNAUD, C., C. DERIEUX and A. POUZET (1980) French satellite broadcasting system, *AIAA, 8th CSSC, Orlando*, Paper 80–0571, 8pp.

ARUGA, T., and T. IGARASHI (1977) A new method for three axis attitude of determination of spacecraft using laser, *IEEE Trans.* **AES 13**(5).

ASSAL, F., R. GUPTA, J. APPLE, and A. LOPATIN (1982) Satellite switching center for SS/TDMA communications, *COMSAT Tech. Rev.*, **12**(1), (Spring) pp. 29–68.

BANGET, J. T. *et al.* (1963) The spacecraft antennas, *Bell System Technical Journal*, July.

BASSNER, H. (1979) Development status and application of the electric propulsion system RIT 10 used for station keeping, *XXXth IAF Congress*, Munich, Paper 79-07, 21pp.

BOURCIER, H., F. DESVIGNES and F. X. DOITTEAU (1980) Les senseurs de terre du satellite Symphonie et leurs développements ulterieurs, *Symphonie Symposium*, Berlin, pp. 867–882.

BREMENSON, C. and J. JAUBERT (1974) Réseau linéariseur pour tube à ondes progressives, *Revue technique THOMSON-CSF*, **6**(2), pp. 529–548.

BRODERSEN, H. and I. RIZOS (1978) Intelsat V solar array, *1st European Symposium Photovoltaic Generators in Space*, Noordwijk, pp. 209–217.

BUGEAT, M. and D. VALENTIAN (1982) Development status of a Cesium field emission thruster, *Acta Astronautica*, **9**(9).

CARIOU, A. (1980) Les mesures en obitographie, *Le mouvement du véhicule spatial en orbite, Cours de Technologie spatiale, CNES*, pp. 179–200.

CARTON, C. (1980) Le contrôle thermique de Symphonie, *Symphonie Symposium*, Berlin, pp. 883–900.

CCIR (1982) Report 453–3, Vol. IV-1, Geneva.

CCIR (1982) Report 676, Vol II. Geneva.

CCIR (1982) Report 558–2 Vol IV-1, Geneva.

CCIR (1982) Report 843 Station keeping techniques for geostationary satellites, Vol II, Geneva.

CHANG, D. C. D. and F. A. TAORMINA (1980) Interference cancellation in multiple spot beam satellite antenna designs, *Compte rendu du symposium international Antennes et Propagation*, Université Laval, Canada.

CHANTERANNE, M. (1979) Use of ultra-light adhesive for the metal honeycomb bonding, *Proceedings of an ESA Symposium on Spacecraft Materials, ESTEC*, pp. 287–291.

CHILDS, W. H., P. A. CARLTON, R. EGRI, C. E. MAHLE and A. E. WILLIAMS (1981) A 14 GHz regenerative receiver for spacecraft application, *5th Int. Conf. on Digital Satellite Comm.*, Genoa, pp. 453–459.

CLARK, R. T. and R. D. McCALLISTER (1982) Development of an LSI maximum likelihood convolutional decoder for advanced forward error correction capability on the NASA 30/20 GHz program, *AIAA, 9th CSSC*, San Diego, Paper 82-0459.

CLARRICOATS, J. B. and G. T. POULTON (1977) High efficiency microwave reflector antennas. A review, *Proc. IEEE*, **65**, (10).

COBAN, E., J. WISNIEWSKI, J. PELOSE, E. BALDERRAMA, N. CHIANG and P. HO (1983) High speed wideband 20 × 20 microwave switch matrix, *ICC83*, Boston, pp. B1.6.1–B1.6.6.

COIRAULT, R. (1980) L band phased array transmitter for communications satellites, *AIAA, 8th CSSC*, Orlando, Paper 80-0556 (6pp.).

COLETTE, C. and B. L. HERDAN (1977) Design problem of spacecraft for communications missions, *Proc. IEEE*, Mar., pp. 342–356.

CZECH, J. (1984) A linearized 4 GHz wideband FET power amplifier for communications satellites, *AIAA, 10th CSSC*, Orlando, Paper 84-0766.

DEBAY, P. *et al.* (1979) Symphonie in flight performance of TM TC, power supply thermal control, and transponder subsystems, *Colloque Espace, télécommunications spatiales*, Toulouse, pp. 715–732.

DONNELLY, E. E. *et al.* (1969) The design of a mechanically despun antenna for Intelsat III communications satellite, *IEEE Trans. Antennas Propagation*, **AP 17**, pp. 407–414.

DRABOWITCH, S. and M. MORLON (1967) Contribution à la technique des antennes contrarotatives, *Signal Processing Arrays, AGARD Conf. Proc.*, No 16 WT Blackband.

DUCHON, P. and M. P. VERMANDE (1980) La mesure d'attitude, *Cours de technologie spatiale*, CNES, pp. 315–373.

DUNLOP, J. D. and J. F. STOCKEL (1980) Nickel hydrogen battery technology. Development and status, *Comsat Tech. Rev.*, **10**(2).

DURRANI, S. H. and F. S. KEBLAWI (1980) Communications satellite system concept band on the AMPA experiment, *J. Spacecraft*, **17**(1), pp. 15–18.

ENGLISH, N. J. (1980) Improving future communications satellite antenna designs, *AIAA, 8th CSSC*, Orlando, Paper 80-554, 17pp.

ESA (1980) Ranging standard, *ESA TTC* A-04(1) (July).

FEARN, D. G. (1977) A review of the UKT 5 electron bombardment mercury ion thruster, *ESTEC Conference on Attitude and Orbit Control System*, Noordwijk, Holland, 18pp.

FEARN, D. G. (1978) Cyclic life tests of T5 thruster hollow cathodes, *AIAA, 13th Int. Electric Propulsion Conference*, San Diego, Paper 78-108.

FOLDES, P. and M. W. DIENEMANN (1980) Large multibeam antennas for space, *J. Spacecraft*, **17**(4), pp. 363–371.

FREE, B. A. (1972) Chemical and electric propulsion tradeoffs for communications satellites, *Comsat Tech. Rev.*, **2**(1) pp. 123–145.

FREE, B. A., W. J. GUMAN, B. G. HERRON, and S. ZAFRAN (1978) Electric propulsion for communications satellites, *AIAA, 7th CSSC*, San Diego, pp. 746–758.

FREELING, M. R. and A. W. WEINRICH (1984) Advanced Satcom: the first all-solid-state communications satellite, *AIAA, 10th CSSC*, Orlando, Paper 84-0715.

FRITZ, D. E., R. L. SACKHEIM, and H. MACKLIS (1983) Trends in propulsion systems for communications satellites, *Space Communications and Broadcasting*, **1**(2) (July) pp. 173–188.

GIBSON, M., D. MADDEN and P. MONIER (1984) Microwave technology developments in the European Space Agency *AIAA, 10th CSSC*, Orlando, Paper 84-0744.

GIRAUBIT, J. N. (1979) Application des matériaux composites aux structures spatiales, *AAAF, Note technique 79-45*, 43pp.

GIRAUBIT, J. N., G. MOUILHAYRAT and G. BARKATS (1979) Performances d'utilisation du generateur solaire rigide allége en fibre de carbon, *Colloque Espace Télécommunications spatiales*, Toulouse, pp. 553–577.

GREENSITE, A. L. (1970) *Control Theory, Vol. II, Analysis and Design of a Space Vehicle Flight Control System*, New York: Spartan Books.

HATCH, G. W. (1969) Communications subsystem design trends for the DSC program, *IEEE Trans. on Aerospace and Elec. Systems*, AES 5, (5).

HO, P., J. WISNIEWSKI, J. PELOSE and H. PERASSO (1982) Dynamic switch matrix for the TDMA satellite switching system, *AIAA, 9th CSSC*, San Diego, Paper 82-0458.

HSING, J. C., A. RAMOS and M. BARETT (1978) Gyro-board attitude reference systems for communications satellites, *AIAA, 7th CSSC*, San Diego, Paper 78-568.

ICHIKAWA, Y. (1978) The results of initial checkup of Japanese broadcasting satellite for experimental purpose, *IEEE Trans. Broadcasting*, Vol. BC 24, No. 4, pp. 74–80.

INUKAI, T., S. J. CAMPANELLA and T. DOBYNS (1983) On board baseband processing: rate conversion, *6th Int. Conf. on Digital Satellite Communications*, Phoenix, pp. XI-24–XI-31.

IZUMISAWA, T., S. KATO and T. KOHRI (1984) Regenerative SCPC satellite communications systems, *AIAA, 10th CSSC* Orlando, Paper 84-0708.

JILG, E. T. (1972) The Intelsat IV spacecraft, *Comsat Tech. Rev.*, 2(2), pp. 271–369.

KAPLAN, M. H. *et al.* (1978) A nuclear-powered communications satellite for the 1980's, *AIAA, 7th CSSC*, San Diego, Paper 78-625.

KATO, H., S. OKASAKA and K. KONDOH (1983) Multibeam satellite communication system for Japanese domestic communications, *SCC83*, Ottawa, pp. 22.2.1–22.2.4.

KOELLE, D. E. and H. V. BASEWITZ (1976) Ultralight solar array (ULP) for future communications satellites, *AIAA/CASI, 6th CSSC*, Montreal, Canada.

KOGA, K., T. MURATANI and A. OGAWA (1977) Onboard regenerative repeaters applied to digital satellite communications, *Proc. IEEE*, 65(3), pp. 401–410.

KUDSIA, C. M., K. R. AINSWORTH, and M. V. O'DONOVAN (1980) Microwave filters and multiplexing networks for communications satellites in the 1980's, *AIAA, 8th CSSC*, Orlando, 1980, Paper 80-0522.

LACOMBE, J. L. and R. HAVAS (1978) Systèmes de contrôle d'attitude et d'orbite de satellites: évolution jusqu'aux années 80-90, *L'Aeronautique et l'Astronautique*, No. 69, pp. 33–56.

LAZENNEC, H. (1966) *Pilotage des missiles et véhicules spatiaux*, Dunod.

LEE, Y. S. (1978) 14 GHz MIC 16 ns delay filter for differentially coherent QPSK regenerative repeater, *IEEE Int. Microwave Symposium Digest*, pp. 37–40.

LEE, Y. S. and W. H. CHILDS (1979) Temperature compensated $BaTi_4O_9$ microstrip delay line, *IEEE Int. Microwave Symposium Digest*, pp. 419–421.

LEGENDRE, P. (1980) Maintien à poste des satellites géostationnaires II Stratégie des corrections d'orbite, *Le mouvement du véhicle spatial en orbite, Cours de Technologie spatiale*, CNES, pp. 611–625.

LOEB, H. W. *et al.* (1979) European electric propulsion activities. *AIAA, 14th International Electric Propulsion Conference*, Princeton, Paper 79-2120.

LOPRIORE, M., A. SAITTO and G. K. SMITH (1982) A unifying concept for future fixed satellite service payloads for Europe, *ESA Journal*, 6(4) pp. 371–396.

MATTHEWS, E. W., W. G. SCOTT and C. C. HAN (1976) Advances in multibeam satellite antenna technology, *Record of the IEEE Eascon*, pp. 132A–132D.

MATTHEWS, E. W., L. F. BROKISH and G. F. WILL (1978) The communications antenna system on the Japanese experimental communications satellite, *AIAA, 7th CSSC*, San Diego, Paper 78-584, pp. 376–382.

MCKINNEY, H. N. and D. C. BRIGGS (1978) Electrical power subsystem for the

Intelsat V satellite, *13th Intersociety Energy Conversion Engineering Conference*, San Diego, pp. 47–53.

MITCHELL, D. H. and M. N. HUBERMAN (1976) Ion propulsion for communications satellites, *AIAA, 6th CSSC*, Montreal, Paper 76-290.

MOAT, D., D. SABOURIN, G. STILWELL, R. MCCALLISTER and M. BOROTA (1982) Baseband processor development for the Advanced Communications Satellite program, *NTC'82*, Galveston, pp. A2.4.1–A2.4.4.

MOBLEY, F. L. (1968) Gravity gradient stabilization result from the Dodge satellite, *AIAA, Paper 68-460*, San Francisco, California.

DE MONTLIVAULT, J. L. and R. HAGENBUCHER (1982) Possible configuration of space communication systems at 20/30 GHz in Europe, *33rd International Astronautical Congress*, Paris, Paper IAF 82-62.

MOSELEY, V. A. (1984) Bipropellant propulsion systems for medium class satellites, *AIAA, 10th CSSC*, Orlando, Paper 84-0726-CP (4 pp).

NYGREN, E. C. (1980) Shaped beam frequency reuse feed arrays for offset fed reflectors, *AIAA, 8th CSSC*, Orlando, Paper 80-0558 (18pp.).

OHM, G. (1981) Experimental 14-11 GHz regenerative repeater for communication satellites, *5th Int. Conf. on Digital Satellite Communications*, Genoa, pp. 445–451.

OHM, G. and M. ALBERTY (1982) 11 GHz QPSK modulator for regenerative satellite repeater, *IEEE Transactions on Microwave Theory and Techniques*, **30**, (11), pp. 1921–1926.

OWENS, J. (1976) Intelsat satellite on board propulsion systems past and future, *AIAA/CASI, 6th CSSC*, Montreal, *AIAA, Paper 76–289*.

PFEFFER, VIELLARD (1970) The Franco-German telecommunications satellite symphonie, *AIAA, 3rd CSSC*, Paper 70-46. Los Angeles.

PFEFFER, H. A., E. SLACHMUYLDERS and C. ROSETTI (1976) The future of European electric propulsion by ESA, *ESA Scientific and Technical Review*, Vol. 2, pp. 256–267.

POWELL, R. V. (1979) A future for large space antennas, *AIAA, 7th CSSC*, San Diego, Paper 78-588.

LA PRADE, J. N. (1983) A solid-state C-band power amplifier for communications satellites, *National Telemetry Conference*, San Diego, pp. 755–763.

QUAGLIONE, G. (1980) Evolution of the Intelsat system from Intelsat IV to Intelsat V, *J. Spacecraft*, **17**(2), pp. 67–74.

RANKIN, J. B. *et al.* (1969) Multifunction single package antenna system for spin-stabilized near synchronous satellite, *IEEE Trans. on antenna and Propagation*, **17**, 435–442.

REISENFELD, S. (1982) Onboard processing for 30/20 GHz communications satellite, *ICC'82*, Philadelphia, pp. 5E.3.1–5E.3.4.

REUDNIK, D. O. and Y. S. YEH (1980) Rapid scan area coverage communications satellite, *J. Spacecraft*, **17**(1), pp. 9–13.

RICARDI, L. J. (1977) Communications satellite antennas, *Proc. IEEE*, **65**(3), pp. 356–369.

RINTALA, W. M. *et al.* (1979) Intelsat V power control electronics system, *IEEE, 4th IECE Conf.*, Boston, pp. 1346–1349.

ROLFO, A. (1981) Le contrôle thermique, *Cours de technologie spatiale*, CNES.

ROSENTHAL, M. *et al.* (1969) VHF antenna systems for spin stabilized satellites, *IEEE Trans. on Ant.*, **AP17**(4), pp. 443–451.

ROZEC, X. and F. ASSAL (1976) Microwave switch matrix for communications satellites, *ICC76*, Philadelphia.

RUDGE, A. N. (1975) Multiple beam antenna offset reflectors with offset feeds, *IEEE Trans. on Ant.*, **23**(3), pp. 317–322.

SACKEIM, P. L., D. E. FRITZ and M. MACKLIS (1980) Performance trends in spacecraft auxiliary propulsion systems, *J. Spacecraft*, **17**(5), 390–395.

SAINT-AUBERT, P., D. VALENTIAN and W. BERRY (1984) Utilization of electric propulsions for communications satellites, *AIAA, 10th CSSC*, Orlando, paper 84.0729, pp. 354–364.

SAINT-ETIENNE, J. (1973) Localisation des véhicules spatiaux, *Rapport CNES*, No. 488/CB/ES.

SCHMIDTBAUER, B., H. SAMUELSON and A. CARLSSON (1973) Satellite attitude control and stabilization using on board computers, *ESRO-CR 100*.

SCOTT, W. G. *et al.* (1976) Development of multiple beam lens antennas, *AIAA, 6th CSSC*, Montreal, Paper 76-250, 13pp.

SCOTT, W. G., H. S. LUH, A. E. SMOLL and E. W. MATTHEWS (1982) 30/20 GHz communications satellite multibeam antenna, *AIAA, 9th CSSC*, San Diego, Paper 82-0449.

SEPP, G. (1975) Earth laser beacon sensor for earth oriented geosynchronous satellites, *Appl. Opt.*, **14**, 1719–1726.

SHIMOSEKO, S. and H. MATSUMOTO (1980) Three axis attitude control acquisition results and regular operational status of BSE, *AIAA*, 8th CSSC, Orlando, Florida, pp. 584–591.

SPARKS, R. H. (1980) Nickel cadmium battery technology advancement for geosynchronous orbit spacecraft, *J. Spacecraft*, **17**(6), pp. 554–557.

SPISZ, E. W. (1983) NASA development of a satellite switched SS/TDMA IF switch matrix, *CECON'83*, Cleveland, pp. 19–27.

STARK, L. (1974) Microwave theory of phased-array antennas, a review, *Proc. IEEE*, **62**(12), pp. 1661–1701.

STILWELL, J. H. (1982) Serial MSK Modem for the Advanced Communications Satellite, *NTC'82*, Galveston, pp. A2.5.1–A2.5.5.

SUSPLUGAS, M. J. (1979) Evolution des performances de la plateforme ECS entre 1980 et 1990, *Colloque Espace, télécommunications spatiales*, Toulouse, pp. 619–635.

TAMMES, J. B. and J. J. BLEIWEIS (1976) An RF monopulse attitude sensing system, *Intern. Telemetering Conf.*, Los Angeles, pp. 46–55.

TEMPLETON, L. and C. L. CUCCIA (1979) Communication satellite evolution during the 1980's, *Colloque Espace, télécommunications spatiales et radiodiffusion par satellite*, Toulouse, pp. 579–605.

THOMPSON P. T. and E. C. JOHNSTON (1983) INTELSAT VI. A new satellite generation for 1986–2000, *Int. Journal of Sat. Comm.*, **1**(1), pp. 3–14.

VALENTINI, R., M. LECOQ and J. MAQUET (1979) Problèmes technologiques de développement des moteurs à hydrazine à surchauffe, *Colloque Espace, télécommunications spatiales et radiodiffusion par satellite*, Toulouse, pp. 519–526.

WAYNE, R. H. and R. C. FINKE (1980) Electric propulsion circa 2000, *AIAA Intern. Meeting on Technical Display*, Global Technology 200, Baltimore, Paper 80-0912.

WERTZ, J. R. (1978) *Spacecraft Attitude Determination and Control*, D. Reidel.

WHEELON, A. D. (1983) Trends in satellite communications, *4th World Telecommunication Forum, Geneva*.

CHAPTER 7   Launching and positioning geostationary satellites

In the previous chapters we examined how the satellite while in its prescribed orbit fulfils its mission. However, even while it is being placed in position, the satellite plays an important role which is the key factor to the whole system becoming operational. This chapter outlines the particular requirements that have to be complied with, and this is followed by a description of some characteristics of specific launchers.

7.1   PLACING THE SATELLITE IN ORBIT

7.1.1   General principles

The procedure is based on the well known *Hohmann transfer* which allows the satellite to be transferred from a circular Earth orbit to another at different altitudes in the same plane with minimum energy consumption. A first *velocity increment* changes the lower circular Earth orbit into an elliptical orbit with perigee at the initial altitude and with apogee at the altitude of the final circular orbit. A second velocity increment at apogee circularizes the transfer orbit into the final one.

Figure 7.1 illustrates the above procedure for the launch of a geostationary satellite. To become geostationary, a satellite must reach a final circular orbit of 35 788 km in the equatorial plane (see Section 5.2.1.3).

According to the type of launch vehicle, this procedure is realized in three ways which differ according to how many of the two required velocity increments are provided by the satellite itself (most often with the assistance of auxilary propulsion stages):

(a) *From a circular low earth orbit* (LEO) as described above. This applies to the Space Transportation System (STS) which orbits at an altitude of about 290 km (parking orbit). The satellite must provide two velocity increments: one for injection into the geosynchronous transfer orbit (GTO) at perigee and another at the apogee (near the 35 788 km final altitude) of the GTO for injection into the geostationary satellite orbit (GEO).

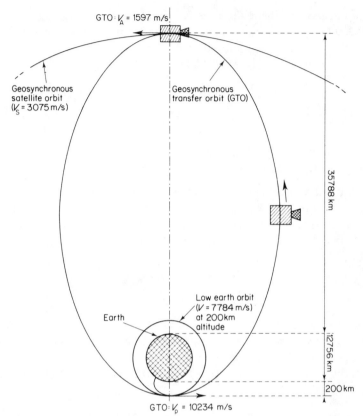

Figure 7.1   Geosynchronous transfer orbit (GTO) from a low Earth orbit to a geosynchronous orbit.

(b) *By injection into GTO* with required velocity at the perigee of the geosynchronous transfer orbit. This geosynchronous transfer orbit has its perigee at an altitude of about 200 km and its apogee near the 35 788 km final altitude. This applies to expendable launch vehicles such as the Ariane, Delta or Atlas-Centaur launchers, where there is no initial circular orbit. Only one velocity increment is required from the satellite at the apogee.

(c) *By direct injection into GEO* with required velocity. This applies to specific expendable launch vehicles: the US Titan IIIC and the USSR Proton launchers.

7.1.2   Velocity increment calculation (orbits in the same plane):

The equation $V^2 = 2\,\mu/r - \mu/a$ gives the satellite velocity at the perigee and the apogee of the transfer orbit (see Section 4.1.3.2) where a is the semi major axis of the ellipse, μ is the gravitational constant of Earth ($\mu = 398\,600$ km^3/s^2), r is the distance from the centre of the Earth to the point at which the velocity V is required to maintain the ellipse.

For instance with a perigee at 200 km,

$$a = \frac{200 + 35\,788}{2} + 6378 \text{ (Earth's radius)} = 24\,372 \text{ km} \qquad (1)$$

and

at the perigee: $r = 6\,578$ km and $V_p = 10\,234$ m/s
at the apogee: $r = 42\,166$ km and $V_a = 1597$ m/s

The required velocity increment at apogee is the difference between the satellite velocity on the geosynchronous satellite orbit and the velocity of the satellite at the apogee of the transfer orbit. Therefore since a geosynchronous satellite is maintained by a velocity $V = 3075$ m/s, the required velocity increment at apogee is:

$$\Delta V = 3075 - 1597 \text{ m/s}$$
$$\Delta V = 1478 \text{ m/s}$$

This velocity increment value corresponds to a minimum and applies to a *circularization manoeuvre* when transfer orbit and circular orbit lie in the same plane.

If any, the velocity increment at perigee can be calculated in the same way.

7.1.3   Orbital corrections

Orbital corrections described above occur in the same plane. To achieve a final orbit in the equatorial plane the launch pad should be in the plane of the Equator. As this is not the case, with present launch sites, an *orbit inclination correction manoeuvre* is mandatory. Indeed, the satellite inclination is a least equal to the latitude of the launch-pad if no inclination correction manoeuvre is performed. For example, for a launch from the Kennedy Space Center (KSC) at Cape Canaveral (*Eastern Test Range: ETR.*, latitude 28.5°), the orbit cannot be inclined less than 28.5°. For a launch from the Centre Spatial Guyanais (CSG) Kourou in Guiana (latitude 5.23°) the inclination cannot be less than 5.23°. This is demonstrated as follows:

7.1.3.1   Minimum inclination of the transfer orbit

The satellite is launched from the launch-pad M with velocity V making an angle a relative to the east. The components of the unit length vectors aligned

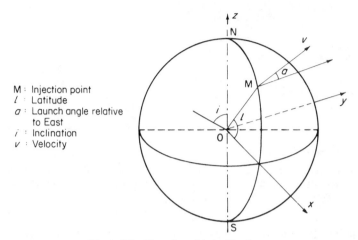

Figure 7.2 Transfer orbit inclination.

M : Injection point
l : Latitude
a : Launch angle relative to East
i : Inclination
v : Velocity

with **0M** and **V** are (Figure 7.2):

	0M	**V**
along 0x:	cos *l*	—
along 0y:	0	cos *a*
along 0z:	sin *l*	—

where *l* is the launch-pad latitude. The unit length vector aligned with $0M \wedge V$ is perpendicular to the orbit plane and has the following component with respect to 0z:

$$\cos i = \cos a . \cos l \qquad (2)$$

The inclination *i* of the orbit plane is then greater or equal than the latitude of the launch-pad *l*.

The minimum inclination is obtained for $a = 0°$ or $180°$. For a geostationary satellite (non-retrograde orbit) the only possible value is $a = 0°$, i.e. a launch towards the east (azimuth $Az = 90°$). Then one benefits from the *speed induced* on the trajectory by the rotation of Earth. A smaller inclination than given by (2) requires that orbit inclination correction manoeuvre be carried out ('dog leg' manoeuvre).

7.1.3.2 Orbit inclination correction strategy

The plane of the transfer orbit is defined by the velocity vector of the spacecraft on the orbit and the centre of the Earth. The inclination of the transfer orbit is the angle between the plane of the transfer orbit and the equatorial plane. Correction of the inclination, i.e. converting the transfer or-

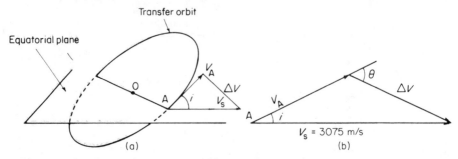

Figure 7.3   Correction of inclination: (a) transfer orbit plane and equatorial plane; (b) required velocity increment (value and orientation) in a plane perpendicular to the line of nodes.

bit plane into the equatorial plane (Figure 7.3(a)), implies a velocity increment applied *at one of the nodes of the orbit* which will cause a resultant velocity vector in the equatorial plane (Figure 7.3(b)). For a given orbital inclination the required velocity increment ΔV depends on the satellite velocity. The lower this velocity is, the more economic the correction manoeuvre is. The correction manoeuvre is therefore best carried out *at the apogee* of the transfer orbit at the same time as the circularization manoeuvre.

It is therefore necessary that:

(1) *The line between the perigee and apogee* (apsidal line) should be the *line of nodes* and therefore the *perigee should be in the equatorial plane*, which implies that the *injection* onto GTO *occurs at equatorial plane crossing.*
(2) The *apogee of the transfer orbit* is at the *altitude* of the *geostationary satellite orbit*;
(3) The *main axis of the satellite*, which is stabilized during all these operations, is *correctly oriented* with an angle θ relative to the satellite velocity in a plane perpendicular to the line of nodes as indicated in Figure 7.3 assurring that the apogee thrust is along the satellite main axis.

Taking into account the fact that V_S is nearly twice V_A, Figure 7.3(b) shows that θ is nearly equal to $2i$. For Cape Canaveral, θ is approximately 56°; for Kourou, the nominal inclination of the transfer orbit Ariane being near to 10°, θ is about 20°. Accurate values for θ can be derived from:

$$\theta = \text{Arc sin} \left(\frac{V_S \sin i}{\Delta V} \right) \tag{3}$$

where ΔV is the total velocity increment to be applied for circularization and inclination correction of value i and given by:

$$\Delta V = \sqrt{V_A^2 + V_S^2 - 2V_A V_S \cos i} \tag{4}$$

or:

$$\Delta V = \sqrt{\frac{\mu}{R_e + h_p}} \sqrt{K} \sqrt{\left(1 + \frac{2K}{1 + K} - 2\sqrt{\frac{2K}{1 + K}} \cos i\right)} \qquad (5)$$

where:

$$K = \frac{R_e + h_p}{R_e + h_a}$$

h_p = perigee altitude
h_a = apogee attitude ($= R_0 = 35\,788$ km)
R_e = Earth radius = 6378 km

μ = gravitational constant = $3.986\ 10^{14}\ \mathrm{m^3/s^2}$

Figure 7.4 shows the curve of ΔV versus the launch-pad latitude assuming a perigee altitude of 200 km. ΔV expresses then as

$$\Delta V = \sqrt{12.006 - 9.822 \cos i} \ (\mathrm{km/s}) \qquad (6)$$

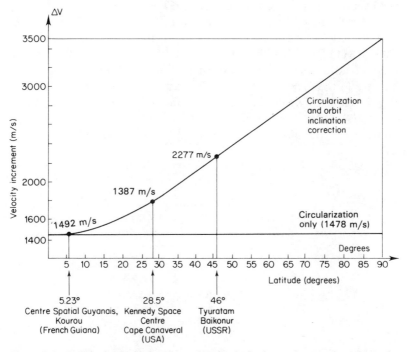

Figure 7.4 Required velocity increment for circularization and correction of transfer orbit inclination versus launch-pad latitude (perigee altitude 200 km); launch towards east ensuring an orbit inclination equal to the latitude of the launch pad.

and i is assumed equal to the launch-pad latitude (launch is towards east with $a = 0°$). Note that for latitudes above $70°$ the inclination correction requires a velocity increment which is greater than that required for circularization only.

7.1.4   Apogee motor

The *apogee* motor, also called *kick* or *boost motor* (AKM or ABM), is required to provide a given velocity increment to the satellite at the apogee of the GTO so that it finally reaches the geostationary satellite orbit. Thus the apogee motor is not required when the satellite is directly injected into GEO by the launch vehicle.

The apogee motor is incorporated in the satellite when current expendable launch vehicles are used. With a launch by the STS the apogee motor may also be part of the upper stage which provides the perigee velocity increment. Up to now, *solid propellent motors* are mainly used to provide the required velocity increment for orbital injection. A solid propellent motor consists of a solid mixture of oxidizer and fuel contained within a titanium or Kevlar-epoxy case ended by a nozzle throat and an exit cone (Figure 7.5). Firing of the propellent grain is performed using an igniter. Propellant formulations are based on high-percentage of carboxy-terminated or hydroxy-terminated polybutadiene (CTPB or HTPB) with additives. Motor *off-loading*, within 10 to 15 per cent, allows for adjustment of the nomimal velocity increment. Specific impulse values are in the order of 290 s. An example of such a motor is the MAGE motor developed by the SEP under contract of the European Space Agency (Isopi, 1979). Other well known motors are the STAR series manufactured by THIOKOL in the USA. Table 7.1 indicates the performance of motors in current use.

Bipropellent motors may benefit from a higher specific impulse (310 s

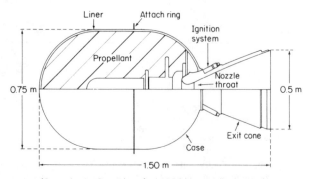

(Approximate dimensions for a 500 kg propellent motor)

Figure 7.5   Construction of a solid propellent motor.

Table 7.1 PERFORMANCES OF VARIOUS SOLID PROPELLENT MOTORS FOR APOGEE OR
PERIGEE BOOST APPLICATIONS.

Name	Total mass (kg)	Propellent load (max.) (kg)	Impulse (10^6 N s)	Max. thrust (vacuum)(N)	I_{sp} * (s)
MAGE 1	368	335	0.767	28 500	287.6
MAGE 1S	447	410	1.168	33 400	290.7
MAGE 2	530	490	1.410	46 700	293.8
STAR 30B	537	505	1.460	26 825	293.1
STAR 30C	620	585	1.645	31 730	285.2
STAR 30E	660	620	1.780	35 365	290.1
STAR 31	1398	1300	3.740	95 675	293.2
STAR 48	2115	1995	5.695	67 820	290.0
STAR 62	2890	2740	7.820	78 320	291.2
U.T.SRM-2	3020	2760	8.100	260 745	303.6
Aerojet 62	3605	3310	9.310	149 965	286.7
STAR 63E	4420	4060	11.865	133 440	298
STAR 75	4800	4560	13.265	143 690	296.3
Aerojet 66	7030	6250	17.595	260 650	286.7
Minuteman III	9085	8390	23.100	206 400	
U.T. SRM-1	10390	9750	28.100	192 685	295.5

*I_{sp} is sometimes expressed in lbf s/lbm or in N s/Kg:
$I_{sp}(s) = I_{sp}(\text{lbf s/lbm}) = I_{sp}(\text{N s/kg}) \times 1/9.807$ (1 lbm = 0.4536 kg; 1 lbf = 4.448 N)

instead of 290 s) when operated in a regulated pressurized mode. When the apogee motor is incorporated in the satellite there is a gradual movement towards the use of bipropellent spacecraft propulsion systems for both geostationary satellite orbit injection and on-orbit control (see Figure 6.20). Such an *unified propulsion system* offers significant *mass savings* over the conventional combination of a solid apogee kick motor and a hydrazine reaction control system. Moreover, any propellent mass not used for orbit injection is still available for on-orbit control, and results in an *increased satellite lifetime*.

Propellent mass required to achieve a given velocity increment can be calculated using formulas of Section 6.4.1.

7.1.4.1 Mass of propellent required for circularization only

As an example this is done below considering the circularization of an equatorial GTO (no inclination correction) with a perigee at 200 km. The required velocity increment at apogee is then $\Delta V = 1478$ m/s (Section 7.1.2). The satellite incorporates a bipropellent apogee motor and the total mass on GTO is 1000 kg. The specific impulse is 310 s with nitrogen tetroxide (N_2O_4) as oxidizer and monomethylhydrazine (MMH) as fuel, considering a regulated pressurized mode of operation.

According to formula (23) of Chapter 6 the required mass of bipropellent: m, is given by:

$$m = M_s\left[1 - \exp\left(-\frac{\Delta V}{gI_{sp}}\right)\right] = 1000\left[1 - \exp\left(-\frac{1478}{9.81 \times 310}\right)\right] = 385 \text{ kg}$$

Taking into account the mass of the tanks, motor and pipes a mass of about 60 kg must be added, so the apogee motor sub-system has a weight of 445 kg, and the useful weight of the satellite becomes 555 kg (Beginning of life mass).

Typical thrust value for a bipropellent apogee motor is 490 N. According to formula (18) in Chapter 6 the rate at which the propellent is consumed is $\rho = (F/I_{sp}g) = 490/310 \times 9.81 = 0.16$ kg/s, and the burn time is therefore:

$$t = \frac{m}{\rho} = \frac{385}{0.16} = 2406 \text{ s} \qquad (7)$$

This burn time is quite large compared with that of a solid propellent motor. This results from the much smaller thrust level (490 N instead of 50 000 N). Hence the velocity increment is *not an impulse* and this *degrades the manoeuvre effectiveness* as the thrust would not remain correctly oriented during the whole *burn time.* In practice the burn time is *split into smaller burn times,* performed at successive apogees. To compensate for the loss in efficiency over impulsive manoeuvres, one benefits from *better orbital injection accuracy* as a longer time is available to refine measurements, and errors caused by thrust misalignment are smaller. This is because the thrusts are lower and the spacecraft attitude can be corrected after each burn. Another advantage of low thrust operations is that solar panels and antennae can be *deployed* prior to apogee firing as the acceleration remains small. Finally, the much lower thrust levels of bipropellent systems make possible *body-fixed stabilization* of the spacecraft attitude by active control during the transfer orbit while for solid propellent motors, with their high thrust level, *spin stabilization* is the most effective way of ensuring stability.

7.1.4.2   Influence of orbit inclination correction on the mass of propellent

Table 7.2 shows the required ΔV and the corresponding propellent mass for

Table 7.2 INFLUENCE OF THE LATITUDE OF THE LAUNCHING BASE.

	Kourou (France)	Cape Canaveral (USA)	Tyuratam (USSR)
Latitude	5.23°	28.5°	46°
ΔV (m/s)	1490	1836	2277
Propellent mass (kg)	387	453	527
Loss with respect to Kourou (kg)	0	66	140
Usable sat. mass (kg)	613	547	473

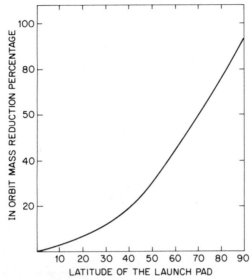

Figure 7.6 Influence of the latitude of the launch-pad on the
satellite mass at the beginning of life (mass at
launch, 1000 kg; specific impulse, 310 s).

various launch-pads assuming a satellite with mass at launch $M_s = 1000$ kg.
Figure 7.6 indicates the mass reduction percentage of the beginning of life
(BOL) satellite mass versus the latitude of the launch-pad. In the case of a
launch-pad at $28.5°$ latitude (Cape Canaveral), one suffers a BOL mass
reduction of 12 per cent compared with the case of an equatorial launch site.
This example underlines the usefulness of a *near equatorial launch site* such
as Kourou.

7.1.5 Orbital injection sequence using an expendable launch vehicle:

The process of injecting a geostationary satellite in orbit by a multistage
expendable launch vehicle can be divided into three phases (Figure 7.7):

7.1.5.1 Launch phase

From takeoff to injection into the transfer orbit, the following operations are
required:

(1) *Ascend in altitude* to reach the nominal altitude of the perigee of the
    transfer orbit.
(2) *Jettison the satellite fairing* once the launch vehicle is above the dense
    layers of the atmosphere.

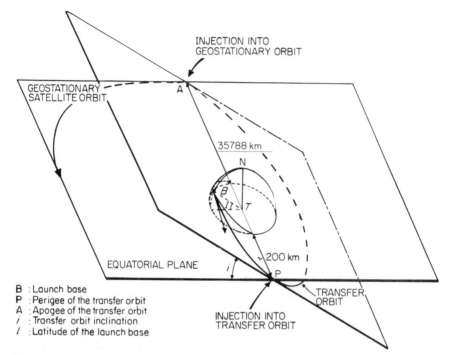

Figure 7.7   Sequence for launch and injection into transfer and geostationary orbit with an expendable launch vehicle.

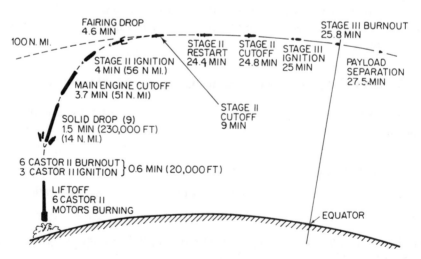

Figure 7.8   Launch sequence of Delta launch vehicle. From Bleviss (1976). *Copyright American Institute of Aeronautics and Astronautics and reprinted with permission of AIAA.*

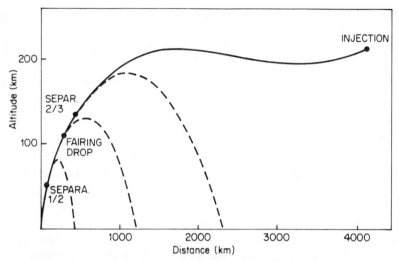

Figure 7.9 Launch sequence of the Ariane launch vehicle. *Reproduced by permission of Centre National d'Etudes Spatiales from Duret (1980).*

(3) *Guide the compound* of the upper stage and the satellite on a curved trajectory which ends *parallel* to the Earth surface with the *required velocity* at the *equatorial plane crossing* (this velocity is that of a satellite at perigee of the transfer orbit).

Figures 7.8 and 7.9 illustrate two possible launch sequences. The first one (Delta 2914, 3914, 3920, and Atlas Centaur) comprises an intermediate coasting phase, during which the upper stage and satellite as a combined unit are oriented and spun in order to keep this orientation when the upper stage is started again. In the second case (Ariane) the thrust is continuous (except during stage separation) and the launch vehicle is guided during the combustion of the upper stage to reach, at an altitude of 200 km, the required velocity at the equatorial plane crossing.

7.1.5.2 Transfer phase

The transfer phase begins with the insertion into the transfer orbit and is completed when the geostationary satellite is placed into the geostationary satellite orbit (i.e. at the apogee of the transfer orbit). During this phase, one has to:

(1) *Monitor* the satellite *trajectory.*
(2) *Measure* the satellite *attitude.*
(3) *Ensure the correct orientation* of the satellite.

During this phase control of the satellite attitude is performed either by spinning the satellite or body-fixed stabilization.

Recall that the nominal transfer orbit requires an *apogee at the geosynchronous orbit* altitude and the *apsidal line must lie in the equatorial plane*. If not, the satellite final orbit will exhibit an abnormal eccentricity and will remain inclined. In such a case the satellite would have to perform orbit correction by its own thrusters and propellent.

7.1.5.3   Positioning phase: drift orbit and station acquisition

The positioning phase begins by *circularization* of the orbit in the equatorial plane by means of the apogee motor, and ends by *positioning the satellite at its nominal station* on the geostationary satellite orbit. The apogee motor must be fired while the satellite is in line of sight of at least two control stations for reliable control during the manoeuvre. Figure 7.10 shows the ground track of the satellite during the transfer orbit (specifically in this case related to the Ariane launcher) and indicates the position of successive apogees. In this way the specific orbit in which the circularization manoeuvre has to take place is determined.

Due to dispersion of the parameters of the transfer orbit and of the apogee manoeuvre, the final orbit is not exactly that of a geostationary satellite: it has a residual eccentricity and inclination which causes the satellite to drift. So this orbit, called *drift orbit*, must be adjusted using the low thrust of the on-board thrusters devoted to station keeping manoeuvres. On the other hand, as the apogee manoeuvre does not necessarily occur at the nominal longitude of the satellite station, the drift allows the satellite to reach the desired longitude.

During the drift orbit, along with orbit corrections, the principal operations are (Figure 7.11):

(1) Sun acquisition
(2) Solar panel deployment.
(3) Earth acquisition.
(4) Station acquisition.

On successful station acquisition the satellite is tested before is starts its operational life.

7.1.6   Orbital injection sequence using the Space System Transportation (Space Shuttle):

The orbital injection sequence differs from that with an expendable launch vehicle as the STS is not able to inject the satellite straight into the transfer orbit. Figure 7.12 illustrates the orbital injection sequence with the Space

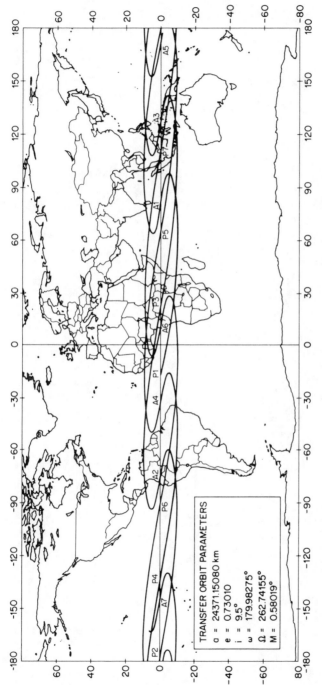

TRANSFER ORBIT PARAMETERS

a = 24371.15080 km
e = 0.73010
i = 9.5°
ω = 179.98275°
Ω = 262.74155°
M = 0.58019°

Figure 7.10 Satellite track during transfer orbit (Ariane launch) showing successive locations of apogee (A) and perigee (P). *Reproduced by permission of Centre National d'Etudes Spatiales from Robert and Foliard (1980).*

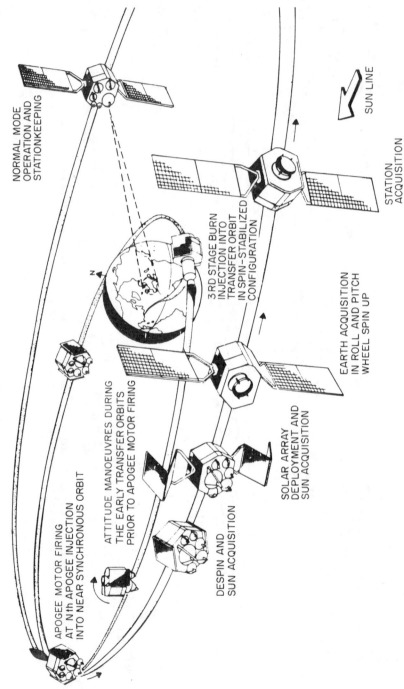

NORMAL MODE
OPERATION AND
STATIONKEEPING

STATION
ACQUISITION

SUN LINE

3RD STAGE BURN
INJECTION INTO
TRANSFER ORBIT
IN SPIN-STABILIZED
CONFIGURATION

EARTH ACQUISITION
IN ROLL AND PITCH
WHEEL SPIN UP

SOLAR ARRAY
DEPLOYMENT AND
SUN ACQUISITION

DESPIN AND
SUN ACQUISITION

ATTITUDE MANOEUVRES DURING
THE EARLY TRANSFER ORBITS
PRIOR TO APOGEE MOTOR FIRING

APOGEE MOTOR FIRING
AT Nth APOGEE INJECTION
INTO NEAR SYNCHRONOUS ORBIT

Figure 7.11 Operations during drift orbit. *Reproduced by permission of Centre National d'Etudes Spatiales from Robert and Foliard (1980).*

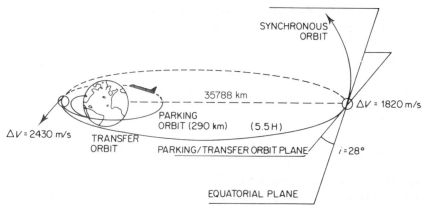

Figure 7.12 Orbital injection sequence with the space transportation system.

System Transportation. The STS orbit is a circular one at an altitude of about 290 km, called parking orbit. After deployment of the satellite from the STS, a velocity increment must be applied to inject the satellite into the transfer orbit. This is done by means of an additional motor: the *perigee motor*. The perigee motor is fired when the spacecraft crosses the equator, so the transfer orbit has its perigee (and hence apogee) in the equatorial plane as discussed above. Owing to the inherent inefficiency of inclination corrections at perigee as underlined above the perigee motor is not used for orbital inclination correction and the transfer orbit inclination is that of the STS orbit (28°). The perigee motor can be either *incorporated* to the satellite (INTELSAT VI, SYNCOM IV) or most often a *specific additional stage*, and uses either solid (see Table 7.1) or liquid propellant. When specific, one finds stages whose primary purpose is to provide only perigee thrust to place their payload onto the geosynchronous transfer orbit (the apogee motor which places the satellite on the geostationary satellite orbit is then combined with the satellite) and those which are capable of providing both perigee and apogee thrust (See section 7.2.1 and Table 7.4).

7.1.7 Launch window

The '*launch window*' is a time period within which the satellite can be launched taking into account the constraints on the satellite. These constraints are as follows:

(1) *Ensure an adequate power supply* (the position of the satellite in relation to the Sun and periods over which eclipses are encountered);
(2) *Guarantee the thermal control*, taking into account the eclipses' duration.
(3) *Permit an accurate determination of the attitude* (angle Sun–satellite and satellite–Earth)

(4) *Ensure visibility from the control stations* at the time of the critical events (position of the apogees of the transfer orbits).

One can determine what are the transfer orbits which would guarantee constraints are complied with. Knowing the perigee of these orbits, one deduces the hours possible for the launching of the satellite, for all days of the year.

7.2   LAUNCH VEHICLES

Launch vehicles operational at the mid-eighties are summarized in Table 7.3. (Gilli, 1984) No satellites are available without the means to launch them. Many nations perceived the situation whereby the US becoming the only country selling launch vehicles as a serious imposition on their freedom of choice and have applied substantial resources towards their own launch vehicle development. The national developments are as follows.

Table 7.3   EXPENDABLE LAUNCH VEHICLES.

Launch vehicle	Country	Total mass	Stages	Propellent	Payload delivery capacity Low Earth orbit (kg)	Payload delivery capacity Geostationary transfer orbit (kg)	Payload delivery capacity Escape (kg)
SLV 3	India	17.5 t	4	4 S	50	—	—
Scout	USA	21.3 t	4	4 S	230	55	35
MU 3S	Japan	54 t	3 + 8 SRB	3 S	300	60	40
C2	USSR	100 t	2	2 L	1 500	—	—
N2	Japan	135 t	3 + 9 SRB	2 L + 1 S	1 900	700	470
Long March 2	China	200 t	2	2 L	2 000	—	—
Delta 3914	USA	190 t	3 + 9 SRB	2 L + 1 S	2 500	950	640
Delta 3920PAM	USA	193 t	3 + 9 SRB	2 L + 1 S	3 000	1247	830
Atlas F	USA	140 t	1½	1½ L	1 400	—	—
Atlas G Centaur	USA	150 t	2½	1½ L + 1 C	5 100	2360	1465
F2	USSR	185 t	3	3 L	7 000	?	?
Ariane 3	EUROPE	237 t	3 + 2 SRB	2 L + 1 C	6 000	2580	1500
A 2-e	USSR	312 t	2½	2½ L	7 500	2400	1800
Titan 34D	USA	635 t	3 + 2 SRB	3 L	12 250	4500*	3200
DI–e	USSR	1000 t	3	3 L	22 000	9000	6350

S = Solid propellent.
L = Liquid propellent.
C = Cryogenic propellent.
SRB = Solid rocket booster.
*1900 kg in geostationary orbit.

7.2.1 US launch vehicles

As an alternative to the Space Transportation System or *Space Shuttle*, the US maintain the launchers *Delta* and *Atlas Centaur*, and have even developed upgraded versions such as the Delta 3920 PAM D, as well as a launcher derived from the Titan 3C, the Titan 34D, which can be used instead of the Shuttle, in particular for the launching of military satellites.

The Shuttle, as shown in Figure 7.13 is a reusable vehicle consisting of an *orbiter* capable of re-entering the atmosphere and making an aircraft-like landing, two recoverable *solid rocket boosters* (SRB), and an expendable LH_2/LO_2 *tank* which furnishes propellent to the three main engines located in the orbiter. The Shuttle can be launched from either the Kennedy Space Center into an easterly trajectory, or from the Vandenberg Air Force Base into polar or near-polar orbits.

The Shuttle is launched vertically with both boosters and the three main engines burning. After booster burn-out, about two minutes into the flight, the boosters separate from the main tank and orbiter, and the orbiter continues its flight into an Earth orbit of about 290 km altitude for a due east launch from the Kennedy Space Center.

The important *dimensions of the cargo bay* (18.3 m long and 4.60 m diameter) and the *launching capabilities* (29 metric tons in circular orbit of 300 km) allow for the placing of heavy payloads in orbit. Therefore, the construction of real 'space stations' for the requirements of telecommunications and other scientific missions becomes a possibility. Another capability of the Shuttle is that of recovering satellites in low Earth orbit.

However, the Shuttle can only reach a low circular orbit ($\simeq 290$ km) additional upper stages are necessary to place payloads into orbits of higher energy than the Shuttle alone is capable of attaining. (See section 7.1.6)

Various upper stages may be used to reach the final geostationary orbit. Some of the upper stages are new developments while others are adaptations of existing stages from expendable launch vehicles. At the current time the following options are available (Mahon and Wild, 1984):

(1) *Perigee stage* which propels the satellite *with its apogee motor*, from the cargo bay, into the *geosynchronous transfer orbit* (SSUS D (Solid Spinning Upper Stage Delta) or PAM D (Payload Assist Module)).

   The PAM-D is carried to orbit in specially designed airborne support equipment (ASE) which restrains the stage and payload during the Shuttle launch, spins the stage and payload at selected rates between 30 and 100 r.p.m. to provide stabilization, and launches it from the Shuttle payload bay by means of ejection springs. (Figure 7.14)

   After sufficient separation between the stage and Orbiter is attained (by manoeuvring the Orbiter and allowing an inactive 45-minute coast period),

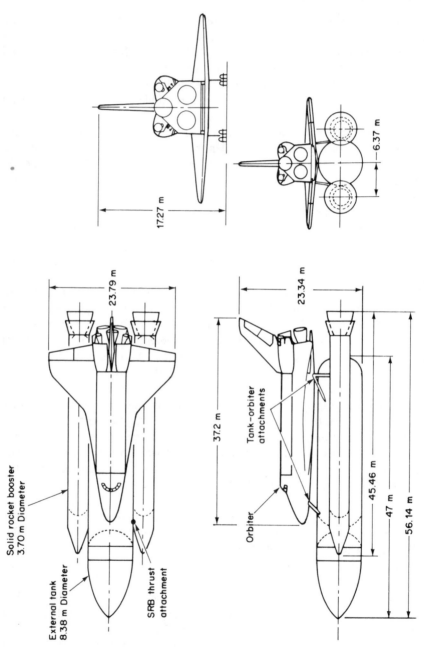

Figure 7.13   Space transportation system.

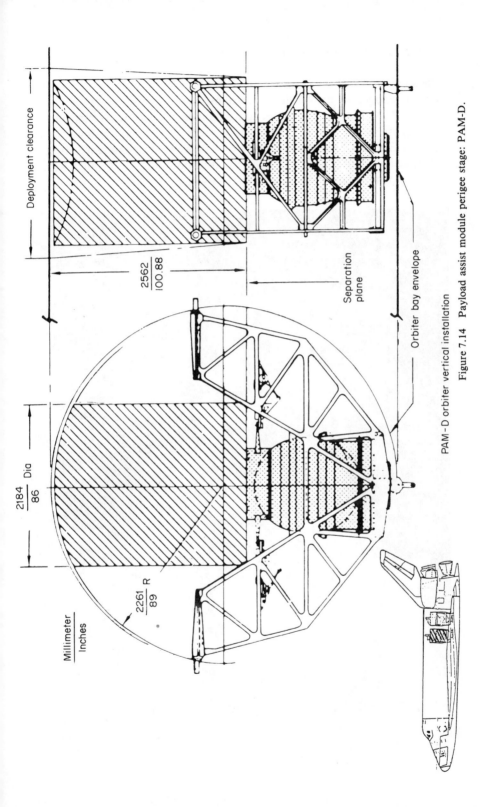

Millimeter / Inches

Deployment clearance

2562 / 100.88

2184 / 86 Dia

2261 / 89 R

Separation plane

Orbiter bay envelope

PAM-D orbiter vertical installation

Figure 7.14 Payload assist module perigee stage: PAM-D.

the solid motor of the stage is ignited by a timer, thus placing the satellite and its apogee motor on geosynchronous transfer orbit (GTO).

The PAM-D system occupies approximately 2 m of the Orbiter payload bay length (up to four PAM-Ds can be accommodated on a single Shuttle flight). Figure 7.15 is an illustration of a PAM-D being launched from the Orbiter bay. Note that the PAM-D is transported in the Orbiter with its axis perpendicular to the longitudinal axis of the Orbiter (Figure 7.17(a)).

Figure 7.15   Spring ejection of the SBS-3 satellite from the STS payload bay. After coasting to a controlled firing point, and once the STS has moved to a safe distance, PAM-D will be fired (NASA photograph).

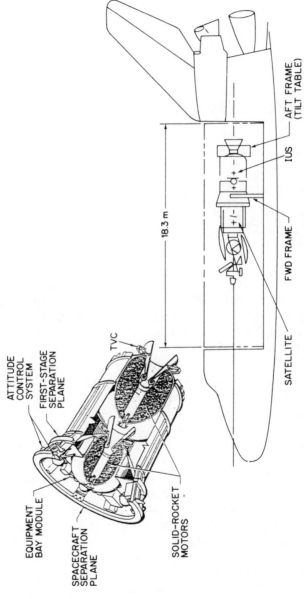

Figure 7.16   Inertial upper perigee apogee stage (IUS).

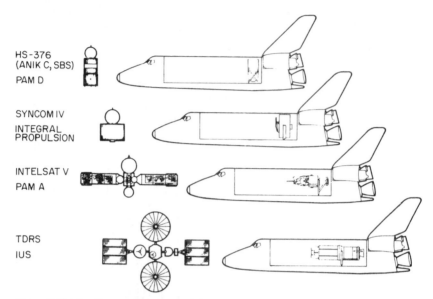

HS-376
(ANIK C, SBS)
PAM D

SYNCOM IV
INTEGRAL
PROPULSION

INTELSAT V
PAM A

TDRS
IUS

Figure 7.17   Loading configurations of the space transportation system (STS). *Reproduced with permission from Rosen (1980). Copyright © 1980 IEEE.*

The performance is 1250 kg on geostationary orbit (equivalent to the performance of a launcher of the Delta and Atlas Centraur class)

(2) *Inertial upper stage (IUS)*, capable of placing a satellite *without an apogee motor* of its own on the *geostationary satellite orbit* (Hanford, 1980) (Figures 7.16 and 7.17 (d)).

The IUS is a two-stage solid propellent vehicle capable of placing 2270 kg on geosynchronous orbit when launched from the Shuttle Orbiter, or 1910 kg on geosynchronous orbit when launched from the Titan-34D expendable launch vehicle. It is a body-fixed stabilized vehicle which employs a highly sophisticated avionics system.

The IUS is launched with its longitudinal axis parallel to that of the Shuttle. It is rotated upwards at an angle of 60° in preparation for deployment. As with the systems previously discussed, deployment is accomplished by springs. The stage maintains attitude control for 45 minutes before ignition of the first stage to place the payload and the IUS second stage on geosynchronous transfer orbit. The second stage is fired to place the 2270 kg payload on geosynchronous orbit.

(3) An *integrated propulsion system on-board the satellite*, including the perigee motor and the apogee motor (SYNCOM IV, (Pasley and Donatelli, 1980) (Figure 7.17 (b)). This is designed to optimize the occupied volume in the cargo bay, and reduce the cost of the launching, using the *frisbee deployment* illustrated on Figure 7.18: a simple spring and latch

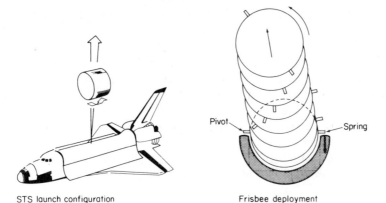

STS launch configuration   Frisbee deployment

Figure 7.18   Frisbee deployment.

Table 7.4   STS UPPER STAGES.

Name	Year	Type[a]	Propellent[b]	Weight[c] (kg)	Length[d] (m)	Performances[e] (kg) GTO	Performances[e] (kg) GEO[f]
IRIS – S	1987	P/S	S	2 290		600	—
IRIS	1987	P/S	S	2 700		950	—
PAM-D	1982	P/S	S	3 320	2.2*	1250	—
PAM-DII	1985	P/S	S	5 360	2.4*	1840	—
PAM-A	—	P/S	S	5 900	2.3	1990	—
AMS	1988	TS/3A	L	6 430	1.65	2550	—
SCOTS	1986	P/S	S	6 000	2.5	2700	—
HPPM			L			—	1400
STV	1987	P/3A	L	8 300		3400	—
TOS-S	1987	P/3A	S	9 560	2.5	4060	—
TOS	1986	P/3A	S	12 950	3.3	6090	—
IUS	1983	TS/3A	S	18 900	5.0	—	2270
TOS/AMS	1988	TS/3A	L	18 320	4.8	8860	2950
CENTAUR-G	1987	TS/3A	C	19 800	5.9	—	4540
CENTAUR-G′	1986	TS/3A	C	23 100	8.9	—	5910
OTV	1995	TS/3A	C				7250
SYNCOM-IV	1985	I/S	S + L				1320
INT-VI	1986	I/S	S + L				2270

[a] P = perigee stage; TS = transfer stage (perigee + apogee); I = satellite integrated stage; S = spin stabilization; 3A = three-axis stabilization with own avionics.
[b] S = solid propellent; L = liquid propellent; C = cryogenic propellents.
[c] Fully loaded stage weight including support equipment.
[d] Length in orbiter cargo bay occupied by stage (spacecraft height must be added except for*).
[e] GTO = on geostationary transfer orbit, $i = 27.5°$
[f] GEO = on geostationary orbit (mass in GEO with perigee stages is about half the mass in GTO provided that an additional apogee stage is needed) (total weight in orbiter cargo bay is [c] + [e]).

mechanism gives an upward motion to the spacecraft, ejecting it from the cargo bay.

Note that, while the PAM-A upper stage (Figure 7.17(c)), with a performance equivalent to that of an Atlas Centaur class launch vehicle, has not up to now found its market, other upper stages are under development, such as the PAM D2, the Transfer Orbit Stage (TOS) and the Centaur. Table 7.4 summarizes the performance of upper stages presently either operational or under development.

Other concepts of reliable launchers are at present under consideration (Reed *et al.*, 1979; Wihilte, 1980; MacConochie *et al.*, 1980).

7.2.2   Europe's expendable launch vehicles

Europe has developed the *Ariane* family of *expendable launch vehicles* (Figure 7.19). Table 7.5 shows the performance of these launch vehicles.

The Ariane launcher is optimized for a *geosynchronous transfer orbit injection* (200 km, $i = 8°$). This orbit is achieved with great accuracy due to the guidance of the third stage. Moreover, an attitude and spin control system (SCAR—*système de controle d' attitude et de roulis*) ensures that the satellite is positioned in the attitude required for the inclination correction, with a

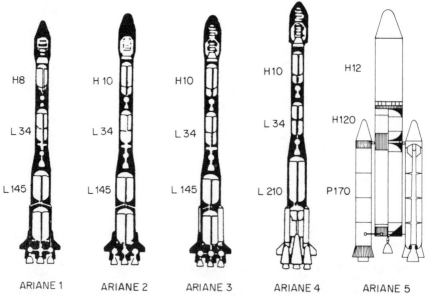

Figure 7.19   The Ariane launch vehicle family.

Table 7.5 ARIANE LAUNCH VEHICLE PERFORMANCES.

	Ariane 1 (max.)	Ariane 2 (max.)	Ariane 3	Ariane 4	Ariane 5
Year of operation	1981	1984	1984	1986	1995
Synchronous transfer orbit	1800 kg	2175 kg	2580 kg	1900–4200 kg	5200–8200 kg
Geostationary orbit[a]	1000 kg	1170 kg	1400 kg	1050–2300 kg	2800–4500 kg
Low Earth orbit (200 Km)	4900 kg	5100 kg	5900 kg	8700 kg	18 000 kg
Sun synchronous orbit[b]	2400 kg	2800 kg	3250 kg	4550 kg	10 000 kg
Escape	1100 kg	1330 kg	1550 kg	2580 kg	—

[a] Approximatively, as it depends on the performance of the apogee motor. Obtained by dividing the GTO capability by 1.8.
[b] 800 km, 98.6° inclination.

possible spin of 10 turns per minute. In this way a saving can be made on the amount of propellent required for the orbital and attitude corrections of the satellite.

Ariane I is a three-stage inertially stabilized launch vehicle which weights 210 metric tons at lift-off. The first stage which has a height of 18.4 m and a diameter of 3.8 m, is powered by four Viking V engines which burn for 146 s. The second stage has a height of 11.6 m and a diameter of 2.6 m. It is equipped with a single Viking IV engine which burns for 136 s. The third stage is a cryogenic LOX/LH$_2$ stage—the first cryogenic stage to be developed in Europe, which burns for 570 s. This stage is 9.1 m high and 2.6 m in diameter. The fairing has a height of 8.65 m and a diameter of 3.2 m. For dual launches, use is made of the 'SYLDA'. (*Système de Lancement Double Ariane*) a secondary structure containing a lower satellite and supporting an upper satellite (see Figure 7.20).

Ariane II is an uprated version of the Ariane I. In this uprating, the thrust of the Viking engines has been increased, and the available mass of the third stage cryogenic propellents has been increased along with an increase of the specific impulse of the engines using these propellents. To accommodate these increases in performance, the volume of the payload fairing was also increased.

Ariane III is the same configuration as Ariane II except for the addition of two solid propellent boosters to the first stage for additional thrust augmentation (see Figure 7.21). Table 7.6 indicates the GTO capabilities of the Ariane II and III.

Ariane IV is a further uprating of Ariane II and Ariane III. The first stage is lengthened by 7 m to increase the propellent capacity. The first stage engines and the second and third stages are identical to those of Ariane II and Ariane III except for structural strengthening of the stages. However, there is a new a larger fairing (diameter: 4 m, height: 11 m), and a larger device for multiple launches (SPELDA).

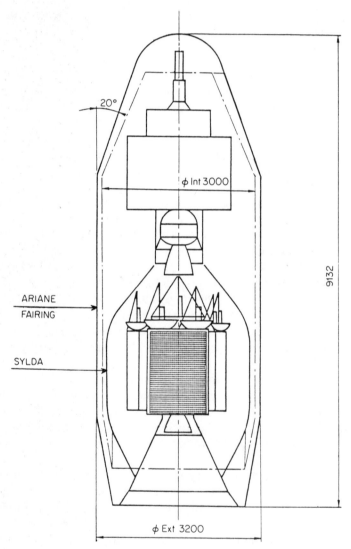

Figure 7.20   Sylda adapter.

Figure 7.21   First Ariane 3 launch, 4 August 1984. *Reproduced by permission of Centre National d'Etudes Spatiales.*

Table 7.6   GEOSYNCHRONOUS TRANSFER ORBIT CAPABILITIES (200/35 786 km, $i = 8°$).

	Payload mass (M_p)				
	Single launch			Dual launch	
	PAF[a] 1497 (27 kg)	PAF 1194 (43 kg)	PAF 937 (48 kg)	SYLDA[b] 3990 (182 kg)	SYLDA 4400 (190 kg)
Ariane 2 (M_p max. = 2175 kg)	2103	2087	2082	1993	1985
Ariane 3 (M_p max. = 2580 kg)	2508	2492		2×1199	2×1195

[a] Payload attach fitting.
[b] *Système de lancement double ariane* (dual launch adapter).

Table 7.7   ARIANE 4: GEOSYNCHRONOUS TRANSFER ORBIT CAPABILITIES
(200/35 786 km, $i = 7.5°$).

	Dedicated launch Spacecraft + Adapter (kg)	Dual launch Spacecraft + Adapters (kg)
AR40	1900	—
AR42P	2600	2225
AR44P	3000	2625
AR42L	3200	2825
AR44LP	3700	3325
AR44L	4200	3825

AR4nP = Solid propellent booster.
AR4nL = Liquid propellent booster.
n: indicates the number of boosters.

There are five versions of Ariane IV as indicated in Table 7.7. These variations are achieved by the addition of various combinations of solid and liquid boosters to the basic vehicle.

A more powerfull new vehicle for the 1990s, *Ariane V*, perhaps carrying a manned reusable vehicle, is under study by the European Space Agency. Ariane V will have a performance of 5200 Kg in GTO.

7.2.3   Japan's expendable launch vehicles

The National Space Development Agency of Japan (NASDA) has been conducting the development and operation of a satellite launcher called *N rocket*. This N series rocket, Japan's first liquid propellent launcher, is a three-stage vehicle capable of launching medium size applications satellites. The N launch vehicle programme, from the basic N, called N-I vehicle, to the current N, called *N-II vehicle*, has been supported by the United States through government level agreements. Hardware and technology of the NASA Thor-Delta have been transferred and applied to both N-I and N-II vehicles.

The N-I rocket project started in 1970 and this launch vehicle became operational in 1975 when it successfully put an 85 kg Engineering Test Satellite into a planned orbit accurately in its first flight. The upgraded N-II vehicle has entered on the operational phase with two successful flights in 1981. It is capable of placing about 350 kg payload into geostationary orbit.

To meet the need for launching large capacity applications satellites in the near future as requested by Japan's satellite users, NASDA has been carrying out study and development work of a future rocket, called *H-I launch vehicle* with high performance capability. This new launch vehicle has an estimated capability of placing about 550 kg payload into geostationary orbit.

7.2.4 USSR's expendable launch vehicle

Among the numerous launch vehicles the USSR has, one is available for commercial launches: the PROTON (or D-1). It is a huge 1 000 metric tons three-stage launch vehicle equipped with six liquid propellent boosters. The basic fairing size is 2.5 m diameter and 3.3 m high, but other fairing configurations are available. It has the capacity of placing a 21 000 kg payload on a low Earth orbit (200 km, $i = 46°$), 9000 kg on the geosynchronous transfer orbit, and 2500 kg on the geostationary orbit with no need for an apogee motor.

7.2.5 China's expendable launch vehicle

China's launch vehicle is the *CZ-3 launcher*, also called the *Long March launch vehicle*. It is a three-stage liquid propellent vehicle, 43 m high, and weighing 202 metric tons at take-off. Its configuration is very similar to that of the ARIANE I launch vehicle, but its performance is not as good: 1400 kg on the geosynchronous transfer orbit.

7.3 ECONOMICS OF LAUNCH

It is difficult to establish the precise *cost of launching a satellite*, as it depends on the launch vehicle used and the servicing. Moreover, the charge does not necessarily reflect the actual cost. Table 7.8 indicates announced costs for various launch vehicles. Note that these launch vehicles do not offer similar performances and the final choice must take into account the compatibility of

Table 7.8 LAUNCH COSTS.

Launch vehicle	Cost (in 1982 dollars)
Delta 3914	$25 m
3920 PAM	$38 m
Atlas Centaur	$55 m
Titan 34D	$110–130 m
STS	$74 m
STS + PAM D ($4m)	$26 m
STS + PAM D II ($7m)	$35 m
STS + PAM A ($10 m)	$49 m
STS + IUS ($55 m)	$150 m
STS + Centaur ($35 m)	$110
STS + TOS ($20 m)	$82 m
Ariane ⎰ 1140 kg s/c	$27 m
⎱ 2500 kg s/c	$49 m
Proton	$25 m
N2 (Japan)	$15 m

the payload with the launcher (mass, size, real estate, level of vibrations, etc.) and the availability of the launcher. With some launch vehicles which offer a large capacity, such as the STS and the ARIANE launcher the cost of the launch vehicle can be shared between users.

With the STS the $74 million charge corresponds to a dedicated launch (use of the whole capacity). When shared between several users, the charge depends on the load factor assuming a filling efficiency of 75 per cent. The load factor is the largest of the two following ratios: (a) relative mass ratio (used mass/29 500 kg), (b) relative bay length occupancy (occupied length in the cargo bay/18.3 m).

For the launch of a geostationary satellite the mass and the length must take into account the upper stages and the associated airborne support equipment (ASE). So the total launch cost is:

$$\text{total cost} = (74/0.75) \times \text{load factor} + \text{upper stages cost} + \text{servicing}$$

The above price policy constitutes a leading factor in the design of satellites and upper stages, as an optimum load factor corresponds to nearly equal mass and length ratios. A typical example of such a concern is the HS 399 spin satellite developed by Hughes Aircraft Company which has been optimized for STS launch with an integrated propulsion sub-system. The spacecraft fits compactly in the cargo bay using only about one-fourteenth of the STS capacity, measured either by mass or bay length occupancy. This 400 kg BOL satellite can be launched by STS into transfer orbit for $10 million. This is about a twofold reduction in launch cost compared with other satellites which points out the consequence of efficient design optimization regarding a STS launch.

Baseline price for a satellite launched on Ariane begins at about $27 million for a 1140 kg satellite as part of a double payload, and increases on the linear scale to $49 million for a satellite weighting 2500 kg.

Concluding, the cost of a launch depends on many factors and is the result of commercial and political negotiations. As a rule of thumb it can be considered that the launch of a 600 kg payload on the geostationary orbit would be charged at about $30 million (in 1982 dollars), that is $50 000 per kilogram.

REFERENCES

BLEVISS, Z. O. (1976) Expendable launch vehicles for synchronous communication satellites, *AIAA, 6th CSSC, Montreal,* Paper 76-274 38pp.

DURET, M. F. (1980) Fonctionnement des fusées porteuses, *Cours CNES,* 'Le mouvement du véhicle spatial en orbite', pp. 417–428.

FREE, B. A. *et al.* (1978) Electric propulsion for communications satellites, *AIAA, 7th CSSC,* San Diego, Paper 78-537.

GILLI, M. (1984) Les systèmes de lancement de la décennie 80-90, *Note technique CNES.*

ISOPI, R. (1979) New solid propellent for European apogee boost motor, *J. Spacecraft and Rockets,* **16**(6), pp. 355–357.

HANFORD, D. R. (1980) IUS. A key transportation element for communications satellites, *AIAA, 8th CSSC,* Orlando, Paper 80-0589, 10pp.

MACCONOCHIE, J. O., R. W. MESSURIER and J. P. BAILEY (1980) Large delta wings for earth-to-orbits transports, *J. Spacecraft,* **17**(5), pp. 453–458.

MAHON, J. and J. WILD (1984) Commercial launch vehicles and upper stages, *Space Communications and Broadcasting,* **2**(4), pp. 339–362.

MARTIN, J. A. (1980) Economy and programmatic considerations for advanced transportation propulsion technology, *J. Spacecraft,* **17**(5), pp. 385–389.

MATSUDA, T., M. MIYAZANA and S. NIO 'Japan expandable launch vehicles'. ICC 82, Philadelphia, pp. 3F.1.1–3F.1.4.

PASLEY, G. F. and P. A. DONATELLI (1980) Leasat liquid apogee motor subsystem design, *J. Spacecraft,* **17**(5), pp. 396–399.

ROBERT, J. M. and J. FOLIARD (1980) Mise à poste des satellites geostatonnaires, *Cours CNES,* 'Le mouvement du véhicle spatial en orbite', pp. 505–582.

REED, D. A. *et al.* (1979) Star Raker, *AIAA/NASA, Conference on Advanced Technology for Future Space Systems, Hampton, USA,* Paper 79-0895, 13pp.

ROSEN, M. A. (1980) Space telecommunications, *IEEE Communications Magazine,* **18**(5), pp. 5–11.

STOCKWELL, B. (1980) Ariane performances and cost for communications satellite launches, *AIAA, 8th CSSC,* Orlando, Paper 80-0558, 4pp.

WIHILTE, A. W. (1980) Optimization of rocket propulsion systems for advanced earth-to-orbit shuttles, *J. Spacecraft,* **17**(2), pp. 99–104.

CHAPTER 8   Earth Station Technologies

The purpose of this chapter is to examine the aspects of earth stations directly related to the satellite communications link.

8.1   ORGANIZATION OF AN EARTH STATION
(Dorian, 1979; Van Trees, 1979)

Figure 8.1 depicts a typical earth station which comprises four major sub-systems: *receiver*, *antenna*, *transmitter* and *tracking equipment*.

Along with those sub-systems directly related to the satellite communications link, one should also consider the *interface equipment* with the terrestrial network and power supply installations. Note that Figure 8.1 applies to large

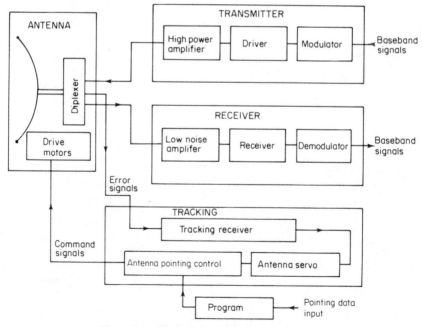

Figure 8.1   Block diagram of an earth station.

328

Figure 8.2
Large Standard A earth station antenna, along with a tripod mounted smaller antenna.

stations where the tracking equipment is required to ensure the precise pointing of the narrow beamwidth antenna, while with small earth stations with large beamwidth antenna no tracking is needed.

As the same antenna is used for transmitting and receiving, a *diplexer* is used to separate the transmitted and received signals. Very often *dual polarization* (circular or linear) on the transmitted and received signals is employed to allow for frequency reuse.

The size and complexity of earth stations depends on the required service and the EIRP of the satellite down-link transmitter. At one extreme, for example, are portable stations with antennae of about 1 m diameter, and at the other are the largest Intelsat Standard A stations with antennae of 32.5 m diameter, as illustrated on Figure 8.2.

8.2   EARTH STATION DESIGN OBJECTIVES

The desired characteristics for most earth stations are:

(1) *High gain* in the direction of *wanted signals.*
(2) *Low gain* in the direction of *unwanted signals.*
(3) *Low effective noise temperature* for the entire receiving system.
(4) *High antenna efficiency.*
(5) *Continuous satellite pointing* with a specified accuracy.
(6) *Minimum performance variations* caused by local wind and weather.
(7) *Minimal variation in illumination* of the satellite by the earth station.
(8) *High discrimination* between orthogonally polarized signals.

8.2.1   Figure of merit of the receiving station

Signal levels received by a satellite earth station are very low. This makes it essential that the receiving antenna is of high gain in the direction of the satellite and that the receiving system introduces as little noise as possible. The efficiency of this combination is usually quoted as the ratio of the gain to the noise temperature and is termed the '*figure of merit*'. This figure indicates the relative capability of the receiving system to receive a signal and is directly related to the *overall carrier power to noise power density ratio* of the down-link at the input of the receiver, according to formula(16) already discussed in Chapter 2:

$$\frac{C}{N_0} = P_T G_T \frac{1}{L} \frac{G_R}{T} \frac{1}{k} \tag{1}$$

where G_R and T are the antenna power gain and the system noise temperature referred to the input of the low noise receiver, respectively. T is dependent on the antenna noise temperature T_A, the noise temperature at the receiver input

T_R and the feed temperature θ_F in the following manner:

$$T = \frac{T_A}{L} + \theta_F\left(1 - \frac{1}{L_F}\right) + T_R \tag{2}$$

where L_F is the feeder loss.

As the noise temperature of the antenna varies with the elevation angle (Figure 2.3) the figure of merit also varies with the elevation angle.

8.2.2 Gain of the antenna major lobe

The simplest way of determining antenna gain is by measurement of the transmitter output power and received level when the free space loss is known. This practical method is not very accurate primarily because of the difficulty in calibrating the transmit and received signal levels. Comparison with the standard antenna method is most widely used. The absolute method which involves receiving the noise from a known celestial source is also used.

The equation for antenna gain

$$G = 4\pi\eta A/\lambda^2 \tag{3}$$

suggests that using a large antenna, i.e. an increased surface area A, is preferable whenever this is practical. But this equation applies only in the case of a perfect reflector; in reality the gain is expressed by:

$$G = \frac{4\pi\eta A}{\lambda^2} \exp\left[-B\left(\frac{4\pi\varepsilon}{\lambda}\right)^2\right] \tag{4}$$

where ε is the surface r.m.s. error between the real reflector profile and the theoretical one caused by irregularities and imperfections of its surface finish and B is the correction factor, less than or equal to 1, whose value depends on the radius of the curve of the reflector. The lower the radius, the higher this factor. For parabolic antennae of focal length f, it varies in terms of the ratio f/D, where D is the antenna diameter as shown in Figure 8.3.

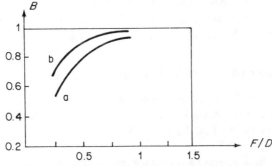

Figure 8.3 Correction factor versus focal length-to-antenna diameter ratio: (a) deformation measured along the paraboloid axis; (b) deformation measured perpendicular to the reflector.

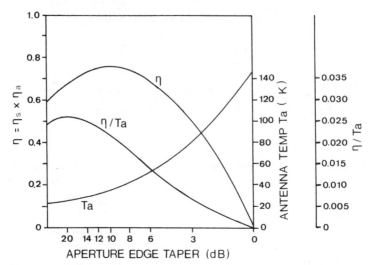

Figure 8.4   Efficiency and noise temperature of a Cassegrain antenna versus the aperture edge taper.

The *efficiency* is $\eta = \eta_s \eta_a$, where η_s is the efficiency related to the *spill-over*, i.e. the radiation emitted by the primary feed which does not illuminate the reflector and η_a is the efficiency dependent on the *uniformity of the radiation emitted by the primary feed*. The more uniform the illumination, the larger is η_a. However, as pursuance of uniform illumination and a minimum spill-over are conflicting objectives, a trade-off involves attenuating the illumination at the periphery of the reflector. This is called the *aperture edge taper*.

In the case of a Cassegrain antenna (Section 8.3.3) the efficiency η is at the maximum for an aperture edge taper of about 10 dB (Figure 8.4). On the other hand, with 18 dB attenuation we achieve a maximum η/T_A ratio and therefore an improved figure of merit, but this improvement is negligible with a medium noise figure receiver (noise temperature of the system is in the order of 200 K); so, the antenna can be designed for maximum gain.

8.2.3   Radiation characteristics outside the main beam

The *sidelobes* of earth station antennae have a direct influence on the level of interference from one network to another. This is particularly important for those antenna sidelobes that lie within a few degrees of the direction of the geostationary orbit.

To minimize these effects, a *reference pattern* has been agreed to WARC-79 (CCIR, Rep.391–4, 1982). This pattern is shown on Figure 8.5 according to the following formulas.

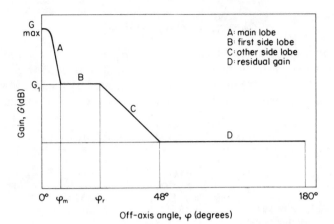

Figure 8.5 Fixed satellite service reference pattern. *Reproduced with permission from CCIR Report 391-4 (1982).*

(1) For antennae with a diameter D larger than 100 λ where λ is the wavelength

$$G(\phi) = G_{max} - 2.5 \times 10^{-3} \left(\frac{D}{\lambda} \phi\right)^2 \quad \text{for} \quad 0 < \phi < \phi_m$$

$$G(\phi) = G_1 = 2 + 15 \log \frac{D}{\lambda} \quad \text{for} \quad \phi_m \leqslant \phi < \phi_r$$

where $G_1 = $ gain of the first side lobe

$$G(\phi) = 32 - 25 \log \phi \quad \text{for} \quad \phi_r \leqslant \phi < 48°$$

$$G(\phi) = -10 \quad \text{for} \quad 48° \leqslant \phi \leqslant 180° \quad (5)$$

$$\phi_m = \frac{20\lambda}{D} \sqrt{(G_{max} - G_1)} \text{(degrees)}$$

$$\phi_r = 15.85 \left(\frac{D}{\lambda}\right)^{-0.6} \text{(degrees)}$$

(2) For antennae with a diameter D smaller than 100 λ

$$G(\phi) = G_{max} - 2.5 \times 10^{-3} \left(\frac{D}{\lambda} \phi\right)^2 \quad \text{for } 0 < \phi < \phi_m$$

$$G(\phi) = G_1 = 2 + 15 \log \frac{D}{\lambda} \quad \text{for } \phi_m \leqslant \phi < 100 \frac{\lambda}{D}$$

$$G(\phi) = 52 - 10 \log \frac{D}{\lambda} - 25 \log \phi \quad \text{for } 100 \frac{\lambda}{D} \leqslant \phi < 48°$$

$$G(\phi) = 10 - 10 \log \frac{D}{\lambda} \quad \text{for } 48° \leqslant \phi \leqslant 180°$$

$$\phi_m = \frac{20\lambda}{D} \sqrt{(G_{max} - G_1)} (\text{degrees})$$

It should be noted that CCIR has defined a *more stringent recommendation* concerning antennae of *new earth stations* installed after 1987 (CCIR, Rec. 580, 1982). This recommendation has been motivated by the fact that *efficient use of the radio spectrum* has become a primary factor in the management of

Table 8.1   (a) INTELSAT BUSINESS SERVICE—STANDARD E TERMINALS
(TDM/QPSK/FDMA) AT 14/11GHz

	E–1	E–2	E–3
Antenna diameter	3.5 m	5.5 m	8 to 10 m
EIRP	57–86 dBW	55–83 dBW	49–77 dBW
G/T	25 dB K^{-1}	29 dB K^{-1}	34 dB K^{-1}
64 Kbps channels equivalent (available to customer)	400	700	1,000
Equivalent number of 1.5 Mbps information rate carriers	16	28	42
Largest available information rate per earth station (72 MHz transponders)	2.048 Mbps	4.096 Mbps	6.144 Mbps
System performance (99% of the time)	1×10^{-6} bit error rate	1×10^{-6} bit error rate	10×10^{-6} bit error rate
Link performance (clear sky conditions)	1×10^{-8} bit error rate	1×10^{-8} bit error rate	1×10^{-8} bit error rate

(b) INTELSAT DOMESTIC LEASE SYSTEM STANDARD Z TERMINALS
(6/4 GHz) (Kelley, 1984).

	'Large'	'Small'	TV/radio 'receive only'
Antenna size (m)	11.0–13.0	6.0–8.0	4.5–5.0
Transmit gain (dBi)	54.5–56.0	49.3–51.7	—
Receive gain (dBi)	51.5–53.0	46.3–48.7	44.0
LNA temperature (K)	45°–80°	100°	100°
G/T at 10° (dB/K)	31.7–33.0	24.5–26.9	22.0
Tracking	Automatic	Manual	Manual
Polarization	Dual/circular	Dual/circular	Dual/circular
Axial ratio (Tx/Rx)	1.06/1.09	1.06/1.09	—/1.4
Communications capability			
No. of channels	More than 12	2–12	1 Video + Audio 1 Radio
Type of HPA	TWT or Klystron	TWT	—
HPA rating	3 kW—TV 1 kW—TPY	50–400 W	—
Types of carriers	FDM/FM, FDM/CFM SCPC/DM TV/FM	SCPC/CFM SCPC/DM	Video/Audio—FM Radio–SCPC/CFM
Power per SCPC channel			
toward large	< 1 W	1 W	—
toward small	1 W	10 W	—

the geostationary satellite orbit. For these future earth station antennae the design objective requires that the gain of 90 per cent of the side lobe peaks will not exceed

$$G = 29 - 25 \log \phi$$

This requirement should be met for any off-axis direction which is within $3°$ of the geostationary satellite orbit for which $1° \leqslant \phi \leqslant 20°$.

8.2.4 INTELSAT earth station standards:

INTELSAT has specified the following standards:

(1) *Standard A* (6/4 GHz, antenna diameter of about 30 m, EIRP = 70–90 dBW, G/T = 40.7 dBK^{-1}) mainly for FDM/FM/FDMA and 120 Mbit/s PSK/TDMA trunking telephony, and TV transmission.
(2) *Standard B* (6/4 GHz, antenna diameter = 11–14 m, EIRP = 60–85 dBW, G/T = 31.7 dBK^{-1}) mainly for SCPC/QPSK and FDM/FM FDMA thin route telephony and data, and transmission of TV signals.
(3) *Standard C* (14/11 GHz, antenna diameter = 14–18 m, EIRP = 72–87 dBW, G/T = 39 dBK^{-1}) mainly for FDM/FM/FDMA and 120 Mbit/s PSK/TDMA trunking telephony, and TV tranmission.
(4) *Standard D* (6/4 GHz, antenna diameter = 5 m (D1) or 11 m (D2), EIRP = 53–57 dBW, G/T = 22.7 dBK^{-1}(D1) or 31.7 dBK^{-1}(D2)) for the VISTA (low density telephone service) using SCPC/FM.
(5) *Standard E* (14/ 11GHz) to provide TDM/QPSK/FDMA for international business services (IBS). See Table 8.1(a) for more details.
(6) *Standard F* (6/4 GHz, antenna diameter = 4.5–5 m (F1), 7.5–8 m (F2), 9–10 m (F3), EIRP = 63–91 dBW (F1), 60–87 dBW (F2), 59–86 dBW (F3), G/T = 22.7 dBK^{-1} (F1), 27 dBK^{-1} (F2), 29 dBK^{-1} (F3)) to provide TDM/QPSK/FDMA for international business services (IBS).
(7) *Standard Z* (6/4 GHz and 14/11 GHz) concerning all stations operating with leased transponders for domestic uses. Transmission characteristics may be selected by the lessee as long as mandatory requirements related to protection against interference are satisfied. Table 8.1(b) indicates example characteristics of standard Z earth stations operating at 6/4 GHz.

8.2.5 EUTELSAT earth station standards:

EUTELSAT uses INTELSAT Standard C for stations supporting trunking telephony and TV transmission and considers two standards for *satellite multiservices* (SMS): Standard 1 relates to a 5 m diameter antenna with G/T = 30.4 dBK^{-1} and Standard 2 to a 3.7 m diameter antenna with G/T = 27.4 dBK^{-1}.

8.3   EARTH STATION EQUIPMENT

8.3.1 Receivers

The total noise temperature at the receiver input has already been examined.

Antennae temperatures are low if the gain is sufficiently high (cold antennae). Feeder attenuation is best reduced by placing the first stage of the receiver close to the antenna feed. Therefore, receivers with low noise first stages and sufficiently high gain should be used (LNA or low noise amplifier). Large stations use cooled or uncooled *parametric amplifiers* while small economic ones use *FET amplifiers*.

Figure 8.6 shows the structure of a two-stage parametric amplifier. The main element is the varactor (diode with non-linear capacitance) which is connected to three ports tuned to the signal frequency F_s, the oscillator or pump frequency F_p and to the idler frequency F_i. When $F_p > F_s$ and when $F_i = F_p - F_s$, signals at frequency F_s are amplified; the power supplied to the output circuit is $P_S = -P_P(F_s/F_p)$, where P_S and $-P_P$ are respectively the powers given by the varactor at the frequencies F_s and F_p. So a power transfer takes place from the pump to the signal.

The circulator, which had initially routed the signal delivered by the antenna from port 1 to port 2, directs the reflected and amplified signal from port 2 to port 3.

The advantage of the varactor is to allow amplification using a reactance theoretically devoid of noisy resistive elements. The noise temperature T is given by:

$$T_N = \theta \left[\frac{1}{Q^2} \frac{F_i}{F_s} + \frac{F_s}{F_i} \right] \qquad (7)$$

where θ is the physical temperature of the diode and Q its quality factor. From

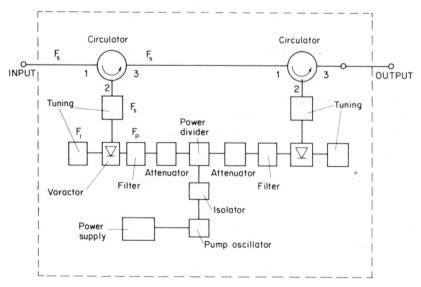

Figure 8.6   Structure of a two-stage parametric amplifier.

equation (7) one sees various means to reduce the noise temperature of a parametric amplifier.

(1) Use of a high Q varactor.
(2) Selection of an appropriate idler frequency.
(3) Cooling the varactor and its associated circuits.

For example, if $\theta = 20$ K, $Q = 10$, $F_i = 28$ GHz and $F_s = 4$ GHz, the temperature T is 4.3 K. In the same conditions, if $\theta = 300$ K, $T = 67$ K.

If B_s and B_i are respectively the bandwidth of the circuits tuned to the signal and to the idler frequency, the amplifier gain G and bandwidth B are roughly linked by:

$$\sqrt{(G)}B = \frac{2}{\dfrac{1}{B_s} + \dfrac{1}{B_i}} \tag{8}$$

Table 8.2 summarizes the performance of usual types of low noise amplifiers used in receivers.

Table 8.2 Low Noise Amplifiers Characteristics.

Amplifier type	Frequency	Noise temp.	Cost	Use
Cryogenic parametric	4 GHz	15 K	High	First-built large stations
Cooled parametric	4 GHz 12 GHz	35 K 100 K	Medium Medium	Standard A and B Standard C
Uncooled parametric	4 GHz 12 Ghz	50 K 120 K	Medium Medium	Large domestic stations EUTELSAT stations
Cooled FET GaAs	4 GHz 12 GHz	50 K 130 K	Low Low	Domestic stations
Ambient FET GaAs	4 GHz 12 GHz	75 K 250 K	Low Low	Small domestic stations

8.3.2 Transmitters

An earth station transmits at high frequency from tens to thousands of watts. Presently *klystrons* and *travelling wave tubes* have been used. Compared with klystrons, TWTs permit high powers over a wide bandwidth. Cooling of these tubes is essential and this is achieved by water circulation using a closed refrigeration system. Table 8.3 summarizes the performance of usual types of power amplifiers used in transmitters.

Table 8.3 POWER AMPLIFIER CHARACTERISTICS.

Table 8.3 POWER AMPLIFIER CHARACTERISTICS.

Amplifier type	Output power	Bandwidth	Cost
Klystron	500–5000 W	Small (40 MHz)	Medium
Travelling wave tube, (TWT)	100–2500 W	Large (500 MHz)	High
FET (6 GHz) (14 GHz)	5–50 W 1–6 W	Large	High

8.3.3 Antennae types

There are several types of antennae: (a) horn antennae, (b) phased array antennae and (c) reflector antennae.

The *horn antenna* has a good performance, but is expensive and very bulky when high gain is required. Hence they are not used for earth stations (the Pleumeur Bodou one is an historical example).

The *phased array antenna* is of practical interest as long as the antenna has small dimensions, which implies low gain value. Large antennae require severe mechanical performance to avoid any deformation. Still, the phased array antenna has advantages when the beam requires steering as in the case of stations on-board mobiles;

Finally *reflector or dish antenna* are most appropriate for satellite communications as they offer a wide range of gain values with affordable mechanical complexity. The three principal implementations are:

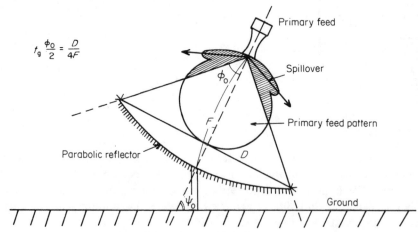

$$tg\,\frac{\phi_0}{2} = \frac{D}{4F}$$

Figure 8.7   Axisymmetric reflector antenna.

(1) *Single parabolic reflector* antenna.
(2) *Single 'offset reflector'* antenna.
(3) *Dual reflector 'Cassegrain'* antenna.

8.3.3.1 Single Parabolic Reflector Antenna

Figure 8.7 illustrates the assembly of an *axisymmetric single parabolic reflector* antenna with primary feed at its focus. With such earth station antennae, the primary feed faces the ground, so the *noise temperature* can be considerable due to the *spill-over* radiated from the ground.

This spill-over can be reduced by *aperture edge tapering*. Obtaining a low noise temperature requires a highly directional primary feed and therefore a large focal length. This makes the antenna more cumbersome and does not facilitate the implementation of electronic hardware immediately behind feed source, as would normally be desirable to reduce feeder losses; indeed, any extra bulk at the axial feed axis contributes to masking or partial *blocking of the aperture plane*.

8.3.3.2 Offset Reflector Antenna

Figure 8.8 illustrates the configuration of an *offset reflector antenna*. 'Offset' mounting makes possible the placing of RF hardware directly behind the

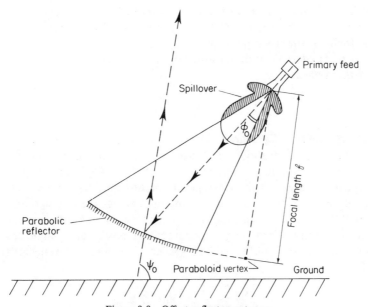

Figure 8.8   Offset reflector antenna.

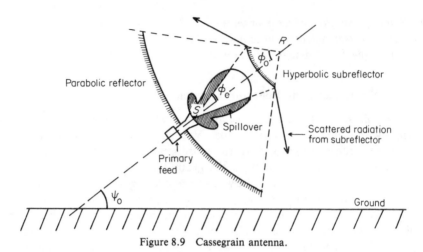

Figure 8.9   Cassegrain antenna.

primary feed without *any aperture blocking effect*. It is nevertheless not often used as the frame which supports the feed and the RF circuits are heavy and costly and in any case it does not solve the problem of the spill-over.

8.3.3.3   Cassegrain Antenna

Figure 8.9 shows an antenna with a *Cassegrain mounting*. The Cassegrain antenna has the phase-centre of the primary feed located at one of the focuses S of a hyperbolic subreflector. The other focus R of the subreflector coincides with the focus of the main parabolic reflector. Therefore if D is the diameter of the aperture of the parabolic reflector, f_d its focal length, the half-apex-angle of this reflector is:

$$\text{tg}\,\frac{\phi_0}{2} = \frac{D}{4f_d} \tag{9}$$

The performance of a Cassegrain antenna can be evaluated by using the concept of 'equivalent parabolic' reflector antenna. An *equivalent parabolic reflector antenna* is defined as an antenna which has a focal length equal to the equivalent focal length of the Cassegrain antenna and diameter equal to that of the main reflector of the Cassegrain antenna. Hence the Cassegrain antenna of Figure 8.9 is equivalent to the prime focus feed antenna of Figure 8.7 with a diameter of D and a focal length f_e characterized by an apex angle $2\phi_e$ which is the angle with apex S subtended by the edge of the auxiliary reflector (Figure 8.10).

$$\text{tg}\,\frac{\phi_e}{2} = \frac{D}{4f_e} \qquad \text{with } f_e \geqslant f_d \tag{10}$$

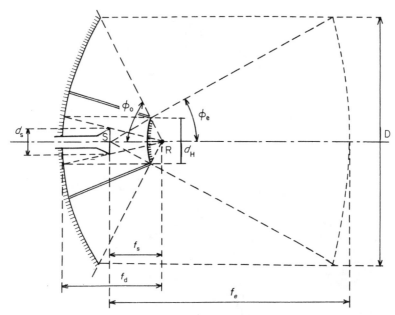

Figure 8.10   Equivalent reflector of a Cassegrain antenna.

The Cassegrain antenna is therefore *less cumbersome* than the equivalent reflector antenna, even though it retains the *advantages of antennas* with a *long focal length*. Cassegrain antennae offer the following advantages:

(1) The *noise temperature* of this antenna can be *very low* for two reasons:
(i) The major portion of the spill-over is not earth but sky generated.
(ii) The spill-over can be made very low as the total equivalent focal length is large allowing the use of very directional primary feeds, and small focal lengths of the parabolic and hyperbolic reflectors f_d and f_s result in a reduction of the residual spill-over.
(2) *Imperfections* in manufacture *do not degrade the antenna* gain very much, as the correction ratio B in Equation (4) is nearly equal to 1 as a result of a high f/D value. (Figure 8.4).

A further advantage is that it is possible to place the RF circuits immediately behind the primary feed and so limit the effect of feeder loss. So as to allow the implementation of RF circuits in a shelter at ground level, and still operate whatever the antenna pointing is, a *beam waveguide system* is often adopted on large earth stations. As illustrated in Figure 8.11 the beam waveguide comprises a set of four reflectors which ensures a proper illumination of the subreflector by the feed located in the shelter. This solution presents the advantage of small losses and permits the rotation of the reflector antenna about two orthogonal axes while the feed and the RF equipment remain fixed.

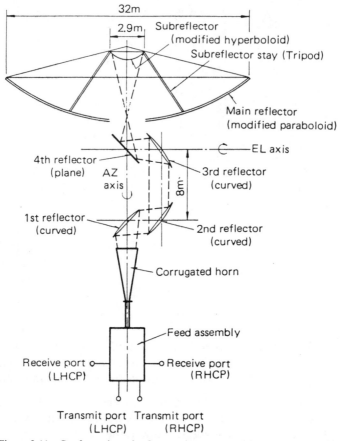

Figure 8.11   Configuration of a Cassegrain antenna with a beam waveguide
system.

All these factors have made the Cassegrain antenna the most widely used on
large earth stations. The remaining problem is the aperture blockage of the
subreflector, and this can be ignored if d_H/D is small (d_H is the subreflector
diameter). For a mean size antenna this effect can be minimized by choosing
the dimensions such that:

$$\frac{f_s}{f_d} = \frac{d_s}{d_H}$$

$$d_s = \left(\frac{2f_d\lambda}{\eta_s}\right)^{1/2}$$

(11)

where the symbols are as indicated in Figure 8.10 and η_s is the primary feed
efficiency.

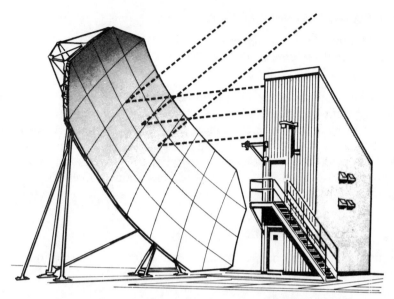

Figure 8.12 Multibeam torus antenna. *Reproduced by permission of Communications Satellite Corporation from Kreutel and Potts (1980).*

Finally, if the Cassegrain offset mounting is chosen blocking is avoided (Combes, 1980).

8.3.3.4 Multibeam Torus Antenna

It is possible to specially design an antenna, of relatively small dimensions, to receive simultaneously several satellites located in one section of the geostationary arc. Such a system developed by COMSAT laboratories (Kreutel and Potts, 1980) called the MBTA (Multibeam Torus Antenna) acts as a parabolic antenna of 9.8 m diameter with a gain in the order of 50 dB at 6/4 GHz and a noise temperature of 30 K for an elevation angle of 20° (Figure 8.12).

8.4 ANTENNA POINTING AND TRACKING

Pointing an antenna at a geostationary satellite requires the knowledge of the azimuth and elevation of the antenna in terms of the geographic location of the station, as calculated in Chapter 4. *Tracking* consists of maintaining the axis of the antenna beam in the direction of the satellite despite the movement of the satellite or the station.

Pointing and tracking depend on:

(1) The *beamwidth* of the antenna beam.
(2) The *apparent motion* of the satellite.
(3) The *type of station,* fixed or mobile.

A further factor affecting orientation equipment performances is the mass associated with larger diameter reflectors. Small reflectors weigh tens to hundreds of kilograms, whereas a large INTELSAT Standard A earth station such as Pleumer Bodou IV or Etham has reflectors that weigh hundreds of tons. The meteorological conditions (wind speed) and the antenna weight, moreover, cause diameter depending deformations which vary with the elevation.

8.4.1 Antenna beamwidth

As a first approximation, the *antenna beamwidth* with tapered illumination assuming a 3 dB gain fall-out is:

$$\theta_{\text{degrees}} = 70 \, \frac{\lambda}{D} \tag{12}$$

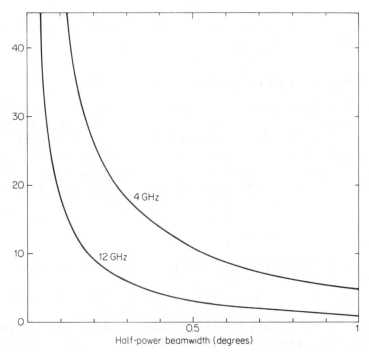

Figure 8.13 Diameter of a reflector antenna versus the half-power beamwidth.

where D is the antenna diameter. (For a uniformly illuminated aperture the half-power beamwidth is 57 λ/D.)

Figure 8.13 shows the diameter of the antenna against the half-power beamwidth angle θ_{3dB} at frequencies of 4 GHz and 12 GHz. The diameter of the antenna beam footprint on a sphere whose radius is that of geostationary orbit is:

$$d_{km} = \frac{(35\ 800\ \theta_{3dB})}{57}$$

where θ_{3dB} is the half-power beamwidth expressed in degrees, assuming the station is located near the subsatellite point. With a station of zero elevation the footprint diameter is increased by approximately 16 per cent.

8.4.2 Apparent motion of a satellite

The apparent motion of a satellite has been studied in Chapter 4, and we have seen that the satellite describes a figure-of-eight pattern, the dimensions of which are determined by the angle of inclination of the orbit. If orbit eccentricities are negligible, then the maximum latitudinal displacement is equal to i and the maximum longitudinal displacement (L_{max}) and its related latitude

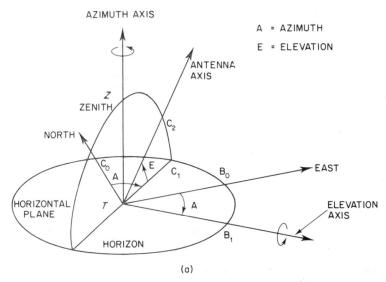

(a)

Figure 8.14   Azimuth-elevation antenna mounting: (a) geometry, TZ = primary fixed vertical axis, TB = secondary axis. Pointing of the antenna axis towards C_2 starting from the north (C_0): (1) rotation A around TA: C_0 becomes C_1; (2) rotation E around TB,: C_1 becomes C_2. (b) Implementation.

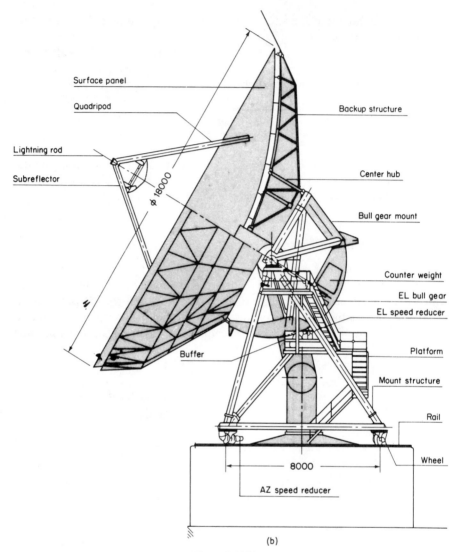

Figure 8.14(b)

l are expressed by:

$$L_{\max} = 4.36 \times 10^{-3}\, i^2$$

and   (13)

$$l = 0.707i$$

all values are expressed in degrees.

The longitudinal displacement is many times less than latitude. The apparent velocity of the satellite is about $2°/h$.

However, it has been shown in Chapter 6 that the combination of drifts and station keeping manoeuvres results in an apparent motion of a geostationary satellite within a window whose dimensions depend on the station keeping accuracy (e.g. $\pm 0.1°$ NS and EW for INTELSAT V and Telecom I). Hence with geostationary satellites the only parameters to be considered are the *window dimensions*.

8.4.3 Antenna mounting

The antenna is moved around two axes, a *primary axis* being fixed relative to the Earth, and a *secondary mobile axis* around the former. Three mountings are possible: azimuth-elevation, X–Y, equatorial. The first one is very frequently adoped, the second one sometimes, and the third one very seldom. In addition, tripod mounting is used when changes in the pointing of the antenna are not required.

8.4.3.1 Azimuth-elevation mounting:

The *'azimuth-elevation' mounting* has a fixed vertical primary axis and the secondary axis is on the horizontal (Figure 8.14). This mounting is currently used for earth stations. A variation of this moutning called 'non-orthogonal azimuth elevation' has the secondary axis not in the horizontal plane but at an angle α with the primary axis. This is advantageous as the volume generated by the antenna contour during its motion is smaller than with the 'azimuth-elevation' mounting (Figure 8.15).

8.4.3.2 X–Y mounting

The *X–Y mounting* has an horizontal fixed primary axis and a secondary axis

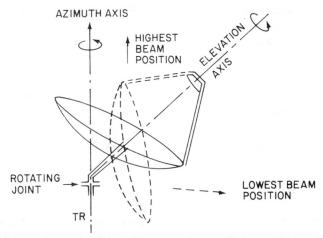

Figure 8.15 Azimuth-elevation antenna mounting with non-orthogonal axes.

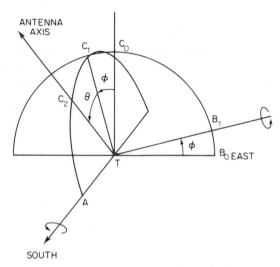

Figure 8.16   X-Y antenna mounting: TA, primary fixed horizontal
axis; TB, secondary axis. Pointing of the antenna axis
towards C_2 starting from the zenith (C_0): (1) rotation ϕ
around TA, C_0 becomes C_1; (2) rotation θ around TB_1,
C_1 becomes C_2.

perpendicular to the first one (see Figure 8.16). It avoids the drawback of the
azimuth-elevation mounting which undergoes high-speed displacements when
the satellite passes near zenith. Hence the X–Y mount antennae are more
suitable to low orbiting satellites than to geostationary satellites.

8.4.3.3   Equatorial mounting

The *'equatorial' mounting* has a fixed primary axis (hour axis) parallel to the
Earth's rotation axis and a declination axis perpendicular to the former
(Figure 8.17). If the orbit inclination is nil, tracking is nearly achieved by
rotation around the primary axis.

8.4.3.4   Tripod mounting

The *tripod mounting* is very suitable for geostationary satellites and small
earth stations. The antenna (Figure 8.18) is fixed to a support by three legs.
The length of one leg is fixed, while the other two are variable. Once positioned
such an antenna allows only for limited amplitude variations of the pointing
(up to $10°$).

8.4.4   Tracking

The *tracking procedure* employed by an earth station antenna depends on the
relative dimensions of the antenna beamwidth and of the window of the

satellite. The worst case is when the station is located at the subsatellite location. If not, the amplitude of the satellite motion as seen from the earth station is smaller.

8.4.4.1 Fixed-pointing of the antenna

Tracking is not necessary when the half-power beamwidth of the antenna is much larger than the satellite window. The depointing angle depends on the satellite station-keeping tolerance and the systematic pointing error during the initial adjustment of the earth station antenna. This adjustment should be co-ordinated with the satellite control centre in order to determine at what time

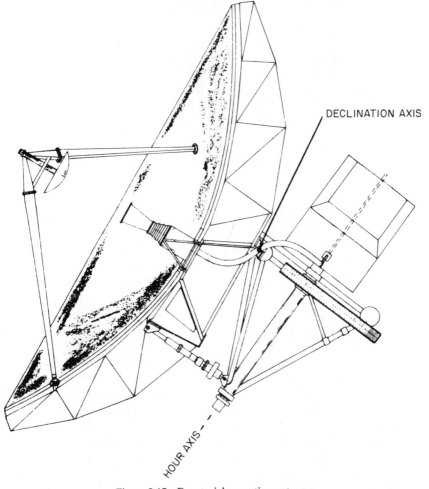

Figure 8.17 Equatorial mounting antenna.

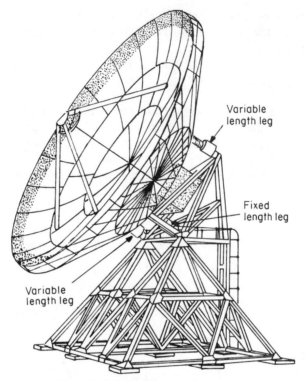

Figure 8.18   Tripod mounting antenna.

the satellite is at nominal station. The procedure for adjustment of the antenna is to offset it in order to measure the gain on either side of the axis and then to set it to the midpoint between equal off-axis gain settings. The *systematic pointing error* is then equal to $0.2\,\theta_{3dB}$. It should be noted that the control centre determines the actual position of the satellite with an accuracy of about $0.02°$.

8.4.4.2   Automatic tracking

When the half-power beamwidth of the antenna is relatively small compared with the dimensions of the satellite window, antenna pointing corrections must be quasi-permanent. This is achieved by *continuous automatic tracking*, using the signal received from the satellite. The depointing error reduces to the systematic pointing error only which depends on the performance of the tracking receiver and of the servo-system. A large signal to noise power ratio is required to ensure that changes in signal level caused by stepping of the antenna are detected. Signal fluctuations owing to propagation impairments and satellite EIRP variations affect the performance severely.

Two techniques are used for automatic tracking:

(1) *Monopulse or multi-horn tracking* involves excitation of the antenna polar response with a zero in its axis which permits the antenna to be directed so that the received signal is at a minimum. The error signals originate by summing and differencing the signals received by four microwave horns arranged around the radioelectric axis of the antenna, or through detection of higher order modes developed in the primary feed by depointing of the antenna. The latter solution offers better sensitivity (Gorton, 1980). Monopulse systems present a *low systematic pointing error*, of the order of $0.1\theta_{3dB}$, but are complex and expensive and need care in installation and maintenance.

(2) *Step by step* (also called step-track or hill climb) *tracking* is achieved by a search for the maximum reception of the received signal. It works by successive displacements of the antenna around the two rotation axes and compares the signal levels received for each of the antenna positions. Pointing is achieved either periodically at regular intervals, or on detection of a drop-off in the signal level at the receiver. The *systematic pointing error* is of the order of $0.2\,\theta_{3dB}$

The use of step-by-step tracking systems is becoming prevalent for medium sized as well as for large stations as they provide cost savings over monopulse. The resetting of the antenna can be achieved by simply moving the subreflectors (Akagawa and Yokoi, 1976).

8.4.4.3 Programmed pointing

An intermediate solution between fixed pointing of the antenna and automatic tracking is the use of ephemerids of the satellite orbit to determine pointing angles at every moment. The pointing angles are computed and stored at the earth station and fed to the antenna drive system.

8.5 MOBILE AND TRANSPORTABLE EARTH STATIONS

Transportable and mobile stations must be able to move freely within a coverage area and it must be possible for the antenna to be pointed in all directions above the horizontal plane.

For mobile stations azimuth-elevation mountings have the disadvantage of needing high angular velocities when the satellite is in the vicinity of the primary vertical axis. This disadvantage can be avoided by giving the pointing system a further degree of liberty.

Finally, for small transportable earth stations, and with geostationary satellites, the tripod mounting is very suitable.

Figure 8.19   INMARSAT Standard—a ship Earth station.

8.5.1   Stations on ships

Small beamwidth antennae require automatic tracking. The difficulties with acquisition and tracking of the satellite by the servo-system may necessitate the use of an inertial stabilized platform. The antenna movement in relation to the ship therefore results from the information derived by the tracking receiver and by the inertial platform.

The INMARSAT organization has issued standards for ship stations. The Standard A ship earth station consists of two portions, *above deck equipment* and *below deck equipment*. The above deck equipment consists of an antenna with stabilization and automatic steering equipment enabling the antenna beam to remain pointed at a satellite regardless of course and ship movements, solid state L-band power amplifier, L-band low noise amplifier diplexer and a low loss protective radome (see Figure 8.19). The below deck equipment consists of an antenna control unit, communications electronics used for transmission, reception, access control and signalling, and telephone and teleprinter equipment. Optional equipment for low speed data, high speed data, facsimile, etc., can be installed with the below deck equipment. *Standard A* ship earth station characteristics are given in Table 8.4.

In addition to the Standard A ship earth stations, INMARSAT has intro-

Table 8.4 INMARSAT STANDARD—A SHIP EARTH STATION CHARACTERISTICS.

Antenna system	Communications
Diameter: typically 1.2 m Gain: typically around 23 dB Beamwidth: typically around 10° Polarization: right-hand circular transmit and receive	Transmit band: 1636.5–1645.0 MHz Receive band: 1535–1543.5 MHz Receive G/T: $\geqslant -4$ dBK^{-1} Transmit EIRP: 37 dBW ± 1 dBW Capacity: Single duplex telephony or telegraphy call at any time

duced two other standards: *Standard B* ($G/T = -12$ dB K^{-1}) and *Standard C* ($G/T = -19$ dB K^{-1}). These standards allow for a significant reduction in size and weight of the associated terminals. Vocoders, possibly in conjunction with forward error correction techniques such as convolutional encoding, can provide a speech quality with Standard B terminals estimated to be fair to good. Standard C terminals provide the same basic services as Standard B (telephony, data transmission, facsimile and telegraphy) but with reduced speech quality. Another standard that might deserve consideration is a -24 to -26 dB K^{-1} G/T terminal that could be installed on very small ships and provide telegraphy or possibly digital speech transmission links using low bit rate vocoders (below 1 Kbit/s).

8.5.2 Stations on aeroplanes

Telecommunication antennae may be placed on aeroplanes, within the framework of communications and navigation satellites (Navstar (Sidford and Owen, 1978), Aerosat (Le Bow *et al.*, 1971; Brown and Swann, 1978)).

REFERENCES

AKAGAWA, M. and H. YOKOI (1976) A new earth station antenna for geostationary satellites. *Progress in Astronautics and Aeronautics,* Vol. 55, pp. 451–467, Montreal.
BROWN, D. L. and G. E. SWANN (1978) A study of aerosat payload configurations, *Inf. Conf. Maritime and Aeronautical Satellite Communication and Navigation,* London, pp. 109–115.
COMBES, P. F. (1980) Ondes métriques et centimetriques, Dunod Université.
DORIAN, C. (1979) The Marisat system, in D. J. Curtin (ed.) *Trends in Communication Satellites,* Pergamon Press.
GORTON, W. D. (1980) Transportable earth stations for multiple applications, *Wescon,* pp. 9.3.1–9.3.7.
KELLEY, T. M. (1984) Leased services on the INTELSAT system: domestic service and international television, *Int. Journ. Satellite Comm.,* 2(1), pp. 29–40.

KREUTEL, R. W. and J. B. POTTS (1980) The multiple-beam torus earth stations antennas, *IEEE, International Conference on Communications, Seattle,* pp. 25.4.1–25.4.3.

LE BROW, J. L. *et al.,* (1971) Satellite Communications to mobile platforms, *Proc. IEEE,* **59**(2) (Feb.), pp. 139–159.

SIDFORD, M. J. and J. I. R. OWEN (1978) Aircraft aerials for the Navstar satellite systems, *Inf. Conf. Maritime and Aeronautical Satellite Communication and Navigation,* London, pp. 101–105.

VAN TREES, H. L. (1979) Earth stations, *Satellite Communications.* IEEE Press, pp. 579–606. (a) E. R. WALTHALL, Earth stations technology, pp. 579–588; (b) R. W. GRUNNER, Intelsat V Earth station technology, pp. 588–594; (c) L. CUCCIA and C. HELLMAN, The low cost capacity earth terminal.

CHAPTER 9 Reliability of Satellite Communication Systems

9.1 INTRODUCTION TO RELIABILITY (Corazza, 1975)

The reliability is the probability that the system will perform satisfactorily over its specified life. It depends on the two principal component parts of the system—the satellite and ground segment.

The availability is the probability that the system is operating satisfactorily at any point in time. Availability moreover depends on the success of the launch, replacement time, number of operational satellites and back-up satellites (in orbit and on Earth).

For earth stations, availability does not solely depend on reliability but also on the maintainability. For a satellite, the availability depends only on its reliability, as no maintenance is possible.

9.1.1 Failure Rate

A complex piece of equipment such as a satellite, displays two modes of failure:

(1) Accidental failure (random failures)
(2) Failure caused by wear out (bearings, solenoid-operated valves, TWT cathodes) or exhaustion of expendables (propellent).

Assuming instantaneous replacement of a failing part, it is possible to determine the instantaneous failure rate (the number of failures per unit time) for a given equipment over the equipment life. The curve of failure rate against time would resemble Figure 9.1 and is called a 'bath tub' curve.

In a satellite the infant mortalities are eliminated in advance of the launch by various methods such as heat cycling and operating the equipment and components (burn-in), and therefore over the duration of the mission, most equipments comprising electronics and mechanical components will exhibit a constant failure rate λ often expressed in Fit (number of failures per 10^9 hours).

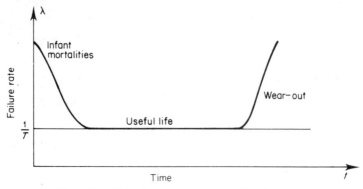

Figure 9.1   Failure rate versus time (bath tub curve).

9.1.2   Probability of Survival or Reliability

If an equipment has a failure rate of λ, its probability of no failure, i.e. its reliability at an instant t, is given by:

$$R = \exp\left(- \int_0^t \lambda \, dt \right) \qquad (1)$$

This is an entirely general expression, and makes no assumption about the way in which λ varies with time.

For the case of a failure rate λ constant, the reliability reduces to

$$R = e^{-\lambda t} \qquad (2)$$

In the case of a satellite, one can define a maximum mission life U at the end of which the service is terminated, typically by exhaustion of propellent. At the end of the mission life U, the probability of survival is nil. Figure 9.2 shows the reliability: the smaller λ is, the better the reliability.

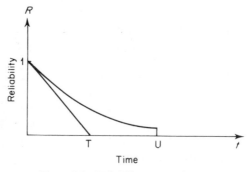

Figure 9.2   Reliability versus time.

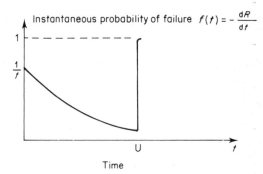

Figure 9.3 Instantaneous probability of failure versus time.

9.1.3 Instantaneous probability of failure

The probability of failure F is complementary to the reliability R. Its derivative is a probability distribution function, defined as:

$$f(t) = \frac{dF}{dt} = -\frac{dR}{dt} \tag{3}$$

The probability of a failure during the period from 0 to time t is

$$F = \int_0^t f(t)\, dt.$$

If λ is a constant, $f(t) = \lambda\, e^{-\lambda t}$ (Figure 9.3).

9.1.4 Mean Time to Failure (MTTF)

This is the mean value of the time to first failure and can be used to predict life of the equipment. It is expressed as the mean of a distribution having a density function $f(t)$:

$$T = \int_0^\infty t\, f(t)\, dt = \int_0^\infty R\, dt \tag{4}$$

For the case of λ constant, this reduces to: $T = 1/\lambda$

In the case of a satellite with a maximum mission life U, the average life τ is given by the sum of two integrals, the second of which is a delta function normalized so that the cumulative probability of failure to infinity is equal to 1.

The average life is:

$$\tau = \int_0^U t\lambda e^{-\lambda t}\, dt + e^{-U/T} \int_U^\infty t\delta(t-U)\, dt \tag{5}$$

$$\tau = T(1 - e^{-U/T}) \tag{6}$$

Table 9.1 AVERAGE LIFE FOR AN MTTF OF 10 YEARS.

Max. Mission Life U(years)	Av. Life τ(years)
$U = T/3 = 3.3$	$\tau = 0.28\ T = 2.8$
$U = T/2 = 5$	$\tau = 0.39\ T = 3.9$
$U = T = 10$	$\tau = 0.63\ T = 6.3$
$U = 2T = 20$	$\tau = 0.86\ T = 8.6$
$U = 3T = 30$	$\tau = 0.95\ T = 9.5$

τ is dependent on the mean time to failure T, as defined for a constant failure rate λ. τ/T is the probability of failure during the mission life U. Table 9.1 shows the average life τ for an MTTF of 10 years in relation to the maximum mission life U.

9.1.5   Wear-Out

Components such as bearings, valves, TWT cathodes wear out in accordance with a normal distribution rather than failing at random. In such cases:

$$f(t) = \frac{1}{\sigma\sqrt{(2\pi)}} \exp\left[-\frac{1}{2}\left(\frac{t-\mu}{\sigma}\right)^2 \right] \tag{7}$$

and

$$R = 1 - \frac{1}{\sigma\sqrt{2\pi}} \int_t^\infty \exp\left[-\frac{1}{2}\left(\frac{t-\mu}{\sigma}\right)^2 \right] \mathrm{d}t \tag{8}$$

where μ is the mean life and σ the standard deviation.

A composite reliability must be determined by multiplying reliability due to wear-out and random failure reliability. Systems are normally designed so that mean wear-out times are long when compared with the mission and design lives. Other distribution functions, like Weibull functions, can be used to model time dependent failure rates.

9.2   MISSION RELIABILITY (Baker and Baker, 1980)

To ensure a service defined by a given availability A during a fixed period L, one must plan the number n of satellites to be launched during the system lifetime L. This is of main importance in the average cost of the service.

The required number of satellites n and the system availability A will be evaluated in two typical cases, in which the time required to replace a satellite in orbit is T_R and the probability of success of the launch is p.

9.2.1   Without spare satellite in orbit

Required number of satellites   As the average life of a satellite is τ, during L

years, on average $S = L/\tau$ satellites have to be placed in orbit. As the probability of success of each launching is p, $n = S/p$ launchings must be attempted, so:

$$n = \frac{L}{pT(1 - e^{-U/T})} \tag{9}$$

System availability Satellites which are near the end of their life U must be replaced sufficiently early so that even in case of a launch failure another launching can be completed in time. The unavailability of the system in this instance is low compared with the unavailability due to accidental failures.

During its life U, the probability of a satellite random failure is $P_f = 1 - e^{-U/T}$. In L years, S replacements have to be made of which P_fS for accidental failure. Each replacement requires a time T_R if successful and, on average, a time T_R/p. The average duration of unavailability during L years is $P_fST_R/p = LT_R/pT$. Hence, the system unavailability is:

$$B = \frac{T_R}{pT} \tag{10}$$

and the system availability $A = 1 - B$ is:

$$A = 1 - \frac{T_R}{pT} \tag{11}$$

9.2.2 With a back-up Satellite (in orbit spare)

Taking the prudent, though pessimistic view that a back-up satellite has a failure rate λ and a life U equal to an active satellite, it would be necessary during a year to launch twice as many satellites as in the preceding case:

$$n = \frac{2L}{pT(1 - e^{-U/T})} \tag{12}$$

But in this case, taking into account that T_R/T is small, the availability of the systems becomes:

$$A = 1 - \frac{2T_R^2}{p^2T^2} \tag{13}$$

9.2.3 Conclusions

Table 9.2 gives three examples in which the time to repair T_R is equal to 0.25 year and the probability of launch success is 0.9.

One must note that the mean time to failure for random failures must be substantially longer than the design life if reasonable system availabilities are to be achieved.

Table 9.2   EXAMPLES OF AVAILABILITY AND NUMBER OF SATELLITES TO BE LAUNCHED ACCORDING TO DESIGN LIFE AND MTTF.

Design life U		5 years	7 years	10 years
MTTF T		10 years	20 years	20 years
Average lifetime τ		3.9 years	5.9 years	7.9 years
Probability of failure during life $P_f = \tau/T$		0.393	0.295	0.395
Time to replace T_R		0.25 year	0.25 year	0.25 year
Probability of launch p success		0.9	0.9	0.9
No spare	Annual launch rate: n/L	0.28	0.19	0.14
	Availability A	0.72	0.986	0.986
One in orbit spare	Annual launch rate: n/L	0.56	0.38	0.28
	Availability A	0.9985	0.9996	0.9996

Without a back-up satellite the service is not ensured on average for 3.4 or 1.7 months in ten years. In order to limit the unavailability to one month, an availability of 99.2 per cent is required, which means one back-up satellite, four to six launchings and an MTTF of at least ten years (10^5 h). Hence satellites have to be designed with a failure rate better than 10^{-5} per hour (10^4 Fit)

9.3   SUB-SYSTYEM RELIABILITY

Calculating the reliability of a system takes into account the reliability of all the elements the system comprises. Unless there are parallel paths contributing towards fulfilling a function, most systems and sub-systems are essentially in series in reliability terms. As reliability is a statistical measure, the composite probability of no failure in the series model is the product of the constituent sub-system reliabilities. With n units, the overall system reliability expresses:

$$R = R_1 \cdot R_2 \cdot R_3, \ldots, R_n \qquad (14)$$

So the reliability of the system with four units each with a reliability of 0.98 would be $0.98^4 = 0.922$. Expressed as a failure rate we simply add the respective failure rates to calculate the failure rate of a serial system.

A telecommunications satellite has about 10 sub-systems (see Chapter 6). It is expected that any sub-system is unlikely to fail in less than 10^{-7} h (10^2 Fit). In the case of a sub-system which is essential for telecommunications, for instance the transponder sub-system, the satellite is equipped with redundant parts to obtain the required reliability.

There are three forms of redundancy:

(1) Parallel—where each item takes part in the function and can continue this function even when one has failed.
(2) Active stand-by—where the item is held in reserve but is operating or powered.
(3) Passive stand-by—As for active but the item is quiescent.

Figure 9.4(a) illustrates the case of a stand-by equipment with one out of two redundancy and Figure 9.4(b) that of a 2/3 redundancy. In Figure 9.4(b) the stand-by equipment may replace any of the two operating equipments after failure of one of these.

Note that some form of switch is required which is activated when failure is expected or actually occurs so as to connect the stand-by item. The reliability of the switch can play an important role in the reliability of that item as it is in series with the component in stand-by and it is a single point failure.

When two equipments operate simultaneously (active parallel redundancy),

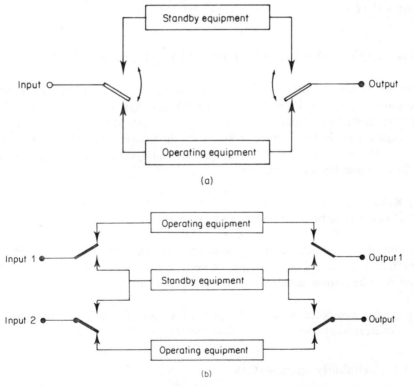

Figure 9.4   Standby equipment: (a) with redundancy ½; (b) with redundancy ⅔.

the probability of no failure is given by $R = 1 - Q_1 Q_2$, where Q_1, Q_2 are the probability of failure of one equipment ($Q_i = 1 - R_i$). Assuming the same failure rate and MTTF T:

$$R = e^{-t/T}(2 - e^{-t/T}) \qquad (15)$$

If only one equipment is operational at a time, and we assume that the back-up will not fail, the probability of survival is as follows:

$$R = e^{-t/T}(1 + t/T) \qquad (16)$$

In reality, even in standby, an equipment can age and actual reliability will be in between the answer provided by these two formulas. To be safe it is advisable to use the first one.

If an equipment without a stand-by has a failure rate λ, with redundancy its MTTF is improved by 50 per cent,

$$T' = 1/\lambda = \frac{3}{2} T = \frac{3}{2\lambda'}.$$

To obtain $\lambda' = 10^{-7}$ per hour (10^2 Fit) the term required is: $\lambda = 1.5 \times 10^{-7}$ per hour = 150 Fit

9.4 COMPONENT RELIABILITY AND PROCUREMENT

Any sub-system consists of hundreds of components. To obtain failure rates for some units of the order of one per 100 000 hours, a failure rate in the order of 1 per ten million hours for each component should not be exceeded.

A satellite project must have an active Product Assurance (PA) project team from the moment the mission is conceived and specified. Project assurance includes a number of disciplines.

(1) Reliability.
(2) Procurement of components and materials.

In practice, however, reliability and procurement are only a part of the Product Assurance. Other disciplines need to be assessed if a successful PA plan is to be implemented. The most important of these are:

(1) Quality assurance which is the plan of action of testing and inspection.
(2) Configuration control of the documentation.

9.4.1 Reliability specifications

Data on the failure rates of components and materials can be obtained from

Table 9.3 TYPICAL FAILURE RATES OF COMPONENTS FOR SPACE APPLICATIONS, EX-
PRESSED IN FIT, (FAILURE RATE AT 75 PER CENT—WHERE APPLICABLE).

Resistors		Germanium tunnel diodes	200
Solid carbon	5		
Metallic Film	5	Transistors	
Wirewound	10	(planar, silicium)	
Potentiometers	200	Standard	10
		Switching	10
Capacitors		HF	20
Solid carbon	3	Power	50
Polycarbonate	3		
Mylar	5	Integrated circuits:	
Paper	20	Digital (bipolar)	10
Solid tantalum	20	Analog	20
Variable	20	FET IC	
High Voltage	100	1–10 gates	100
		11–50 gates	500
Silicon Diodes			
Switching	4	TWT	
Standard	10	Transformers	200
Power	20	Power	30
Zener	50	Signal	10
Detector/mixer	100		
		Inductors	
Filter sections		Power	20
Hybrid	25	Signal	10
Passband	10		
		Quartz crystals	80
Couplers	10	Relays	400
Circulators	10		
Connectors	1		

GIDEP (Government and Industries Data Exchange Program) and from data of parts with flight experience.

Table 9.3 gives the failure rates which have been specified in a study on the reliability of a satellite (Dawson, 1969). In this connection MIL-HDBK-217C (1979) can also be consulted. The components with lower reliability are the electronic tubes, components with moving parts (relays, motors and potentiometers). The most reliable components are the carbon or metal film resistors, ceramic and glass capacitors, the rectifiers and connectors.

The failure rate of components is improved by derating. This can be in the form of reducing their maximum power consumption, voltage, current, temperature or some combination of all. Derating reduces the wear-out by a geometric relationship to the amount of derating applied and is of concern for meeting specified failure rates.

9.4.2 Procurement of components

Essentially every project develops a 'Preferred Components List' which is especially developed from a general 'Qualified Components List'. Several lists can be used as references and are available from space agencies such as the NASA, the Centre National d'Etudes Spatiales (CNES), the Royal Aircraft Establishment (RAE) and the European Space Agency (ESA). Another source of specification and components lists is the US JAN (Joint Army and Navy) which has specifications (e.g. MIL STD 38510 for microcircuits, MIL STD 19 500 for diode and transistors, etc.) associated with a components list.

REFERENCES

BAKER, J. C. and G. A. BAKER (1980) Impact of the space environment on spacecraft lifetimes, *J. Spacecraft,* **17**(5), 479–480.

BAZOVSKY, I. Fiabilité théorie et pratique de la sûreté de fonctionnement, Dunod.

CORAZZO, M. (1975) Techniques mathématiques de la fiabilité previsionnelle, Cépadues Editions, Toulouse.

DAWSOW, G. (1969) *Revue des télécommunications,* ITT no. 44/9.

MIL-HDBK (1979) 217C—Reliability prediction of electronic equipment, Department of Defense, USA.

Appendices

1 Error function tabulation

$$\mathrm{erf}(x) = \frac{2}{\sqrt{\pi}} \int_0^x e^{-u^2}\, du \quad \mathrm{erfc}(x) = 1 - \mathrm{erf}(x) = \frac{2}{\sqrt{\pi}} \int_x^\infty e^{-u^2}\, du$$

x	$\mathrm{erf}(x)$	$\mathrm{erfc}(x)$	x	$\mathrm{erf}(x)$	$\mathrm{erfc}(x)$
0.0	0.000	1.000			
0.1	0.112	0.888	2.1	0.997	$0.298\ 10^{-2}$
0.2	0.223	0.777	2.2	0.998	$0.186\ 10^{-2}$
0.3	0.329	0.671	2.3	0999	$0.114\ 10^{-2}$
0.4	0.428	0.572	2.4	—	$0.689\ 10^{-3}$
0.5	0.520	0.480	2.5	—	$0.407\ 10^{-3}$
0.6	0.604	0.396	2.6	—	$0.236\ 10^{-3}$
0.7	0.678	0.322	2.7	—	$0.134\ 10^{-3}$
0.8	0.742	0.258	2.8	—	$0.750\ 10^{-4}$
0.9	0.797	0.203	2.9	—	$0.411\ 10^{-4}$
1.0	0.843	0.157	3.0	—	$0.221\ 10^{-4}$
1.1	0.880	0.120	3.1	—	$0.116\ 10^{-4}$
1.2	0.910	$0.897\ 10^{-1}$	3.2	—	$0.603\ 10^{-5}$
1.3	0.934	$0.660\ 10^{-1}$	3.3	—	$0.306\ 10^{-5}$
1.4	0.952	$0.477\ 10^{-1}$	3.4	—	$0.152\ 10^{-5}$
1.5	0.966	$0.339\ 10^{-1}$	3.5	—	$0.743\ 10^{-6}$
1.6	0.976	$0.237\ 10^{-1}$	3.6	—	$0.356\ 10^{-6}$
1.7	0.984	$0.162\ 10^{-1}$	3.7	—	$0.167\ 10^{-6}$
1.8	0.989	$0.109\ 10^{-1}$	3.8	—	$0.769\ 10^{-7}$
1.9	0.993	$0.721\ 10^{-2}$	3.9	—	$0.348\ 10^{-7}$
2.0	0.995	$0.468\ 10^{-2}$	4.0	—	$0.155\ 10^{-7}$
			5.0	—	$0.15\ \ 10^{-11}$

2 Terms and definitions

ITU RADIO REGULATIONS—ARTICLE 1

Introduction

For the purposes of these Regulations, the following terms shall have the meanings defined below. These terms and definitions do not, however, necessarily apply for other purposes. Definitions identical to those contained in the International Telecommunication Convention (Malaga–Torremolinos, 1973) are marked '(CONV.)'.

Note: If, in the text of a definition below, a term is printed in italics, this means that the term itself is defined in this Article.

Section I. General Terms

1.1 *Administration:* Any governmental department or service responsible for discharging the obligations undertaken in the Convention of the International Telecommunication Union and the Regulations (CONV.).

1.2 *Telecommunication:* Any transmission, *emission* or reception of signs, signals, writing, images and sounds or intelligence of any nature by wire, *radio*, optical or other electromagnetic systems (CONV.).

1.3 *Radio:* A general term applied to the use of *radio waves* (CONV.).

1.4 *Radio Waves* or *Hertzian Waves:* Electromagnetic waves of frequencies arbitrarily lower than 3 000 GHz, propagated in space without artificial guide.

1.5 *Radiocommunication: Telecommunication* by means of *radio waves* (CONV.).

1.6 *Terrestrial Radiocommunication:* Any *radiocommunication* other than *space radiocommunication* or *radio astronomy.*

1.7   *Space Radiocommunication:* Any *radiocommunication* involving the use of one or more *space stations* or the the use of one or more *reflecting satellites* or other objects in space.

1.8   *Radiodetermination:* The determination of the position, velocity and/or other characteristics of an object, or the obtaining of information relating to these parameters, by means of the propagation properties of *radio waves*.

1.9   *Radionavigation: Radiodetermination* used for the purposes of navigation, including obstruction warning.

1.10   *Radiolocation: Radiodetermination* used for purposes other than those of *radionavigation*.

1.11   *Radio Direction-Finding: Radiodetermination* using the reception of *radio waves* for the purpose of determining the direction of a *station* or object.

1.12   *Radio Astronomy:* Astronomy based on the reception of *radio waves* of cosmic origin.

1.13   *Coordinated Universal Time (UTC):* Time scale, based on the second (SI), as defined and recommended by the CCIR[1], and maintained by the International Time Bureau (BIH).
       For most practical purposes associated with the Radio Regulations, UTC is equivalent to mean solar time at the prime meridian (0° longitude), formerly expressed in GMT.

1.14   *Industrial, Scientific and Medical (ISM) Applications* (of radio frequency energy): Operation of equipment or appliances designed to generate and use locally radio frequency energy for industrial, scientific, medical, domestic or similar purposes, excluding applications in the field of *telecommunications*.

[1]The full definition is contained in CCIR Recommendation 460-2.

Section II. Specific Terms Related to Frequency Management

2.1   *Allocation* (of a frequency band): Entry in the Table of Frequency Allocations of a given frequency band for the purpose of its use by one or more terrestrial or space *radiocommunication services* or the *radio astronomy service* under specified conditions. This term shall also be applied to the frequency band concerned.

2.2 *Allotment* (of a radio frequency or radio frequency channel): Entry of a designated frequency channel in an agreed plan, adopted by a competent conference, for use by one or more administrations for a terrestrial or space *radiocommunication service* in one or more identified countries or geographical areas and under specified conditions.

2.3 *Assignment* (of a radio frequency or radio frequency channel): Authorization given by an administration for a radio *station* to use a radio frequency or radio frequency channel under specified conditions.

Section III. Radio Services

3.1 *Radiocommunication Service:* A service as defined in this Section involving the transmission, *emission* and/or reception of *radio waves* for specific *telecommunication* purposes.

In these Regulations, unless otherwise stated, any radiocommunication service relates to *terrestrial radiocommunication.*

3.2 *Fixed Service:* A *radiocommunication service* between specified fixed points.

3.3 *Fixed-Satellite Service:* A *radiocommunication service* between *earth stations* at specified fixed points when one or more *satellites* are used; in some cases this service includes satellite-to-satellite links, which may also be effected in the *inter-satellite service*; the fixed-satellite service may also include *feeder links* for other *space radiocommunication* services.

3.4 *Aeronautical Fixed Service:* A *radiocommunication service* between specified fixed points provided primarily for the safety of air navigation and for the regular, efficient and economical operation of air transport.

3.5 *Inter-Satellite Service:* A *radiocommunication service* providing links between artificial earth *satellites.*

3.6 *Space Operation Service:* A *radiocommunication service* concerned exclusively with the operation of *spacecraft*, in particular *space tracking, space telemetry* and *space telecommand.*

These functions will normally be provided within the service in which the *space station* is operating.

3.7 *Mobile Service:* A *radiocommuniation service* between *mobile* and *land stations*, or between *mobile stations* (CONV.).

3.8 *Mobile-Satellite Service:* A *radiocommunication service:*

between *mobile earth stations* and one or more *space stations*, or between *space stations* used by this service: or

between *mobile earth stations* by means of one or more *space stations*.

This service may also include *feeder links* necessary for its operation.

3.9 *Land Mobile Service:* A *mobile service* between *base stations* and *land mobile stations*, or between *land mobile stations*.

3.10 *Land Mobile-Satellite Service:* A *mobile-satellite service* in which *mobile earth stations* are located on land.

3.11 *Maritime Mobile Service:* A *mobile service* between *coast stations* and *ship stations*, or between *ship stations*, or between associated *on-board communication stations*; *survival craft stations* and *emergency position-indicating radiobeacon stations* may also participate in this service.

3.12 *Martime Mobile-Satellite Service:* A *mobile-satellite service* in which *mobile earth stations* are located on board ships; *survival craft stations* and *emergency position-indicating radiobeacon stations* may also participate in this service.

3.13 *Port Operations Service:* A *maritime mobile service* in or near a port, between *coast stations* and *ship stations*, or between *ship stations*, in which messages are restricted to those relating to the operational handling, the movement and the safety of ships and, in emergency, to the safety of persons.

Messages which are of a *public correspondence* nature shall be excluded from this service.

3.14 *Ship Movement Service:* A *safety service* in the *maritime mobile service* other than a *port operations service*, between *cost stations* and *ship stations*, or between *ship stations*, in which messages are restricted to those relating to the movement of ships.

Messages which are of a *public correspondence* nature shall be excluded from this service.

3.15 *Aeronautical Mobile Service:* A *mobile service* between *aeronautical stations* and *aircraft stations*, or between *aircraft stations*, in which *survival craft stations* may participate; *emergency position-indicating radiobeacon stations* may also participate in this service on designated distress and emergency frequencies.

3.16 *Aeronautical Mobile-Satellite Service:* A *mobile-satellite service* in which *mobile earth stations* are located on board aircraft; *survival craft stations* and *emergency position-indicating radiobeacon stations* may also participate in this service.

3.17 *Broadcasting Service:* A *radiocommunication service* in which the transmission are intended for direct reception by the general public. This service may include sound transmissions, *television* transmissions or other types of transmission (CONV.).

3.18 *Broadcasting-Satellite Service:* A *radiocommunication service* in which signals transmitted or retransmitted by *space stations* are intended for direct reception by the general public.

In the broadcasting-satellite service, the term 'direct reception' shall encompass both *individual reception* and *community reception*.

3.19 *Radiodetermination Service:* A *radiocommunication service* for the purpose of *radiodetermination*.

3.20 *Radiodetermination-Satellite Service:* A *radiocommunication service* for the purpose of *radiodetermination* involving the use of one or more *space stations*.

3.21 *Radionavigation Service:* A *radiodetermination service* for the purpose of *radionavigation*.

3.22 *Radionavigation-Satellite Service:* a *radiodetermination-satellite service* used for the purpose of *radionavigation*.

This service may also include *feeder links* necessary for its operation.

3.23 *Maritime Radionavigation Service:* A *radionavigation service* intended for the benefit and for the safe operation of ships.

3.24 *Maritime Radionavigation-Satellite Service:* A *radionavigation-satellite service* in which *earth stations* are located on board ships.

3.25 *Aeronautical Radionavigation Service:* A *radionavigation service* intended for the benefit and for the safe operation of aircraft.

3.26 *Aeronautical Radionavigation-Satellite Service:* A *radionavigation-satellite service* in which *earth stations* are located on board aircraft.

3.27 *Radiolocation Service:* A *radiodetermination service* for the purpose of *radiolocation*.

3.28 *Meteorological Aids Service:* A *radiocommunication service* used for meteorological, including hydrological, observations and exploration.

3.29 *Earth Exploration-Satellite Service:* A *radiocommunication service* between *earth stations* and one or more *space stations*, which may include links between *space stations*, in which:

> information relating to the characteristics of the Earth and its natural phenomena is obtained from *active sensors* or *passive sensors* on earth *satellites*;
>
> similar information is collected from airborne or Earth-based platforms;
>
> such information may be distributed to *earth stations* within the system concerned;
>
> platform interrogation may be included.

This service may also include *feeder links* necessary for its operation.

3.30 *Meteorological-Satellite Service:* An *earth exploration-satellite service* for meteorological purposes.

3.31 *Standard Frequency and Time Signal Service:* A *radiocommunication service* for scientific, technical and other purposes, providing the transmission of specified frequencies, time signals, or both, of stated high precision, intended for general reception.

3.32 *Standard Frequency and Time Signal-Satellite Service:* A *radiocommunication service* using *space stations* on earth *satellites* for the same purposes as those of the *standard frequency and time signal service*.

3.33 *Space Research Service:* A *radiocommunication service* in which *spacecraft* or other objects in space are used for scientific or technological research purposes.

3.34 *Amateur Service:* A *radiocommunication service* for the purpose of self-training, intercommunication and technical investigations carried out by amateurs, that is, by duly authorized persons interested in radio technique solely with a personal aim and without pecuniary interest.

3.35 *Amateur-Satellite Service:* A *radiocommunication service* using *space stations* on earth *satellites* for the same purposes as those of the *amateur service*.

3.36 *Radio Astronomy Service:* A service involving the use of *radio astronomy*.

3.37 *Safety Service:* Any *radiocommunication service* used permanently or temporarily for the safeguarding of human life and property (CONV.).

3.38 *Special Service:* A *radiocommunication service*, not otherwise defined in this Section, carried on exclusively for specific needs of general utility, and not open to *public correspondence*.

Section IV. Radio Stations and Systems

4.1 *Station:* One or more transmitters or receivers or a combination of transmitters and receivers, including the accessory equipment, necessary at one location for carrying on a *radiocommunication service*, or the *radio astronomy service*.

Each station shall be classified by the service in which it operates permanently or temporarily.

4.2 *Terrestrial Station:* A station effecting *terrestrial radiocommunication*.

In these Regulations, unless otherwise stated, any *station* is a terrestrial station.

4.3 *Earth Station:* A *station* located either on the Earth's surface or within the major portion of the Earth's atmosphere and intended for communication:

with one or more *space stations*; or

with one or more *stations* of the same kind by means of one or more reflecting *satellites* or other objects in space.

4.4 *Space Station:* A *station* located on an object which is beyond, is intended to go beyond, or has been beyond, the major portion of the Earth's atmosphere.

4.5 *Survival Craft Stations:* A *mobile station* in the *maritime mobile service* or the *aeronautical mobile service* intended solely for survival purposes and located on any lifeboat, life-raft or other survival equipment.

4.6 *Fixed Station:* A *station* in the *fixed service*.

4.7   *Aeronautical Fixed Station:* A *station* in the *aeronautical fixed service*.

4.8   *Mobile Station:* A *station* in the *mobile service* intended to be used while in motion or during halts at unspecified points.

4.9   *Mobile Earth Station:* An *earth station* in the *mobile-satellite service* intended to be used while in motion or during halts at unspecified points.

4.10   *Land Station:* A *station* in the *mobile service* not intended to be used while in motion.

4.11   *Base Station:* A *land station* in the *land mobile service*.

4.12   *Land Mobile Station:* A *mobile station* in the *land mobile service* capable of surface movement within the geographical limits of a country or continent.

4.13   *Coast Station:* A *land station* in the *maritime mobile service*.

4.14   *Coast Earth Station:* An *earth station* in the *fixed-satellite service* or, in some cases, in the *maritime mobile-satellite service*, located at a specified fixed point on land to provide a *feeder link* for the *maritime mobile-satellite service*.

4.15   *Ship Station:* A *mobile station* in the *maritime mobile service* located on board a vessel which is not permanently moored, other than a *survival craft station*.

4.16   *Ship Earth Station:* A *mobile earth station* in the *maritime mobile-satellite service* located on board ship.

4.17   *On-Board Communication Station:* A low-powered *mobile station* in the *maritime mobile service* intended for use for internal communications on board a ship, or between a ship and its lifeboats and life-rafts during lifeboat drills or operations, or for communication within a group of vessels being towed or pushed, as well as for line handling and mooring instructions.

4.18   *Port Station:* A *coast station* in the *port operations service*.

4.19   *Aeronautical Station:* A *land station* in the *aeronautical mobile service*.

In certain instances, an aeronautical station may be located, for example, on board ship or on a platform at sea.

4.20 *Aeronautical Earth Station:* An *earth station* in the *fixed-satellite service,* or, in some cases, in the *aeronautical mobile-satellite service,* located at a specified fixed point on land to provide a *feeder link* for the *aeronautical mobile-satellite service.*

4.21 *Aircraft Station:* A *mobile station* in the *aeronautical mobile service,* other than a *survival craft station,* located on board an aircraft.

4.22 *Aircraft Earth Station:* A *mobile earth station* in the *aeronautical mobile-satellite service* located on board an aircraft.

4.23 *Broadcasting Station:* A *station* in the *broadcasting service.*

4.24 *Radiodetermination Station:* A *station* in the *radiodetermination service.*

4.25 *Radionavigation Mobile Station:* A *station* in the *radionavigation service* intended to be used while in motion or during halts at unspecified points.

4.26 *Radionavigation Land Station:* A *station* in the *radionavigation service* not intended to be used while in motion.

4.27 *Radiolocation Mobile Station:* A *station* in the *radiolocation service* intended to be used while in motion or during halts at unspecified points.

4.28 *Radiolocation Land Station:* A *station* in the *radiolocation service* not intended to be used while in motion.

4.29 *Radio Direction-Finding Station:* A *radiodetermination station* using *radio direction-finding.*

4.30 *Radiobeacon Station:* A *station* in the *radionavigation service* the *emissions* of which are intended to enable a *mobile station* to determine its bearing or direction in relation to the radiobeacon station.

4.31 *Emergency Position-Indicating Radiobeacon Station:* A *station* in the *mobile service* the *emissions* of which are intended to facilitate search and rescue operations.

4.32 *Standard Frequency and Time Signal Station:* A *station* in the *standard frequency and time signal service.*

4.33   *Amateur Station:* A *station* in the *amateur service.*

4.34   *Radio Astronomy Station:* A *station* in the *radio astronomy service.*

4.35   *Experimental Station:* A *station* utilizing *radio waves* in experiments with a view to the development of science or technique.

This definition does not include *amateur stations.*

4.36   *Ship's Emergency Transmitter:* A ship's transmitter to be used exclusively on a distress frequency for distress, urgency or safety purposes.

4.37   *Radar:* A *radiodetermination* system based on the comparison of reference signals with radio signals reflected, or retransmitted, from the position to be determined.

4.38   *Primary Radar:* A *radiodetermination* system based on the comparison of reference signals with radio signals reflected from the position to be determined.

4.39   *Secondary Radar:* A *radiodetermination* system based on the comparison of reference signals with radio signals retransmitted from the position to be determined.

4.40   *Radar Beacon (racon):* A transmitter-receiver associated with a fixed navigational mark which, when triggered by a *radar*, automatically returns a distinctive signal which can appear on the display of the triggering *radar*, providing range, bearing and identification information.

4.41   *Instrument Landing System (ILS):* A *radionavigation* system which provides aircraft with horizontal and vertical guidance just before and during landing and, at certain fixed points, indicates the distance to the reference point of landing.

4.42   *Instrument Landing System Localizer:* A system of horizontal guidance embodied in the *instrument landing system* which indicates the horizontal deviation of the aircraft from its optimum path of descent along the axis of the runway.

4.43   *Instrument Landing System Glide Path:* A system of vertical guidance embodied in the *instrument landing system* which indicates the vertical deviation of the aircraft from its optimum path of descent.

4.44   *Marker Beacon:* A transmitter in the *aeronautical radionavigation service* which radiates vertically a distinctive pattern for providing position information to aircraft.

4.45   *Radio Altimeter: Radionavigation* equipment, on board an aircraft or *spacecraft,* used to determine the height of the aircraft or the *spacecraft* above the Earth's surface or another surface.

4.46   *Radiosonde:* An automatic radio transmitter in the *meteorological aids service* usually carried on an aircraft, free balloon, kite or parachute, and which transmits meteorological data.

4.47   *Space System:* Any group of cooperating *earth stations* and/or *space stations* employing *space radiocommunication* for specific purposes.

4.48   *Satellite System:* A *space system* using one or more artificial earth *satellites.*

4.49   *Satellite Network:* A *satellite system* or a part of a *satellite system,* consisting of only one *satellite* and the cooperating *earth stations.*

4.50   *Satellite Link:* A radio link between a transmitting *earth station* and a receiving *earth station* through one *satellite.*

A satellite link comprises one up-link and one down-link.

4.51   *Multi-Satellite Link:* A radio link between a transmitting *earth station* and a receiving *earth station* through two or more *satellites,* without any intermediate *earth station.*

A multi-satellite link comprises one up-link, one or more satellite-to-satellite links and one down-link.

4.52   *Feeder Link:* A radio link from an *earth station* at a specified fixed point to a *space station,* or vice versa, conveying information for a *space radiocommunication service* other than for the *fixed-satellite service.*

Section V. Operational Terms

5.1   *Public Correspondence:* Any *telecommunication* which the offices and *stations* must, by reason of their being at the disposal of the public, accept for transmission (CONV.).

5.2   *Telegraphy:* A form of *telecommunication* which is concerned in any process providing transmission and reproduction at a distance of documentary matter, such as written or printed matter or fixed images, or the reproduction at a distance of any kind of information in such a form. For the purposes of the Radio Regulations, unless otherwise specified therein, telegraphy shall mean a form of *telecommunication* for the transmission of written matter by the use of a signal code.

5.3   *Telegram:* Written matter intended to be transmitted by *telegraphy* for delivery to the addressee. This term also includes *radiotelegrams* unless otherwise specified (CONV.).

In this definition the term *telegraphy* has the same general meaning as defined in the Convention.

5.4   *Radiotelegram:* A *telegram*, originating in or intended for a *mobile station* or a *mobile earth station* transmitted on all or part of its route over the *radiocommunication* channels of the *mobile service* or of the *mobile-satellite service*.

5.5   *Radiotelex Call:* A telex call, originating in or intended for a *mobile station* or a *mobile earth station*, transmitted on all or part of its route over the *radiocommunication* channels of the *mobile service* or the *mobile-satellite service*.

5.6   *Frequency-Shift Telegraphy: Telegraphy* by frequency modulation in which the telegraph signal shifts the frequency of the carrier between predetermined values.

5.7   *Facsimile:* A form of *telegraphy* for the transmission of fixed images, with or without half-tones, with a view to their reproduction in a permanent form.

In this definition the term *telegraphy* has the same general meaning as defined in the Convention.

5.8   Telephony: A form of *telecommunication* set up for the transmission of speech or, in some cases, other sounds.

5.9   *Radiotelephone Call:* A telephone call, originating in or intended for a *mobile station* or a *mobile earth station*, transmitted on all or part of its route over the *radiocommunication* channels of the *mobile service* or of the *mobile-satellite service*.

5.10   *Simplex Operation:* Operating method in which transmission is made possible alternately in each direction of a *telecommunication* channel, for example, by means of manual control[1].

5.11   *Duplex Operation:* Operating method in which transmission is possible simultaneously in both directions of a *telecommunication* channel[1].

[1]In general, *duplex operation* and *semi-duplex operation* require two frequencies in *radiocommunication; simlex operation* may use either one or two.

5.12   *Semi-Duplex Operation:* A method which is *simplex operation* at one end of the circuit and *duplex operation* at the other[1].

[1]In general, *duplex operation* and *semi-duplex operation* require two frequencies in *radiocommunication; simplex operation* may use either one or two.

5.13   *Television:* A form of *telecommunication* for the transmission of transient images of fixed or moving objects.

5.14   *Individual Reception* (in the broadcasting-satellite service): The reception of *emissions* from a *space station* in the *broadcasting-satellite service* by simple domestic installations and in particular those possessing small antennae.

5.15   *Community Reception* (in the broadcasting-satellite service): The reception of *emissions* from a *space station* in the *broadcasting-satellite service* by receiving equipment, which in some cases may be complex and have antennae larger than those used for *individual reception*, and intended for use:

> by a group of the general public at one location; or
>
> through a distribution system covering a limited area.

5.16   *Telemetry:* The use of *telecommunication* for automatically indicating or recording measurements at a distance from the measuring instrument.

5.17   *Radiotelemetry: Telemetry* by means of *radio waves.*

5.18   *Space Telemetry:* The use of *telemetry* for the transmission from a *space station* of results of measurements made in a *spacecraft*, including those relating to the functioning of *spacecraft.*

5.19   *Telecommand:* The use of *telecommunication* for the transmission of signals to initiate, modify or terminate functions of equipment at a distance.

5.20   *Space Telecommand:* The use of *radiocommunication* for the transmission of signals to a *space station* to initiate, modify or terminate functions of equipment on an associated space object, including the *space station*.

5.21 *Space Tracking:* Determination of the *orbit*, velocity or instantaneous position of an object in space by means of *radiodetermination*, excluding *primary radar*, for the purpose of following the movement of the object.

Section VI. Characteristics of Emissions and Radio Equipment

6.1 *Radiation:* The outward flow of energy from any source in the form of *radio waves*.

6.2 *Emission: Radiation* produced, or the production of *radiation*, by a radio transmitting *station*.

For example, the energy radiated by the local oscillator of a radio receiver would not be an emission but a *radiation*.

6.3 *Class of Emission:* The set of characteristics of an *emission*, designated by standard symbols, e.g. type of modulation of the main carrier, modulating signal, type of information to be transmitted, and also, if appropriate, any additional signal characteristics.

6.4 *Single-Sideband Emission:* An amplitude modulated *emission* with one sideband only.

6.5 *Full Carrier Single-Sideband Emission:* A *single-sideband emission* without reduction of the carrier.

6.6 *Reduced Carrier Single-Sideband Emission:* A *single-sideband emission* in which the degree of carrier suppression enables the carrier to be reconstituted and to be used for demodulation.

6.7 *Suppressed Carrier Single-Sideband Emission:* A *single-sideband emission* in which the carrier is virtually suppressed and not intended to be used for demodulation.

6.8 *Out-of-band Emission:* Emission on a frequency or frequencies immediately outside the *necessary bandwidth* which results from the modulation process, but excluding *spurious emissions*.

6.9 *Spurious Emission: Emission* on a frequency or frequencies which are outside the *necessary bandwidth* and the level of which may be reduced without affecting the corresponding transmission of information. Spurious emissions include harmonic *emissions*, parasitic *emissions*, intermodulation products and frequency conversion products, but exclude *out-of-band emissions*.

6.10   *Unwanted Emissions:* Consist of *spurious emissions* and *out-of-band emissions*.

6.11   *Assigned Frequency Band:* The frequency band within which the *emission* of a *station* is authorized; the width of the band equals the *necessary bandwidth* plus twice the absolute value of the *frequency tolerance*. Where *space stations* are concerned, the assigned frequency band includes twice the maximum Doppler shift that may occur in relation to any point of the Earth's surface.

6.12   *Assigned Frequency:* The centre of the frequency band assigned to a *station*.

6.13   *Characteristic Frequency:* A frequency which can be easily identified and measured in a given *emission*.

A carrier frequency may, for example, be designated as the characteristic frequency.

6.14   *Reference Frequency:* A frequency having a fixed and specified position with respect to the *assigned frequency*. The displacement of this frequency with respect to the *assigned frequency* has the same absolute value and sign that the displacement of the *characteristic frequency* has with respect to the centre of the frequency band occupied by the *emission*.

6.15   *Frequency Tolerance:* The maximum permissible departure by the centre frequency of the frequency band occupied by an *emission* from the *assigned frequency* or, by the *characteristic frequency* of an *emission* from the *reference frequency*.

The frequency tolerance is expressed in parts in 10^6 or in hertz.

6.16   *Necessary Bandwidth:* For a given *class of emission*, the width of the frequency band which is just sufficient to ensure the transmission of information at the rate and with the quality required under specified conditions.

6.17   *Occupied Bandwidth:* The width of a frequency band such that, below the lower and above the upper frequency limits, the *mean powers* emitted are each equal to a specified percentage $\beta/2$ of the total *mean power* of a given *emission*.

Unless otherwise specified by the CCIR for the appropriate *class of emission*, the value of $\beta/2$ should be taken as 0.5%.

6.18 *Right-Hand* (clockwise) *Polarized Wave:* An elliptically- or circularly-polarized wave, in which the electric field vector, observed in any fixed plane, normal to the direction of propagation, whilst looking in the direction of propagation, rotates with time in a right-hand or clockwise direction.

6.19 *Left-Hand* (anticlockwise) *Polarized Wave:* An elliptically- or circularly-polarized wave in which the electric field vector, observed in any fixed plane, normal to the direction of propagation, whilst looking in the direction of propagation, rotates with time in a left-hand or anticlockwise direction.

6.20 *Power:* Whenever the power of a radio transmitter etc. is referred to it shall be expressed in one of the following forms, according to the *class of emission*, using the arbitrary symbols indicated:

> *peak envelope power* (*PX* or *p*X);
> *mean power* (*PY* or *p*Y);
> *carrier power* (*PZ* or *p*Z).

For different *classes of emission*, the relationships between *peak envelope power*, *mean power* and *carrier power*, under the conditions of normal operation and of no modulation, are contained in CCIR Recommendations which may be used as a guide.

For use in formulae, the symbol p denotes power expressed in watts and the symbol P denotes power expressed in decibels relative to a reference level.

6.21 *Peak Envelope Power* (of a radio transmitter). The average power supplied to the antenna transmission line by a transmitter during one radio frequency cycle at the crest of the modulation envelope taken under normal operating conditions.

6.22 *Mean Power* (of a radio transmitter): The average power supplied to the antenna transmission line by a transmitter during an interval of time sufficiently long compared with the lowest frequency encountered in the modulation taken under normal operating conditions.

6.23 *Carrier Power* (of a radio transmitter): The average power supplied to the antenna transmission line by a transmitter during one radio frequency cycle taken under the condition of no modulation.

6.24 *Gain of an Antenna:* The ratio, usually expressed in decibels, of the power required at the input of a loss-free reference antenna to the power supplied to the input of the given antenna to produce, in a given direction, the same field strength or the same power flux-density at the same distance. When not specified otherwise, the gain refers to the direction of maximum *radiation*. The gain may be considered for a specified polarization.

Depending on the choice of the reference antenna a distinction is made between:

(a) *absolute or isotropic gain* (G_i), when the reference antenna is an isotropic antenna isolated in space;

(b) *gain relative to a half-wave dipole* (G_d), when the reference antenna is a half-wave dipole isolated in space whose equatorial plane contains the given direction;

(c) *gain relative to a short vertical antenna* (G_v), when the reference antenna is a linear conductor, much shorter than one quarter of the wavelength, normal to the surface of a perfectly conducting plane which contains the given direction.

6.25 *Equivalent Isotropically Radiated Power (e.i.r.p.):* The product of the power supplied to the antenna and the antenna gain in a given direction relative to an isotropic antenna (*absolute or isotropic gain*).

6.26 *Effective Radiated Power (e.r.p.)* (in a given direction): The product of the power suplied to the antenna and its *gain relative to a half-wave dipole* in a given direction.

6.27 *Effective Monopole Radiated Power (e.m.r.p.)* (in a given direction): The product of the power supplied to the antenna and its *gain relative to a short vertical antenna* in a given direction.

6.28 *Tropospheric Scatter:* The propagation of *radio waves* by scattering as a result of irregularities or discontinuities in the physical properties of the troposphere.

6.29 *Ionospheric Scatter:* The propagation of *radio waves* by scattering as a result of irregularities or discontinuities in the ionization of the ionosphere.

Section VII. Frequency Sharing

7.1 *Interference:* The effect of unwanted energy due to one or a combination of *emissions, radiations,* or inductions upon reception in a *radiocommunication* system, manifested by any performance degradation, misinterpretation, or loss of information which could be extracted in the absence of such unwanted energy.

7.2 *Permissible Interference*[1]*:* Observed or predicted *interference* which complies with quantitative *interference* and sharing criteria contained in these Regulations or in CCIR Recommendations or in special agreements as provided for in these Regulations.

[1]The terms 'permissible interference' and 'accepted interference' are used in the coordination of frequency assignments between administrations.

7.3 *Accepted Interference*[1]*: Interference* at a higher level than that defined as *permissible interference* and which has been agreed upon between two or more administrations without prejudice to other administrations.

7.4 *Harmful Interference: Interference* which endangers the functioning of a *radionavigation service* or of other *safety services* or seriously degrades, obstructs, or repeatedly interrupts a *radiocommunication service* operating in accordance with these Regulations.

7.5 *Protection Ratio* (R.F.): The minimum value of the wanted-to-unwanted signal ratio, usually expressed in decibels, at the receiver input, determined under specified conditions such that a specified reception quality of the wanted signal is achieved at the receiver output.

7.6 *Coordination Area:* The area associated with an *earth station* outside of which a *terrestrial station* sharing the same frequency band neither causes nor is subject to interfering *emissions* greater than a permissible level.

7.7 *Coordination Contour:* The line enclosing the *coordination area.*

7.8 *Coordination Distance:* Distance on a given azimuth from an *earth station* beyond which a *terrestrial station* sharing the same frequency band neither causes nor is subject to interfering *emissions* greater than a permissible level.

7.9 *Equivalent Satellite Link Noise Temperature:* The noise temperature referred to the output of the receiving antenna of the *earth station* corresponding to the radio frequency noise power which produces the total observed noise at the ouput of the *satellite link* excluding noise due to *interference* coming from *satellite links* using other *satellites* and from terrestrial systems.

Section VIII. Technical Terms Relating to Space

8.1   *Deep Space:* Space at distances from the Earth approximately equal to, or greater than, the distance between the Earth and the Moon.

8.2   *Spacecraft:* A man-made vehicle which is intended to go beyond the major portion of the Earth's atmosphere.

8.3   *Satellite:* A body which revolves around another body of preponderant mass and which has a motion primarily and permanently determined by the force of attraction of that other body.

8.4   *Active Satellite:* A *satellite* carrying a *station* intended to transmit or retransmit radiocommunication signals.

8.5   *Reflecting Satellite:* A *satellite* intended to reflect radiocommunication signals.

8.6   *Active Sensor:* A measuring instrument in the *earth exploration-satellite service* or in the *space research service* by means of which information is obtained by transmission and reception of *radio waves*.

8.7   *Passive Sensor:* A measuring instrument in the *earth exploration-satellite service* or in the *space research service* by means of which information is obtained by reception of *radio waves* of natural origin.

8.8   *Orbit:* The path, relative to a specified frame of reference, described by the centre of mass of a *satellite* or other object in space subjected primarily to natural forces, mainly the force of gravity.

8.9   *Inclination of an Orbit* (of an earth satellite): The angle determined by the plane containing the *orbit* and the plane of the Earth's equator.

8.10   *Period* (of a satellite): The time elapsing between two consecutive passages of a *satellite* through a characteristic point on its *orbit*.

8.11   *Altitude of the Apogee* or *of the Perigee:* The altitude of the apogee or perigree above a specified reference surface serving to represent the surface of the Earth.

8.12   *Geosynchronous Satellite:* An earth *satellite* whose period of revolution is equal to the period of rotation of the Earth about its axis.

8.13   *Geostationary Satellite:* A *geosynchronous satellite* whose circular and direct *orbit* lies in the plane of the Earth's equator and which thus remains fixed relative to the Earth; by extension, a *satellite* which remains approxi--mately fixed relative to the Earth.

8.14   *Geostationary-satellite orbit:* The *orbit* in which a *satellite* must be placed to be a *geostationary satellite*.

3 Nomenclature of the frequency and wavelength bands used in radiocommunication

ITU RADIO REGULATIONS—ARTICLE 2

The radio spectrum shall be subdivided into nine frequency bands, which shall be designated by progressive whole numbers in accordance with the following table. As the unit of frequency is the hertz (Hz), frequencies shall be expressed:

in kilohertz (kHz), up to and including 3000 kHz;

in megahertz (MHz), above 3 MHz, up to and including 3000 MHz;

in gigahertz (GHz), above 3 GHz, up to and including 3000 GHz.

For bands above 3000 GHz, i.e. centimillimetric waves, micrometric waves and decimicrometric waves, it would be appropriate to use terahertz (THz).

However, where adherence to these provisions would introduce serious difficulties, for example in connection with the notification and registration of frequencies, the lists of frequencies and related matters, reasonable departures may be made.

Band number	Symbols	Frequency range (lower limit exclusive, upper limit inclusive)	Corresponding metric subdivision	Metric abbreviations for the bands
4	VLF	3 to 30 kHz	Myriametric waves	B.Mam
5	LF	30 to 300 kHz	Kilometric waves	B.km
6	MF	300 to 3 000 kHz	Hectometric waves	B.hm
7	HF	3 to 30 MHz	Decametric waves	B.dam
8	VHF	30 to 300 MHz	Metric waves	B.m
9	UHF	300 to 3 000 MHz	Decimetric waves	B.dm
10	SHF	3 to 30 GHz	Centimetric waves	B.cm
11	EHF	30 to 300 GHz	Millimetric waves	B.mm
12		300 to 3 000 GHz	Decimillimetric waves	

Note 1: 'Band Number N' (N = band number) extends from 0.3×10^N Hz to 3×10^N Hz.

Note 2: Prefix: k = kilo (10^3), M = mega (10^6), G = giga (10^9), T = tera (10^{12}).

IEEE RADAR STANDARD 521

Band symbol	Frequency range
L	1000–2000 MHz
S	2000–4000 MHz
C	4000–8000 MHz
X	8000–12500 MHz
K_u	12.5–18 GHz
K	18–26.5 GHz
K_a	26.5–40 GHz

4 List of geostationary space stations notified or under coordination, by orbital positions

IFRB Circular No. 1644
30.10.1984

Frequency bands			GHz	<1	<3	4	6	7	11	12	14	>15
171	W	USA	TDRS WEST		3						14	15
170	W	URS	GALS-4					7				
170	W *	URS	LOUTCH P4						11		14	
170	W	URS	STATSIONAR-10			4	6					
170	W *	URS	VOLNA-7	1	3							
168	W	URS	POTOK-3			4						
160	W	URS	ESDRN						11		14	15
149	W	USA	ATS-1	1		4	6					
143	W *	USA	US SATCOM V			4	6					
139	W *	USA	US SATCOM I-R			4	6					
136	W	USA	US SATCOM-1			4	6					
135	W	USA	GOES WEST	1	3							
135	W	USA	USGCSS 2 E PAC					7				
135	W	USA	USGCSS 3 E PAC		3 *			7				
131	W *	USA	US SATCOM III-R			4	6					
128	W	USA	COMSTAR D1			4	6					
123.5	W	USA	WESTAR-2			4	6					
123	W *	USA	WESTAR 5			4	6					
119	W	USA	US SATCOM-2			4	6					
117.5	W	CAN	ANIK-C3						11		14	
116.5	W	MEX	MORELOS II		3 *	4	6		11 *	12	14	
113.5	W	MEX	MORELOS I		3 *	4	6		11 *	12	14	
114	W	CAN	ANIK-A3			4	6					
112.5	W	CAN	ANIK-C2						11		14	
109	W	CAN	ANIK-B1			4	6		11		14	
106	W	USA	USASAT-6B						11		14	
105	W	USA	ATS-5	1	3							
104.5	W	CAN	ANIK-D1			4	6					
100	W	USA	FLTSAT E PAC	1				7				
99	W	USA	WESTAR-1			4	6					
99	W *	USA	WESTAR 4			4	6					
97	W *	USA	USASAT 6A							12	14	
95	W	USA	COMSTAR D2			4	6					
95	W *	USA	USASAT 6C							12	14	
91	W	USA	WESTAR-3			4	6					
87	W	USA	COMSTAR D3			4	6					
86	W	USA	ATS-3	1								
86	W *	USA	USASAT 3C			4	6					

Frequency bands	GHz	<1	<3	4	6	7	11	12	14	>15
83 W * USA	USASAT 7B				4	6				
81 W * USA	USASAT 7D			4	6		11		14	
79 W USA	TDRS CENTRAL		3						14	15
75.4 W CLM	SATCOL 1A			4	6					
75.4 W CLM	SATCOL 1B			4	6					
75 W CLM	SATCOL 2			4	6					
75 W USA	GOES EAST	1	3							
72 W * USA	USASAT 8B			4	6					
70 W * B	SBTS A-1			4	6					
67 W * USA	USASAT 8A			4	6					
65 W * B	SBTS A-2			4	6					
53 W * USA/IT	INT 4 ATL 5			4	6					
53 W * USA/IT	INT 4A ATL 3			4	6					
50 W * USA/IT	INT 4A ATL 2			4	6					
50 W * USA/IT	INT 4 ATL 1			4	6					
41 W USA	TDRS EAST		3						14	15

FLTSAT = FLTSATCOM;   INT = INTELSAT

Frequency bands	6Hz	<1	<3	4	6	7	11	12	14	>15
34.5 W USA/IT	INT 4 ATL 5			4	6					
34.5 W USA/IT	INT 4A ATL 4			4	6					
34.5 W*USA/IT	INT MCS ATL E		3	4	6					
34.5 W USA/IT	INT 5 ATL 4			4	6		11		14	
31 W USA/IT	INT 4A ATL 4			4	6					
31 W*G	UNISAT I ATL						11	12	14	
31 W*G	UNISAT I							12	14	17
29.5 W USA/IT	INT 4 ATL 2			4	6					
29.5 W USA/IT	INT 4A ATL 3			4	6					
29.5 W USA/IT	INT 5 ATL 3			4	6		11		14	
27.5 W*USA/IT	INT MCS ATL B		3	4	6					
27.5 W*USA/IT	INT 5 ATL 3			4	6		11		14	
27.5 W*USA/IT	INT 5A ATL 2			4	6		11		14	
26.5 W URS	GALS-1					7				
26 W*F/MRS	MARECS ATL 1	1	3	4	6					
25 W*URS	VOLNA-1	1	3							
25 W URS	STATSIONAR-8			4	6					
25 W*URS	LOUTCH P1						11		14	
25 W F/SIR	SIRIO-2	1	3							
24.5 W*USA/IT	INT MCS ATL D		3	4	6					
24.5 W*USA/IT	INT 5A ATL 1			4	6		11		14	
24.5 W USA/IT	INT 5 ATL 1			4	6		11		14	
24.5 W USA/IT	INT 4A ATL 1			4	6					
24 W URS	PROGNOZ-1		3	4						
23 W*F/MRS	MARECS ATL 2	1	3	4	6					
23 W USA	FLTSAT ATL	1				7				
21.5 W*USA/IT	INT 5 ATL 5			4	6		11		14	
21.5 W*USA/IT	INT 4A ATL 1			4	6					
21.5 W*USA/IT	INT MCS ATL C		3	4	6					
19.5 W USA/IT	INT 4 ATL 3			4	6					
19.5 W USA/IT	INT 4A ATL 2			4	6					
19 W F	TDF-1		3				11			17
19 W*F/LST	L-SAT		3					12	14	20/30 17
19 W D	TV-SAT		3							17/18*
18.5 W*USA/IT	INT 4A ATL 2			4	6					
18.5 W USA/IT	INT 5 ATL 2			4	6		11		14	
18.5 W USA/IT	INT MCS ATL A		3	4	6					
18 W BEL	SATCOM-II					7				
18 W BEL	SATCOM 3					7				
16 W URS	WSDRN						11		14	15
15 W USA	MARISAT-ATL	1	3	4	6					
14 W URS	LUTCH-1						11		14	
14 W URS/IK	STATSIONAR-4			4	6					
14 W URS	VOLNA-2		3							
13.5 W*URS	POTOK-1			4						
12 W USA	USGCSS 3 ATL		3*			7				
12 W USA	USGCSS 2 ATL					7				
11.5 W F/SYM	SYMPHONIE-2	1		4	6					
11.5 W F/SYM	SYMPHONIE-3	1		4	6					
11 W*URS	STATSIONAR-11			4	6					
8 W F	TELECOM-1A		3	4	6	7		12	14	
5 W*F	TELECOM-1B		3	4	6	7		12	14	
4 W*USA/IT	INT 4A ATL 3			4	6					

FLTSAT = FLTSATCOM;   INT = INTELSAT

Frequency bands		GHz	<1	<3	4	6	7	11	12	14	>15
4	W * USA/IT	INT 4A ATL 1			4	6					
4	W USA/IT	INT 4 ATL 1			4	6					
1	W * USA/IT	INT 4A ATL 2			4	6					
1	W USA/IT	INT 4 ATL 4			4	6					
1	W * G	SKYNET 4A	1				7				44
0	E F/GEO	GEOS-2	1	3							
0	E F/MET	METEOSAT	1	3							
4	E * F	TELECOM 1C		3			7		12	14	
5	E F/OTS	OTS	1					11		14	
6	E * G	SKYNET 4B	1				7				44
7	E * F	EUTELSAT I-3						11	12	14	
10	E * F	APEX			34	6	7				
10	E F	EUTELSAT I	1					11	12*	14	
12	E URS	PROGNOZ-2		3	4						
13	E F	EUTELSAT I-2	1					11	12*	14	
17	E * ARS	SABS						11		14	
19	E ARS	ARABSAT I		3	4	6					
20	E F/SIR	SIRIO-2	1	3							
26	E ARS	ARABSAT II		3	4	6					
26	E * IRN	ZOHREH-2						11		14	
29	E F/GEO	GEOS-2	1	3							
34	E * IRN	ZOHREH-1						11	12	14	
35	E URS	GALS-6					7				
35	E URS	PROGNOZ-3		3	4						
35	E URS	STATSIONAR-2			4	6					
40	E URS	STATSIONAR-12			4	6					
45	E URS	GAS-2					7				
45	E * URS	LOUTCH P2						11		14	
45	E URS	STATSIONAR-9			4	6					
45	E * URS	VOLNA-3	1	3							
47	E * IRN	ZOHREH-3						11		14	
53	E URS	LOUTCH-2						11		14	
53	E URS/IK	STATSIONAR-5			4	6					
53	E URS	VOLNA-4		3							
57	E USA/IT	INT 5 IND 3			4	6		11		14	
57	E * USA/IT	INT 4A IND 2			4	6					
57	E * USA/IT	INT MCS IND C		3	4	6					
60	E USA/IT	INT 4A IND 2			4	6					
60	E USA/IT	INT 5 IND 2				4	6		11		14
60	E USA/IT	INT MCS IND B		3	4	6					
60	E USA	USGCSS 2 IND					7				
60	E USA	USGCSS 3 IND		3*			7				
63	E USA/IT	INT 4A IND 1			4	6					
63	E USA/IT	INT 5 IND 1			4	6		11		14	
63	E USA/IT	INT MCS IND A		3	4	6					
64.5	E * F/MRS	MARECS IND 1	1	3	4	6					
66	E * USA/IT	INT 4A IND 1			4	6					
66	E * USA/IT	INT 5 IND 4			4	6		11		14	
66	E * USA/IT	INT MCS IND D		3	4	6					
72.5	E USA	MARISAT-IND	1	3*	4*	6*					
73	E * F/MRS	MARECS IND 2	1	3	4	6					
74	E IND	INSAT-1A	1	3	4	6					
75	E USA	FLTSAT IND	1				7				
76	E * URS	GOMSS	1	3							

*Presently being coordinated under RR 1060

Frequency bands			GHz	<1	<3	4	6	7	11	12	14	>15
77	E	INS	PALAPA-2			4	6					
80	E	URS	POTOK-2			4						
80	E	URS	PROGNOZ-4		3	4						
80	E	URS	STATSIONAR-1			4	6					
80	E*	URS	STATSIONAR-13			4	6					
83	E	INS	PALAPA-1			4	6					
85	E	URS	GALS-3					7				
85	E*	URS	LOUTCH P3						11		14	
85	E	URS	STATSIONAR-3			4	6					
85	E*	URS	VOLNA-5	1	3							
90	E	URS	LOUTCH-3						11		14	
90	E	URS	STATSIONAR-6			4	6					
90	E*	URS	VOLNA-8		3							
94	E	IND	INSAT-1B	1	3	4	6					
95	E	URS	CSDRN						11		14	15
95	E*	URS	STATSIONAR-14			4	6					
99	E	URS	STATSIONAR-T	1			6					
99	E*	URS	STATSIONAR-T2	1			6					
108	E	INS	PALAPA-B1			4	6					
110	E	J	BSE		3						14	
110	E	J	BS-2		3					12	14	
113	E	INS	PALAPA-B2			4	6					
118	E	INS	PALAPA-B3			4	6					
125	E	CHN	STW-1			4	6					
128	E	URS	STATSIONAR-15			4	6					
130	E	J	ETS-2	1	3				11			34
130	E	URS	GALS-5					7				
132	E	J	CS-2A		3	4	6					20/30
135	E	J	CSE		3	4	6					18/29
136	E	J	CS-2B		3	4	6					20/340
140	E	J	GMS	1	3							
140	E	J	GMS-2	1	3							
140	E*	J	GMS-3	1	3							
140	E	URS	LOUTCH-4						11		14	
140	E	URS	STATSIONAR-7			4	6					
140	E	URS	VOLNA-6		3							
150	E*	J	CSE			4	6					
156	E	AUS	AUSSAT I							12	14	
160	E	AUS	AUSSAT II							12	14	
160	E	J	GMS	1	3							
164	E	AUS	AUSSAT III							12	14	
172	E	USA	FLTSAT W PAC	1				7				
173	E*	USA/IT	INT 5 PAC 1			4	6		11		14	
173	E*	USA/IT	INT 4A PAC 1			4	6					
174	E	USA/IT	INT 4A PAC 1			4	6					
174	E*	USA/IT	INT 5 PAC 1			4	6		11		14	
175	E	USA	USGCSS 2 W PAC					7				
175	E	USA	USGCSS 3 W PAC					7				
176	E*	USA/IT	INT 4A PAC 2			4	6					
176.5	E	USA	MARISAT-PAC	1	3	4	6					
177.5	E*	F/MRS	MARECS PAC 1	1	3	4	6					
179	E	USA/IT	INT 4A PAC 2			4	6					
179	E*	USA/IT	INT 5 PAC 2			4	6		11		14	
179	E*	USA/IT	INT MCS PAC A		3	4	6					

*Presently being coordinated under RR 1060

Index